T0314194

Introduction to Fuzzy Logic

Introduction to Fuzzy Logic

James K. Peckol

Principal Lecturer Emeritus, Department of Electrical and Computer Engineering, University of Washington, Seattle, WA, USA
and
President, Oxford Consulting, Ltd, Seattle, WA, USA

This edition first published 2021

Registered Office(s)
John Wiley & Sons, Inc., 111 River Street, Hoboken, NJ 07030, USA
John Wiley & Sons Ltd, The Atrium, Southern Gate, Chichester, West Sussex, PO19 8SQ, UK

Editorial Office
The Atrium, Southern Gate, Chichester, West Sussex, PO19 8SQ, UK

For details of our global editorial offices, customer services, and more information about Wiley products visit us at www.wiley.com.

Wiley also publishes its books in a variety of electronic formats and by print-on-demand. Some content that appears in standard print versions of this book may not be available in other formats.

Library of Congress Cataloging-in-Publication Data

Name: Peckol, James K., author. | John Wiley & Sons, publisher.
Title: Introduction to fuzzy logic / James K. Peckol.
Description: Hoboken, NJ : Wiley, 2021. | Includes bibliographical references and index.
Identifiers: LCCN 2021011123 (print) | LCCN 2021011124 (ebook) | ISBN 9781119772613 (cloth) | ISBN 9781119772620 (adobe pdf) | ISBN 9781119772637 (epub)
Subjects: LCSH: Fuzzy logic. | Fuzzy sets. | Logic, Symbolic and mathematical.
Classification: LCC QA9.64 .P43 2021 (print) | LCC QA9.64 (ebook) | DDC 511.3/13–dc23
LC record available at https://lccn.loc.gov/2021011123
LC ebook record available at https://lccn.loc.gov/2021011124

Cover Design: Wiley
Cover Image: © Rasi Bhadramani/iStock/Getty Images

Set in 9.5/12.5pt STIXTwoText by Straive, Pondicherry, India
Printed and bound by CPI Group (UK) Ltd, Croydon, CR0 4YY

C9781119772613_160721

Dedication

To my family: Near and Extended, Close and Distant,
Present and Departed, So Similar,
So Different, So Known, So Surprising . . .
especially to our youngest brother Karl,
taken from us out of season during the last voyage
of the Edmund Fitzgerald.

Contents

Preface

Starting to Think Fuzzy and Beyond

Let's begin with these questions: "Exactly what is fuzzy logic?" "Why is the logic called fuzzy?" "Who might use fuzzy logic?" These are very good questions. People may have heard something about fuzzy logic and other kinds of logic but may not be quite sure what these terms mean or quite understand the applications.

Does fuzzy logic mean that someone's comment in a discussion is very confused? Let's try to answer that question and several of the other more common ones over the course of this text by starting with some simple fuzzy examples.

Our daily language is often routinely fuzzy; yet most of the time we easily understand it. Let's start by looking at some familiar expressions from our everyday exchanges.

Where did you park the car?
  I parked up *close* to the front door of the building.
Please put the box in the trunk of the car.
  I can't lift it. It's *very heavy*.
Are we close to the city yet?
  We're *roughly about* thirty minutes away.
Is that shower warm?
  It's *very, very* hot.
Is he tall?
  Yes, he's *very, very* tall.
Is she smart?
  Trust me, she's *incredibly* smart.

Each of the responses to the questions above is somewhat vague and imprecise yet, for the most part, each provides a reasonable answer that is probably well understood. Each expression in italics is called a fuzzy *linguistic variable* rather than a crisp real number or a simple "yes" or "no." The expressions give a high-level view of fuzzy logic or fuzzy reasoning. Accompanying such reasoning we also find threshold logic and perceptrons, which model the brain.

In daily life, we find that there are two kinds of imprecision: statistical and nonstatistical. Statistical imprecision is that which arises from such events as the outcome of a coin toss or card game. Nonstatistical imprecision, on the other hand, is that which we find in expressions such as "*We're roughly about thirty minutes away*." This latter type of imprecision is what we call fuzzy.

Children learn to understand and to manipulate such instructions at an early age. They quite easily understand phrases such as "Be home by around 5:00." Perhaps children understand too well. They are adept at turning such a fuzzy expression into one that is also fuzzy. When they arrive home shortly after 6:00, they argue that 6:00 is about 5:00.

As we note, humans are quite facile at understanding fuzzy expressions and linguistic variables. For a computer, however, the opposite is true. With fuzzy logic, threshold logic, and perceptrons, increasingly both computer hardware and software are evolving to more challenging and interesting areas of logic such as neural networks, machine learning, and artificial intelligence.

Despite its amusing and seemingly contradictory name, fuzzy logic is not a logic that is fuzzy. On the contrary, fuzzy logic is a way of capturing the vagueness and imprecision that are so common in everyday human language. This capturing of vagueness and imprecision is also found in threshold logic and has significant application in artificial neurons called perceptrons. Capturing and representing the vagueness and imprecision of everyday language in terms that a computer can understand and work with is one of the objectives of fuzzy logic.

The computers we are all so familiar with operate using classical or crisp logic. Classical logic, around since Aristotle, divides the world into precise, nonoverlapping groups such as: yes–no, up–down, true–false, black–white, etc. Like a light bulb that can only be on or off, a classical logic statement can only be true or false. Those of you who have just said, "Wait a minute, what if the light's on a dimmer?" have just taken the first step to understanding fuzzy logic, threshold logic, and perceptrons. Like the light on a dimmer, a fuzzy logic statement can also be completely true or completely false, but it can also be partially true or partially false.

Fuzzy logic is simply a flexible variation and extension of classical logic. Fuzzy logic can represent statements that are completely true or false, and it can also represent those that are partially true. Classical logic lives in a black-and-white world. Fuzzy logic, threshold logic, and perceptrons, like humans, admit shades of gray. This ability to represent degrees of truth makes such tools very powerful for representing vague or imprecise ideas. We can now say, for example, that the tolerance on one capacitor is tighter than that on another or one program runs faster than another and not be concerned about specific values.

Organizing the Book

It is often all too easy to hack together a one-off crisp logic application that appears to work. Trying to replicate a million or more copies of such a design (with elastic timing constraints, variable path impedance, or flexible data values) very quickly runs into the real-world gremlins that are waiting for us. A solid, secure, robust, reliable design must always be based on the proven underlying theory, a thorough problem analysis, and a disciplined development approach. Such methods are growing increasingly important as we continue to push the envelope of designs that are impacting the daily lives of an ever-increasing number of people.

This book takes a developer's perspective to first refreshing the basics of classic or crisp logic, teaching the concept of fuzzy logic, then applying such concepts to approximate reasoning systems such as threshold logic and perceptrons. This book examines, in detail, each of the important theoretical and practical aspects that one must consider when designing today's applications.

These applications must include the following:

1) The formal hardware and software development process (stressing safety, security, and reliability)
2) The digital and software architecture of the system
3) The physical world interface to external analog and digital signals
4) The debug and test throughout the development cycle and finally
5) Improving the system performance

The Chapters

Introduction and Background

The Introduction gives an overview of the topics covered in the book. These topics include some of the vocabulary that is part of the fuzzy logic, threshold logic, and perceptron worlds. The Introduction also includes a bit of background and history, applications where such tools can be used, and a few contemporary examples.

History and Infrastructure

With the preliminary background set, the next two chapters introduce some of the early work that provided the foundation for fuzzy logic, the reasoning process for solving problems, and a brief review of the essentials of classic or crisp logic.

Chapter 1 presents some of the early views on reality, learning, logic, and reasoning that founded the first classic laws of thought that ultimately laid the foundations for fuzzy logic. Working from these fundamentals, the chapter introduces and discusses the basic mathematics and set theory underlying crisp and fuzzy logic and examines the similarities and differences between the two forms of logic. The chapter concludes with the introduction and study of fuzzy membership functions.

Chapter 2 opens with an introduction of the fundamental concepts of crisp logic underlying a classic algebra or algebraic system. The study follows with a review of the basics of Boolean algebra. We then introduce the concept and purpose of a truth table and demonstrate algebraic proofs using such tables. We then learn that the entries in such a table are called minterms and that a minterm is a binary aggregate of logical 0s and 1s that sets the logical value, true or false, of single cell entries in truth tables.

Next the K-Map is introduced and reviewed as a pictorial tool for grouping logical expressions with shared or common factors. Such sharing enables the elimination of unwanted variables thereby simplifying a logical expression. These studies introduce and teach the groundwork for relaxing the precision of classic logic and the concepts and tools similar to those that we'll apply and work with in the worlds of fuzzy logic, threshold logic, and perceptrons.

Sets, Sets, and More Sets

Building on the work of those who opened the path and set the trail for us, the next two chapters introduce and study the fundamental concepts, properties, and operations of sets and set membership first for classic sets and then for fuzzy sets.

Chapter 3 introduces the fundamental concept of sets, focusing on what are known as classical or crisp sets. The chapter begins with an introduction of some of the elementary vocabulary and terminology and then reviews the principle definitions and concepts of the theory of ordinary or classical sets. The concepts of *subsets* and *set membership* are then presented and explored. *Set membership* naturally leads to the concept of *membership functions.*

With the fundamentals of sets and set membership established, we study the basic theory of classic or crisp logic. We then move to the details of the properties and logical operations of using crisp sets and of developing crisp membership applications. Crisp sets and crisp membership applications are a prelude to the introduction of *fuzzy logic*, *fuzzy sets*, fuzzy *set membership,* and threshold logic.

Chapter 4 moves to the fuzzy world introducing and focusing on what are termed fuzzy sets. The chapter reviews some of the principle definitions and concepts of the theory of ordinary or classical sets and illustrates how these are identical to fuzzy subsets when the degree of membership in the subset is expanded to include all real numbers in the interval [0.0, 1.0]. We learned that vagueness and imprecision are common in everyday life. Very often, the kind of information we encounter may be placed into two major categories: statistical and nonstatistical.

The fundamental fuzzy terminology is presented and followed by the introduction of the basic fuzzy set properties and applications. With properties and applications understood, the focus shifts to membership functions and the grade of membership. Up to this point, data has been expressed in numerical form. Often a graphical presentation is a more effective and convenient tool. Such graphs can be expressed in both linear and curved graphical format.

Linguistic Variables and Hedges

Chapter 5, we begin this chapter with a look at early symbols and sounds and their evolution to language and knowledge. Building on such origins, we introduce the concept of *sets* and move into the worlds of formal *set theory*, *Boolean algebra* and introduce *crisp variables.*

From the crisp world, we migrate into the fuzzy world and introduce the concept and term *linguistic variable* as a variable whose values are words or phrases in a natural (or synthetic) language rather than real numbers. Such words are interpreted as representing labels on fuzzy subsets within a universe of discourse, which refers to a collection of entities that are currently being discussed, analyzed, or examined.

We learn that the values for a linguistic variable are generated from a set of primary terms, a collection of modifiers called *hedges*, and a collection of *connectives*. Hedges affect the value of a linguistic variable by either concentrating or diluting the membership distribution of the primary terms. We learn that concentrating or diluting a membership distribution can be very clearly represented graphically as discussed in Chapter 4.

We conclude with a discussion of the purpose, creation and use, and manipulation of hedges.

Fuzzy Inference and Approximate Reasoning

Chapter 6 introduces the concepts of fuzzy inference and approximate reasoning. As part of developing an effective reasoning methodology, we introduce and demonstrate the fundamental concepts and various relationships of equality, containment and entailment, conjunction and disjunction, and union and intersection among sets and subsets.

We stress that for engineers, reasoning is an essential and significant element of effective design and the successful solution to problems. We also introduce and demonstrate a variety of relationships between and among fuzzy sets and subsets as an important part of that whole process.

We conclude with the inference rules, modus ponens and modus tollens, which govern deduction and abduction, the forward and backward reasoning processes.

Chapter 7 introduces and extends the crisp world development process into the fuzzy world. The first step effectively applies the fuzzification process to extend crisp world data and basic knowledge of fuzzy world fundamentals into the domain of fuzzy system design and development. The fuzzy design process is then illustrated through its application in the design and development of several basic systems. The closing step, called defuzzification, is then applied to return crisp data to the real world.

Doing the Work

Chapter 8, we open the chapter with a summary of the growing strengths and major attractions of fuzzy logic. We identify two of the major attractions. The first attraction is fuzzy logic's ability to facilitate expressing problems in linguistic terms. The second is its ability to support applications in the areas where a designer can encounter vagueness and where the numerical mathematical model of a system may be too complex or impossible to build using conventional techniques.

We then introduce and present a formal approach to fuzzy logic design. We point out that a fundamental and essential key to any successful design process is that knowing and understanding the problem and recognizing potential hazards that might occur during the design process. We stress that without such knowledge, one cannot and should not proceed with a design until such issues are rectified. In addition, we recommend periodic design reviews throughout the design process and stress the performance of *failure modes*, *effects*, and *criticality analysis* (FMECA) testing.

We then introduce and walk through the execution of the major steps in a formal design methodology. In particular, we refer to the requirements and design documentation in Appendix A and the Unified Modeling Language tools and touch on text in Appendix B.

Introducing Threshold Logic

Chapter 9, as we now move forward, we introduce and explore two tools that build on and extend the knowledge gained from crisp and fuzzy logic: threshold logic and perceptrons. The incorporated features ultimately make important contributions to the foundation of advanced tools called neural networks, machine learning, and artificial intelligence.

Threshold logic brings into the world of digital logic design the ability to alter the value of input signals using what are designated as weights. The logic also introduces the capability to set a threshold that the summed weighted inputs need to exceed to assert a true output. Although effective with most logical operations, the threshold devices fail at the implementation of the exclusive OR.

Moving to Perceptron Logic

Chapter 10, we open the chapter with an introduction to the architecture of the threshold logic device, which is a basic building block at the heart of what is called a *perceptron*. The perceptron is also known as an *artificial neuron*. Starting with a high-level model of the basic biological neuron, we introduce the vocabulary describing the elementary components of the device.

From there, we move to the *McCulloch–Pitts* (MCP) artificial neuron and illustrate possibilities of implementing the fundamental logic devices using the MCP model. We then bring in the concept of weights from threshold logic and discuss, develop, and implement an MCP neuron network. The next step is to introduce and present the basic perceptron. We walk through each of the major functional blocks from inputs to output in the basic perceptron and also point out potential implementation and operational concerns and possible solutions of which to be aware.

We then present and describe perceptron learning and develop the learning rule. The chapter concludes with a presentation of the essential steps for testing the perceptron.

The Appendices

Two supporting appendices are included. The first provides an introduction to the preparation of formal Requirements Specifications and Design Specifications. The second provides a tutorial of the Unified Modeling Language (UML) and associated tools and some important issues to consider in the testing process.

Appendix A introduces the traditional product development life cycle and stresses the need to thoroughly understand both the operating environment and the system being designed. Further stressed is the difference between a *requirements specification* and a *design specification*. Design is only one element of product development. Each design must also be thoroughly tested to confirm and ensure that it meets specified requirements and national and international standards within its operating environment.

Appendix B introduces and provides an overview of the Unified Modeling Language and associated diagrams. The major static diagrams and the utility of each are presented and discussed. Of particular importance are the *class* and *use case* diagrams. The need for dynamic modeling is also presented. In addition, important test considerations are introduced.

This appendix also restresses the need for serious system test and the test process. It also provides a brief outline and summary of the key elements and components of a design that potentially can cause (serious) problems during operation and therefore should be considered and closely and carefully examined during the test process.

The Audience

The book is intended for students with a broad range of background and experience and also serves as a reference text for those working in the fields of electrical engineering and computer science. The core audience should have at least one quarter to one semester of study in Boolean algebra and crisp logic design, facility with a high-level programming language such as C, C++, or Java, and some knowledge of logic devices and operating systems, and should be an upper-level junior or senior or lower-level graduate student. Some background in formal system design, test, embedded systems, and analog fields would also be helpful.

Notes to the Instructor

This book can be a valuable tool for students in the traditional undergraduate electrical engineering, computer engineering, or computer science programs as well as for practicing engineers who also wish to review the basic concepts in these programs. Here, students may study the essential aspects of the development of contemporary fuzzy, neural, and approximate reasoning systems.

Students are also given a solid presentation of hardware and software architecture fundamentals, a good introduction to the design process and the formal methods used therein (including safety, security, and reliability). They are also given a comprehensive presentation of the interface to local and distributed external world devices and guidance on how to debug and test their designs.

Key to the presentation is a substantial number of worked examples illustrating fundamental ideas as well as how some of the subtleties in application go beyond basic concepts. Each chapter opens with a list of *Things to Look For* that highlights the more important material in the chapter and concludes with review questions and thought questions.

The review questions are based directly on material covered in the chapter and mirror and expand on the *Things to Look For* list. The questions provide students a self-assessment of their understanding and recall of the material covered. Though based on the material covered in the chapter, the thought questions extend the concepts as well as provide a forum in which students can explore, discuss, and synthesize new ideas based on those concepts with colleagues.

The text is written and organized much as one would develop a new system, i.e. from the top-down, building on the basics. Ideas are introduced and then revisited throughout the text, each time to a greater depth or in a new context.

Safety, security, and reliability are absolutely essential components in the development of any kind of system and system design process today. Such material is not an integral component in this text but should be stressed as companion technology.

As we stated in the opening of this Preface, finding a good balance between depth and breadth in today's approximate reasoning systems is a challenge. To that end, a couple of decisions were made at the outset. First, the text is not written around a specific microprocessor or software language. Rather, the material is intended to be relevant to (and has been used to develop) a wide variety of applications running on many different kinds of processors. Second, the artificial intelligence field and approaches to approximate reasoning are rapidly changing even as this sentence is being typed and read. In lieu of trying to pursue and include today's latest technologies, the focus is on the basics that apply to any of the technologies. The underlying philosophy of this book is that the student is well-grounded in the fundamentals will be comfortable working with and developing state-of-the-art systems using the newest ideas. Ohm's law hasn't changed for many years; the fields of computer science and electrical and computer engineering have and are.

The core material has been taught as a one-quarter, senior-level course in fuzzy logic development in several universities around the world. Based on student background, the text is sufficiently rich to provide material for a two-to-three-quarter or two-semester course in fuzzy logic and approximate reasoning systems development at the junior to senior level in a traditional four-year college or university engineering program.

Beyond the core audience, the sections covering the assumed foundation topics can provide a basis on which the student with a limited hardware or software background can progress through the remainder of the material. The logic and software sections are not sufficiently deep to replace the corresponding one- or two-quarter courses in the topics. For those with adequate background in such areas, the material can either be skipped or serve as a brief refresher.

Acknowledgments

Over the years, as I've collected the knowledge and experiences to bring this book together, there have been many, many people with whom I have studied, worked, and interacted. Our discussions, debates, and collaborations have led to the ideas and approach to design presented on the pages that follow.

While there are far too many to whom I owe a debt of thanks to try to list each here, I do want to give particular thanks to David L. Johnson, Corrine Johnson, Greg Zick, Tom Anderson, David Wright, Gary Anderson, Patrick Donahoo, Dave Bezold, Steve Swift, Paul Lantz, Mary Kay Winter, Steve Jones, Kasi Bhaskar, Brigette Huang, Jean-Paul Calvez, Gary Whittington, Olivier Pasquier, Charles Staloff, Gary Ball, John Davis, Patrick F. Kelly, Margaret Bustard, William and Betty Peckol, and Donna Karschney for all they've done over the years. William Hippe, Alex Talpalatskiy, and my brother William Peckol, who have all spent many hours proofreading, commenting, and making valuable suggestions to improve the early, working versions of the text, deserve a special thank you. From John Wiley, I want to thank Sandra Grayson who supported the original idea of publishing this text and especially Louis Manoharan (Project Editor) and Kanchana Kathirvelu (Production Editor), both of whom helped to guide the manuscript through the editing and production phases, and the unknown copyeditors, compositors, and others whose efforts on and contributions to this project have been invaluable.

In any project, design reviews are an essential part of producing a quality product. I wish to express my appreciation and thanks to this project's many reviewers for their evaluations and constructive comments, which helped guide its development.

I want to extend special thanks to my family William, Suzanne, Joe, Paulette, Karl, and Lori, and to my daughters Erin and Robyn, and grandchildren Kyleen, Jordan, and Tara.

Finally, I extend a thank you to my many teachers, friends, colleagues, and students who I've had the pleasure of knowing and working with over the years.

About the Author

James K. Peckol, Ph.D., is a Principal Lecturer Emeritus in the Department of Electrical and Computer Engineering at the University of Washington, Seattle, USA, where he has been named Teacher of the Year three times and Outstanding Faculty twice. He is also the founder of Oxford Consulting, Ltd., a product design and development consulting firm. The author is a member of *Tau Beta Pi, Who's Who in the World, Who's Who in Science and Engineering*, and has been presented with the Marquis *Who's Who* Lifetime Achievement Award.

His background spans over 50 years as an engineer and educator in the fields of software, digital, medical, and embedded systems design and development. Also, the author has published first and second editions of the book *Embedded Systems: A Contemporary Design Tool*.

As an engineer in the aerospace, commercial, and medical electronics industries, the author has worked on and contributed to the design and development of test systems for military aircraft navigation systems and radar systems, the *Apollo* color camera, various weather satellites, the Mars *Viking Lander*, flight control systems for a number of commercial aircraft, production of high-quality electronic test instruments and measurement systems, and several defibrillation systems. Academic experience spans more than 25 years of developing and teaching software, digital design, fuzzy logic, approximate reasoning, networking, and embedded systems design courses for students from academe and industry with experience ranging from limited hardware or software backgrounds to those at the junior, senior, graduate, and industrial levels.

Introduction

THINGS TO LOOK FOR...
• The topics that will be covered in the book. • Important first steps when beginning a new design. • Some of the strengths, applications, and advantages of fuzzy logic. • The differences between crisp and fuzzy logic. • Is fuzzy logic a newly developed technology? • Who is using fuzzy logic? • Should fuzzy logic be used for all designs? • Should a fuzzy system be implemented in hardware or software? • Where might tools called perceptrons and threshold logic be used? • What should we do after we have designed and built the hardware and firmware for our system?

I.1 Introducing Fuzzy Logic, Fuzzy Systems, and • • • • •

This section begins with some personal philosophy about fuzzy logic, fuzzy systems, and other devices. It begins with what fuzzy logic is and presents some introductory questions and answers about the field. It compares and contrasts fuzzy logic with the traditional classic or crisp logic and concludes with a high-level view of the basic design and development process and potential applications for both approaches. In addition to fuzzy logic, this text provides a brief review of classical logic and presents and discusses the related areas of threshold logic and artificial neurons also termed perceptrons.

I.2 Philosophy

The chapters ahead bring us into the interesting world of threshold logic, fuzzy logic, and perceptrons. Welcome. The approach and views on solving engineering problems presented in this work reflect things that I have learned throughout my career. Yours will probably be a bit different, particularly as you learn and develop your skills and as the technology evolves. My approach has been augmented by the views and approaches of many very creative engineers, scientists, mathematicians, and philosophers dating back centuries.

We see here the two main themes that will be interwoven through each of the chapters ahead. With each new design, our first look should be from the outside. What are we designing? How will people use it – *what is its behavior*? What effect will it have on its operating environment – *what are the outputs*? What will be the effect of its operating environment on it – *what are its inputs*? How well do we have to do the job – *what are the constraints*?

We want to look at the high-level details first and then go on to the lower. We can borrow the idea of the public interface (an outside view) to our system from our colleagues working on object-centered designs. Our first view should be of the public interface to our system – we should view it from the outside and then move to the details inside.

I.3 Starting to Think Fuzzy – Fuzzy Logic Q&A

We'll open this book by introducing fuzzy logic. Along our path, we use, design, and develop tools, techniques, and knowledge from just about every other discipline in electrical engineering and computing science.

OK, let's start. Other than an interesting name, exactly what is fuzzy logic? Many people have heard something about fuzzy logic but are not quite sure what it is, what it means, or what's fuzzy about it. Let's try to answer those questions and several of the other more common ones.

Despite its amusing and seemingly contradictory name, fuzzy logic is not a logic that is fuzzy. Exactly the opposite is true. It is a way of capturing the vagueness and imprecision that are so common in our everyday languages and thinking. Capturing and representing such vagueness and imprecision in terms that a computer or learning system can understand and work with becomes the challenge.

Fuzzy logic can represent statements that are completely true or false, and it can also represent those that are partially true and/or partially false. Classical crisp logic lives in a black-and-white, "yes" and "no" world. Fuzzy logic admits shades of gray. Such an ability to represent degrees of truth using what are called hedges and linguistic variables makes fuzzy logic very powerful for representing vague or imprecise ideas.

I.4 Is Fuzzy Logic a Relatively New Technology?

Not really. Although fuzzy logic has been generating a lot of interest in this country recently, it is far from new. Lotfi Zadeh (1965), of the University of California at Berkeley, proposed many of the original ideas when he published his first famous research paper on fuzzy sets in 1965. Japanese companies have been using fuzzy logic for over 50 years. They have been granted over 2000 patents and have designed fuzzy logic into hundreds of products ranging from elevator and traffic control systems to video cameras and refrigerators.

One frequently cited example is a one-button washing machine. This machine senses the size of the load of clothes, the amount and type of dirt and then selects the proper quantity of soap, water level, water temperature, and washing time. Fuzzy logic has also been applied to the classic driverless truck-backer-upper problem and automatic flight control for helicopters.

I.5 Who Is Using Fuzzy Logic in the United States?

Companies in the United States utilizing fuzzy logic in contemporary designs include Eaton Industrial Controls, Motorola, NCR, Intel, Rockwell, Togai InfraLogic, NASA, Gensym, Allen-Bradley Co., General Electric, and General Motors. Some of the fields using fuzzy logic–related technologies include linear and nonlinear control, data analysis, pattern recognition, operations research, and financial systems.

The list of companies outside of the United States becoming involved in developing products that use fuzzy logic or in producing tools for designing fuzzy logic systems is growing at rapid rate.

I.6 What Are Some Advantages of Fuzzy Logic?

Fuzzy logic works very well in conjunction with other technologies. In particular, it provides accurate responses to ambiguous, imprecise, or vague data. Because fuzzy logic allows ideas to be expressed in linguistic terms, it offers a formal mathematical system for representing problems using familiar words.

As a result, fuzzy logic has proven to be a powerful and effective tool for modeling systems with uncertainties in their inputs or outputs or for use when precise models of a system are either unknown or extremely complex.

I.7 Can I Use Fuzzy Logic to Solve All My Design Problems?

Perhaps, however, you should not use fuzzy logic in those systems for which you already have a good or optimal solution using traditional methods. If there is a simple and clearly defined mathematical model for the system, use it. Fuzzy logic, like any other tool, must be used properly and carefully.

Fuzzy logic has been found to give excellent results in several general areas. The most common usage today is in systems for which complete or adequate models are difficult to define or develop and in systems or tasks that use human observations as input, control rules, or decision rules.

The Hitachi's control system for the Sendai subway near Tokyo and Matsushita's one-button washing machine are very good examples. A fuzzy logic approach also works well in systems that are continuous and complex and that have a nonlinear transfer function or in which vagueness is common.

I.8 What's Wrong with the Tools I'm Using Now?

Nothing. Fuzzy logic does not replace your existing tools; it gives you an additional one. Fuzzy logic simplifies the task of representing and working with vague, imprecise, or ambiguous information often common in human speech, ideas, or reasoning. It provides a means for solving a set of problems that have been difficult or impossible to solve using traditional methods. Consider working examples such as automatic flight control for a helicopter, or the precise management of the freezing of fish in a home freezer.

I.9 Should I Implement My Fuzzy System in Hardware or Software?

Either one. Good tools and solutions are available for hardware and software. However, today fuzzy logic is essentially a "software type" technology and should probably be considered and evaluated that way. This is good news to us as designers. It means that we should be able to take advantage of all that we've learned about developing good software and hardware systems and apply it to developing good fuzzy logic solutions.

As with software, continuous technological improvements will ensure a migration of fuzzy logic into hardware. Hardware solutions may certainly run faster, but software solutions are more flexible and do offer a hedge against the possible unavailability of parts. Such systems, as we'll see, also have the ability to learn. Whichever approach is finally selected, one still needs to go through the rigorous process of analyzing the product under development, making the same trade-offs we have always made, and thoroughly testing the resulting design and product for safety and reliability.

I.10 Introducing Threshold Logic

As we now move forward, we introduce and explore two tools that build on and extend the knowledge gained from crisp and fuzzy logic. These tools are called *threshold logic* and *perceptrons*. Threshold logic builds on and extends the capability of the traditional combinational logic gates. The incorporated features ultimately make important contributions to the foundation of advanced tools called *neural networks*, *machine learning*, and *artificial intelligence*.

I.11 Moving to Perceptron Logic

We move next to a very fascinating device called a *perceptron*. This device is also known as an *artificial neuron*. The perceptron incorporates capabilities from both fuzzy logic and threshold logic and includes the capability to learn. Starting with a high-level model, we introduce the vocabulary describing the elementary components and capabilities of the device. From there, we move to the *McCulloch–Pitts* (MCP) artificial neuron, which we will examine and then implement a basic model. We also illustrate implementing fundamental classic logic devices using the MCP model.

I.12 Testing and Debugging

Once we design and build the hardware and firmware for our system, we move to confirming that it works. To support that process from its beginning, we include two appendices that introduce and present how to write solid Requirements and Design Specifications and to outline the fundamental functionality of your design.

I.13 Summary

We introduced the design and development tools called fuzzy logic, threshold logic, and perceptrons and presented a brief high-level overview of how fuzzy logic compares with the traditional crisp logic and other possible sources of additional information. The topics discussed are contributing to and pushing the limits of several very interesting technologies.

The full book can provide a powerful tool for the student in the traditional undergraduate electrical engineering, computer engineering, or computing science programs as well as the practicing engineer.

Our goal in introducing the technologies and designing systems based upon those technologies covered in the book is to help people solve today's and tomorrow's interesting and challenging problems. We stress very strongly that the traditional design, test processes, and formal methods (particularly including safety and reliability) do not go away; rather they become more relevant as we move to ever-increasingly complex systems.

As you work through the book, try to remember a couple of things. People use our products – our designs can affect people's lives. Once again, always do your best to make your designs as safe and as reliable as you can for each application. Remember too that the cost of a product isn't limited to the cost of the parts that make it up. We also have to consider costs of building, selling, supporting, and adding new features to your design.

Remember that good system designers and designs proceed using a minimum of six steps:

✓ Requirements Definition
✓ System Specification
✓ Functional Design
✓ Architectural Design
✓ Prototyping
✓ Testing

Finally, remember that our responsibility for a design doesn't end with design release. Also, we stress that a good, solid, and reliable design always begins with a firm foundation. Without that, everything we add later is fragile. Good luck, have fun, and learn from each design.

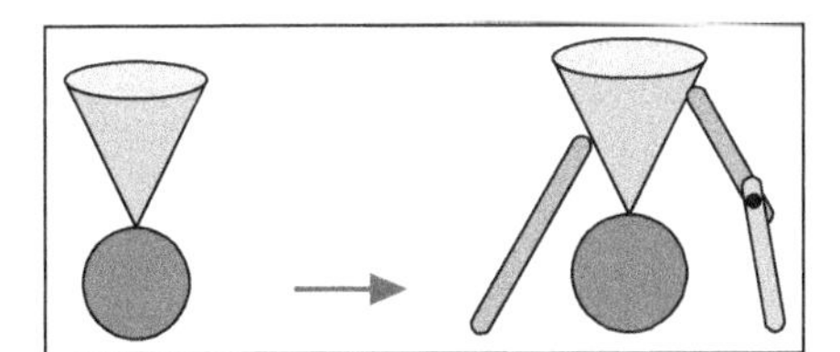

Review Questions

Fuzzy Logic

I.1 What is fuzzy logic and is it a new technology?

I.2 What are the differences between crisp or classic logic and fuzzy logic?

I.3 Can you site several applications where fuzzy logic is used?

I.4 What are some of the advantages of fuzzy logic?

I.5 What is a linguistic hedge and where would it be used?

Threshold Logic

I.6 What is threshold logic?

I.7 Where might threshold logic be used?

I.8 Can you explain what these applications are?

Perceptrons

I.9 What are perceptrons?

I.10 It is claimed that a perceptron can learn. Can you propose what it should be learning and how that might be done?

Thought Questions

I.1 Give two examples of systems that would benefit from fuzzy logic over crisp logic.

I.2 What are some of the more difficult problems that a fuzzy system might confront?

I.3 What criteria might you use to set values for a threshold in a threshold logic circuit?

I.4 Would the same criteria apply for setting a threshold in a perceptron-based design?

I.5 What would be the advantage of a system that could learn?

I.6 What would some of your first steps be in starting a fuzzy logic design, a threshold logic design, or a perceptron logic design?

I.7 What do you think should be the criteria for specifying tests for a fuzzy logic design, a threshold logic design, or a perceptron logic design?

I.8 How would you propose debugging systems developed using such technologies?

I.9 What are the major categories of signals that the described systems would interface with in the external world?

I.10 What are some of the more difficult problems that designers of the described systems might face? Consider examples such as a very popular consumer product, an intelligent robot system, a mission to Mars, or an automatic loading system on a commercial jet airliner.

I.11 What do you think might be some of the more important performance considerations that one should take into account when designing or using any of the systems described?

1

A Brief Introduction and History

THINGS TO LOOK FOR...

- Early views on reality, learning, logic, and reasoning.
- The early classic laws of thought.
- Foundations of fuzzy logic.
- A learning and reasoning taxonomy.
- The mathematics underlying crisp and fuzzy logic.
- Similarities and differences between crisp and fuzzy logic.
- Fuzzy logic and approximate reasoning.
- Fuzzy sets and membership functions.

1.1 Introduction

We open this text with a challenge and a foundation. Whether crisp or fuzzy, whether involving animals, humans, or machines, philosophers, scientists, and educators have studied, debated, and analyzed terms such as *think*, *ponder*, *logic*, *reason*, *philosophize*, or *learn* for centuries. Yet today, our understanding of these processes still has the opportunity to grow. Given such a history, what do we know?

Let us start with learning. Learning is a process that starts (at least) immediately after birth and continues, often unobtrusively, through the remaining years of life. Recent research, however, has found that learning may actually begin months earlier. Nevertheless, the term itself generally evokes childhood memories of old books, pedagogical teachers, and stuffy classrooms on warm spring afternoons when we would rather be outside playing. If we pause and reflect for a moment, we realize that learning is not limited to that proffered by the pendants of previous days but is a natural part of our daily existence. Each time that we encounter a fresh idea, make a new discovery, or solve one of life's many challenges, we are learning; we are growing and enriching our perceived model of our world or potentially what lies in space beyond.

Understanding the concepts of learning and reasoning is playing an increasingly significant role in the modern high-tech design, development, and implementation of perceptrons, neural networks, artificial intelligence, machine learning, and the primary topic of this text, i.e., fuzzy systems. We mentioned two terms: "crisp" and "fuzzy." We now introduce and explore thought and reasoning in such systems.

Introduction to Fuzzy Logic, First Edition. James K. Peckol.

1.2 Models of Human Reasoning

We now move from animals and humans to raise the question that has perplexed for eons. Can a computing machine be designed to think, to reason, to learn, to create, that is, to self-modify? Can such machines learn and operate like a human being? To be able to design and implement such tools and machines, we must first fully understand their intended applications and what these terms actually mean in such contexts.

Throughout history, as an outgrowth of the work of researchers in both the information processing and the epistemological schools, many different models of human reasoning have been explored, proposed, and tried. In each instance, the hypothetical model tries to capture the dynamic nature, inexactness, or intuitive nature of the underlying process.

Frequently, the heuristic character of human reasoning is quantified numerically. This is seen in Lotfi Zadeh's characterization of notions such as *young* or *old* on a mathematical scale or Ted Shortliffe's measures of *belief* and *disbelief*. Herb Simon has countered that people do not reason numerically. Perhaps they do not. However, mathematics is a reasonable first-order approach when attempting to capture the essence of the intuitive inexactness humans so readily accept but which computers have difficulty accommodating. Three or four simple words illustrate the essence of the two philosophies:

$$|(yes, no) \rightarrow (maybe) \rightarrow (\text{maybe not})| \ldots |(\text{crisp}) \rightarrow (\text{fuzzy})|$$

Let's begin our study by looking at some of the early works. This work is rooted mainly in the studies, writings, and teachings of early Greek philosophers including Socrates, Plato, Aristotle, Parmenides, and Heraclitus. Philosophy in the early days often included mathematics and related reasoning.

1.2.1 The Early Foundation

Socrates was one of the early classic Greek philosophers and is often regarded as the founding father of Western philosophy. He is particularly noted for his creation of the *Socratic Method* in which the teacher repeatedly poses clarifying questions until the student grasps and understands the concept(s) being taught.

Plato, a student of Socrates, formed the first institution of higher learning in the Western world. Reflecting the *Socratic Method*, he wrote dialogues in which the participants discuss, analyze, and dissect a topic from various perspectives. He is also considered the developer of the concept of *forms* in which an ideal world of *forms* exists in contrast to a false world of phenomena.

Aristotle, mentored by Plato, brought symbolic logic and the notion of scientific thinking to Western philosophy. In doing so, he contradicted Plato's ideal world of forms with a more pragmatic view and contributed advances in the branch of philosophy known as *metaphysics*. Among Aristotle's fundamental assertions was that it is impossible to both be something and not be the same thing at the same time. Today, that assertion falls apart in the field of fuzzy logic.

Parmenides was a pre-Socratic philosopher known as the *Philosopher of Changeless Being*. He believed that there are two views of reality. One is the way of truth or fact and the other is that change is impossible and existence is unchanging. Over the course of his career, he also had a significant influence on a young Plato.

Heraclitus views contradicted those of Parmenides with his insistence on ever-present change as fundamental to the universe. Such a view was reflected in his saying: "*No man ever steps in the same river twice*." His beliefs continued to identify him as one of the founders of the branch of metaphysics referred to as *ontology*, which deals with the nature of being.

From these early philosophers and their different views on reality, learning, and reasoning, we have the following classic laws of thought:

1.2.1.1 Three Laws of Thought

The *laws of thought* are stipulated to be the rules by which rational discourse can be considered to be based. Thus, they are rules that apply, without exception, to any subject matter of thought. The three laws are as follows:

- The Law of Identity
- The Law of Excluded Middle
- The Law of Non-contradiction

- Law of Identity
  The *law of identity* simply states that each entity or thing is identical to itself.
- Law of Non-contradiction
  The *law of non-contradiction* is given by Aristotle. Among his other assertions, he contends that when trying to determine the nature of reality, the following principle applies. A substance cannot have a quality and yet simultaneously not have that same quality.
- Law of Excluded Middle
  The fundamental *law of excluded middle*, which originated with Plato, states that for any proposition, that proposition is either true or its negation is true.

  Such a statement is an essential component of Boolean algebra and can be written as the classic exclusive OR: $A = (M \vee \sim M)$. A is equal to M or not M. The symbol $\vee$ is OR and the symbol ~ is the term "not."

1.3 Building on the Past – From Those Who Laid the Foundation

As the centuries have rolled forward and technology has advanced, with *thinking*, *reasoning*, and *learning* yet at its roots, the works and ideas of the early philosophers laid the foundation for contemporary concepts and ideas. Yet, as we move forward, if we pause and reflect for a moment, we realize that learning is not limited to that proffered by the pendants of previous days but is a natural part of our daily existence. Each time that we encounter a fresh idea, make a new discovery, or solve one of life's many problems, we are revisiting our store of knowledge, we are reasoning, and, hopefully, we are learning and collecting new knowledge to add to that store. We are growing and enriching our perceived model of the world.

As we have seen, educators, scientists, and philosophers have debated, studied, and analyzed learning for centuries. Our deep understanding of this process remains embryonic, yet our learning continues. Following in the footsteps of early thinkers are Feigenbaum, Hegle, Newell, Minsky, Papert, Winston, McCulloch, Pitts, Rosenblatt, Hebb, Lukasiewicz, Hopfield, Knuth, and Zadeh. We will examine the roles that each has played and how they have contributed. The next few paragraphs present a small portion of their works.

1.4 A Learning and Reasoning Taxonomy

One can distinguish a number of different forms or methods of learning ranging from the most elementary rote learning to more complex processes of learning by analogy and discovery. Variations on this taxonomy have appeared in much of the recent literature. In the following paragraphs, the term "student" is frequently used. Consider that a student or learner could be a human person or, potentially, a machine.

In some of his earlier works, Edward Feigenbaum proposed a five-phase learning process:

1) Request information.
2) Interpret the information.
3) Convert the information into a useable form.
4) Integrate the information into the existing knowledge store.
5) Apply the knowledge and evaluate the results.

The learning situation is composed of two parties, the *learner* and the *teacher* or *environment*, and a body of knowledge to be transferred from the environment to the learner. Based on the five criteria, a six-level learning taxonomy was proposed. The taxonomy considers two extremes: no active learner participation and complete active learner participation. Examining the taxonomy, one can easily see the influence of Socrates.

1) *Rote learning* – A memorization process that requires little thought of meaning by the learner.
2) *Learning with a teacher* – Most of the information is provided by the teacher. Missing details must be inferred by the learner.
3) *Learning by example* – Specific conceptual instances are given; however, generalization must be achieved by the learner.
4) *Learning by analogy* or *metaphor* – Related conceptual instances are given. The learner must recognize the relation and apply it to the task at hand.
5) *Learning by problem solving* – Knowledge embedded in the problem may be gained by the learner through solving the problem.
6) *Learning by discovery* – Knowledge exists but must be hypothesized by the learner through theory formation and extracted by experiment.

1.4.1 Rote Learning

Habit or *rote learning* is the simplest learning process. The environment supplies all of the knowledge, and the learner merely accepts and stores it with no thought to meaning or content. Despite its elementary nature, the rote acquisition of knowledge is essential to all higher forms of learning. The learner must retain base information to be able to apply it to future problems.

B.F. Skinner took a somewhat different view. He suggested that no clear connection had been demonstrated in education between ends and means. He contended that the educational process should be reduced to defining goals or acts that the learner was able to perform. Based on the "present" state of knowledge, a sequence of acts could be created to move the learner from the present to the desired state. Often, the teacher would not be necessary.

One of the most familiar and perhaps best early instances of mechanized rote learning is Arthur Samuel's program designed to play the game of checkers. The program was initially equipped with a number of suggested procedures for playing the game correctly. The intent was to have the program learn by memorizing successful (deemed significant) board positions as it encountered them and then to use them properly and effectively in future games. Ultimately, the program progressed to the level of skilled novice.

1.4.2 Learning with a Teacher

Learning with a teacher is the first level of increased complexity in the learning hierarchy. Here, the learner is beginning to take an active role in several phases of the process, specifically she or he may request information from the teacher. In this situation, abstracted or general information is presented in an integrated manner by the teacher. The learner must accept the information and then complete the store of knowledge by inferring the missing details.

Many successful programs have been written using such a paradigm. In these, the program played the role of the student or learner. Several programs, including Mostow's FOO program for playing the card game "hearts" and Waterman's poker player, were oriented toward game playing. Davis's TEIRESIAS program presented an interesting variation on this scheme.

Rather than being autonomous, TEIRESIAS was designed to sit in front of the MYCIN program written earlier by E.H. Shortliffe. MYCIN was a large rule-based system designed to assist physicians in the diagnosis and therapy of infectious diseases. The design and development process for any such large-scale system is both iterative and refining. If the system makes a misdiagnosis or offers advice contrary to the physician's diagnosis, the knowledge base must be modified. Under such a condition, TEIRESIAS would interact with the user to correct the difficulty. Such a situation reduces to a two-part task: first, explaining to the user the line of reasoning that led to the conclusion and then second, asking what additional or different information is needed to alter the result.

1.4.3 Learning by Example

Learning by example or induction increases the level of participation by the learner in the learning process. Unlike the previous example in which the teacher abstracted and then presented the material, here the student must assume the responsibility for the task. In such a context, specific conceptual instances are presented, and the student must recognize the significant or key features of the examples and then form the desired generalizations.

An early classic example of such an approach is Patrick Winston's work on "*Learning Structural Descriptions from Examples*." The goal of Winston's program was to learn elementary geometrical constructs such as those one might build using toy blocks. The program was presented with training instances from which it evolved an internal description of the concept it was to be learning. The knowledge acquired was incorporated into a semantic network where all of the interrelationships among the constituent elements were described.

Critical to the effective use of Winston's algorithm are the ideas that positive training instances are evolutionary rather than revolutionary. In Winston's algorithm, negative training instances are those that reflect only minimal differences from the concept being investigated; thus, no learning occurs.

1.4.4 Analogical or Metaphorical Learning

Analogical or metaphorical learning is probably one of the more common methods by which human beings acquire new knowledge. As Winston points out in his work *Learning and Reasoning by Analogy*, with such an approach, once again, the learner's contribution to the process is increased. When *learning by example*, the learner is presented with positive and negative instances of the concept to be learned. With an *analogy*, the student has only closely related instances from which to extract the desired concept.

Jaime Carbonell identifies *transformational* and *derivational* as the two principal methods of reasoning by analogy. When learning by transformational analogy, the line of reasoning proceeds incrementally from some old or known solution to the new or desired solution through a series of mappings *means-ends* called *transform operators*. The operators are applied using a *means-ends paradigm* until the desired transformation is achieved.

Knowledge acquisition by *derivational analogy* achieves learning by recreating the line of reasoning that resulted in the solution to the problem. The reconstruction includes both decision sequences and attendant justifications.

1.4.5 Learning by Problem Solving

Learning by problem solving can easily be viewed as subsuming all other forms of learning discussed. However, such a technique has sufficient merit in its own right that it deserves individual consideration. With this approach, the knowledge to be imparted is embedded in a problem or sequence of problems. The objective is for the learner to acquire that knowledge by solving the problem. The most serious difficulty with such an approach is the intolerance of individual method.

Consider the question: "What is the sum of ¼ and ¼?" Although a response of 2/4 would be completely correct, that answer may be considered wrong since it did not match the "correct" answer of ½.

1.4.6 Learning by Discovery

Learning by discovery is the antithesis of rote learning. In this paradigm, the learner is the initiator in all five phases of learning discussed earlier. There is no new knowledge in the world since all knowledge already exists and is merely waiting some clever individual to discover it.

According to Carbonell, two basic methods of acquiring knowledge by discovery are available: *observation* and *experimentation*. Observation is considered to be a passive approach because the learner collects information by watching a particular event and then later forms a theory to explain the phenomenon. In contrast, experimentation is viewed as active. Here, the process generally involves the learner postulating a new theory about the existence of a particular piece of knowledge and verifying that theory by experiment. In neither case is the possibility of serendipitous discovery excluded.

Discovery, Carbonell proposes, is a three-step process begun by hypothesis formation. The hypothesis may be either data driven, as is the case for observation, or theory driven as with experimentation. The initial step is followed by a refinement process in which partial theories are merged and boundary conditions are established. Finally, what has been learned and created is extended to new instances.

Clearly, any theory or model of human (or machine) learning must include aspects of each member of the established taxonomy. Today, no comprehensive and unified theory of human learning exists. Only partial theories that attempt to explain portions of the whole of human learning have been developed. Looking back over each of the ideas, we can take away the notion that learning and problem solving are effectively used interchangeably.

1.5 Crisp and Fuzzy Logic

As we move forward, we will explore, study, and learn two forms of logic and reasoning called *crisp logic* or reasoning and *fuzzy logic* or reasoning. The two graphical diagrams shown in Figure 1.1 suggest the difference between these two forms.

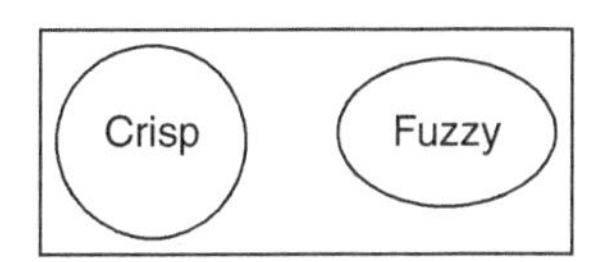

Figure 1.1 Crisp and fuzzy circles.

On the left is a crisp, precise circle and on the right is one in which the shape is less precise but can still be viewed in many contexts as a kind of circle.

1.6 Starting to Think Fuzzy

Over the years, fuzzy logic has been found to be extremely beneficial and useful to people involved in research and development in numerous fields including engineers, computer software developers, mathematicians, medical researchers, and natural scientists. As we begin, with all those people involved, we raise the question: What is fuzzy?

Originally, the word fuzz described the soft feathers that cover baby chicks. In English, the word means indistinct, imprecise, blurred, not focused, or not sharp. In French, the word is flou and in Japanese, it is pronounced "aimai." In academic or technical worlds, the word fuzz or fuzzy is used in an attempt to describe the sense of ambiguity, imprecision, or vagueness often associated with human concepts.

Revisiting an earlier example, trying to teach someone to drive a car is a typical example of real-life fuzzy teaching and fuzzy learning. As the student approaches a red light or intersection, what do you tell him or her? Do you say, "Begin to brake 25 m or 75 ft from the intersection?" Probably not. More likely, we would say something more like "Apply the brakes soon" or "Start to slow down in a little bit." The first case is clearly too precise to be implemented or executed by the driver. How can one determine exactly when one is 25 m or 75 ft from an intersection? Streets and roads generally do not have clearly visible and accurate millimeter- or inch-embedded gradations. The second vague instruction is the kind of expression that is common in everyday language.

Children learn to understand and to manipulate fuzzy instructions at an early age. They quite easily understand phrases such as "Go to bed about 10:00." Perhaps with children, they understand too well. They are adept at turning such a fuzzy expression into one that is very precise. At 9:56, determined to stay up longer, they declare, "It's not 10:00 yet."

In daily life, we find that there are two kinds of imprecision: statistical and nonstatistical. Statistical imprecision is that which arises from such events as the outcome of a coin toss or card game. Nonstatistical imprecision, on the other hand, is that which we find in instructions such as "Begin to apply the brakes soon." This latter type of imprecision is called fuzzy, and qualifiers such as *very*, *quickly*, *slowly* or others on such expressions are called *hedges* in the fuzzy world.

Another important concept to grasp is the linguistic form of variables. *Linguistic variables* are variables with more qualitative rather than numerical values, comprising words or phrases in a natural or potentially an artificial language. That is, whether simple or complex, such variables are *linguistic* rather than *numeric*. Simple examples of such variables are *very*, *slightly*, *quickly*, and *slowly*. Other examples may be generated from a set of primary terms such as *young* or its antonym *old* or *tall* with its antonym *short*.

The first practical noteworthy applications of fuzzy logic and fuzzy set theory began to appear in the 1970s and 1980s. To effectively design modern everyday systems, one must be able to recognize, represent, interpret, and manipulate statistical and nonstatistical uncertainties. One should also learn to work with *hedges* and *linguistic variables*. One should use statistical models to capture and quantify random imprecision and fuzzy models to capture and to quantify nonrandom imprecision.

1.7 History Revisited – Early Mathematics

Fuzzy logic, with roots in early Greek philosophy, finds a wide variety of contemporary applications ranging from the manufacture of cement to the control of high-speed trains, auto focus cameras, and potentially self-driving automobiles. Yet, early mathematics began by emphasizing precision. The central theme in the philosophy of Aristotle and many others was the search for perfect numbers or golden ratios. Pythagoras and his followers kept the discovery of irrational numbers a secret. Their mere existence was also counter to many fundamental religious teachings of the time.

Later mathematicians continued the search for precision and were driven toward the goal of developing a concise theory of mathematics. One such effort was *The Laws of Thought* published by Stephan Korner in 1967 in the *Encyclopedia of Philosophy*. Korner's work included a contemporary version of *The Law of Excluded Middle* which stated that every proposition could only be TRUE or FALSE – there could be no in between. An earlier version of this law, proposed by Parmenides in approximately 400 BC, met with immediate and strong objections. Heraclitus, a fellow philosopher, countered that propositions could simultaneously be both TRUE and NOT TRUE. Plato, the student, made the same arguments to his teacher Socrates.

1.7.1 Foundations of Fuzzy Logic

Plato was among the first to attempt to quantify an alternative possible state of existence. He proposed the existence of a third region, beyond TRUE and FALSE, in which "opposites tumbled about." Many modern philosophers such as Bertrand Russell, Kurt Gödel, G.W. Leibniz, and Hermann Lotze have supported Plato's early ideas.

The first formal steps away from classical logic were taken by the Polish mathematician Lukasiewicz (also the inventor of Reverse Polish Notation, RPN). He proposed a three-valued logic in which the third value, called *possible*, was to be assigned a numeric value somewhere between TRUE and FALSE. Lukasiewicz also developed an entire set of notations and an axiomatic system for his logic. His intention was to derive modern mathematics.

In later works, he also explored four- and five-valued logics before declaring that there was nothing to prevent the development of infinite-valued logics. Donald Knuth proposed a similar three-valued logic and suggested using the values of −1, 0, 1. The idea never received much support.

1.7.2 Fuzzy Logic and Approximate Reasoning

The birth of modern fuzzy logic is usually traced to the seminal paper *Fuzzy Sets* published in 1965 by Lotfi A. Zadeh. In his paper, Zadeh described the mathematics of fuzzy subsets and, by extension, the mathematics of fuzzy logic. The concept of the fuzzy event was introduced by Zadeh (1968) and has been used in various ways since early attempts to model inexact concepts were prevalent in human reasoning. The initial work led to the development of the branch of mathematics called *fuzzy logic*. This logic, actually a superset of classical binary-valued logic, does not restrict set membership to absolutes (Yes or No) but tolerates varying degrees of membership.

Using these criteria, an element is assigned a grade of membership in a parent set. The domain of this attribute is the closed interval [0, 1]. If the grade of membership values is restricted to the two extrema, then fuzzy logic reduces to two-valued or crisp logic.

In his work, Zadeh proposed that people often base their thinking and decisions on imprecise or nonnumerical information. He further believed that the membership of an element in a set need not be restricted to the values 0 and 1 (corresponding to FALSE and TRUE) but could easily be extended to include all real numbers in the interval 0.0–1.0 including the endpoints. He further felt that such a concept should not be considered in isolation but rather viewed as a methodology that moves from a discrete world to a continuous one. To augment such thinking, he proposed a collection of operations supporting his new logic.

Zadeh introduced his ideas as a new way of representing the vagueness common in everyday thinking and language. His fuzzy sets are a natural generalization or superset of classical sets or Boolean logic that are one of the basic structures underlying contemporary mathematics. Under Boolean algebra, a proposition takes a narrow view that a value is either completely true or completely false. In contrast, fuzzy logic introduces the concept of partial truth under which values are expressed anywhere within, and including, the two extremes of TRUE and FALSE.

Based on the idea of the fuzzy variable, Zadeh (1979) further proposed a theory of *approximate reasoning*. This theory postulates the notion of a possibility distribution on a linguistic variable. Using this concept, he was able to reason using vague concepts such as *young, old, tall,* or *short*. Zadeh also introduced the ideas of *semantic equivalence* and *semantic entailment* on the possibility distributions of linguistic propositions. Using these concepts, he was able to determine that a *statement* and its *double negative* are equivalent and that *very small* is more restrictive than *small*. Such conclusions derive from either the *equality* or *containment* of corresponding distributions.

Zadeh's theory is generally effective in reconciling ambiguous natural language expressions. The scope of the work was initially limited to laboratory sentences comparing hair color, age, or height between various people. Zadeh's work provided a good tool for future efforts, particularly in combination with or to enhance other forms of reasoning.

As often follows the introduction of a new concept or idea, questions arise: Why does that thing do this? Why doesn't it do that? Can you make it do another thing? An early criticism of Zadeh's fuzzy sets was: "Why can't your fuzzy set members have an uncertainty associated with them?" Zadeh eventually dealt with the issue by proposing more sophisticated kinds of fuzzy sets. New criteria evolved the original concept into numbered types of fuzzy subsets. His initial work became type-1 fuzzy sets. Additional concepts grew from type-2 fuzzy sets to ultimately type-*n* in a 1976 paper to incorporate greater uncertainty into set membership. Naturally, if a type-2 or higher set has no uncertainty in its members, it reduces to a type-1. In this text, we will work primarily with type-1 fuzzy subsets.

1.7.3 Non-monotonic Reasoning

Non-monotonic reasoning is an attempt to duplicate the human ability to reason with incomplete knowledge and to make default assumptions when insufficient evidence exists to empirically support a hypothesis. This proposed method of reasoning may be contrasted with *monotonic* reasoning in the following way.

A *monotonic logic* states that if a conclusion can be derived from a set of premises X, and if X is a subset of some larger set of premises Y, then x, a member of X, may also be derived from Y. This does not hold true for non-monotonic logic since Y may contain statements that may prevent the earlier conclusion from being derived.

Consider the following example scenario: The objective is to cross a river, and at the edge of the river there is a row boat and a set of oars. Using monotonic logic, one can conclude that it is possible to cross the river by rowing the boat across. If the new information that the boat is painted red is added, this will not alter the conclusion. On the other hand, using non-monotonic logic, the same initial conclusion may be drawn; however, if the new information that the boat has a hole in it arises, the original conclusion can no longer be drawn.

An English philosopher, William of Ockham (or Occam) (1280–1348) held a number of beliefs that foreshadowed the development of non-monotonic logic. In particular, there is an element of his philosophy called *Occam's Razor* that provides a succinct description of this form of logic. Occam stated "...that for the purposes of explaining, things not known to exist should not, unless it is absolutely necessary, be postulated as existing" (1280–1348); this has also been called the *Law of Parsimony*.

This belief may be reformulated slightly as "in the absence of any information to the contrary, assume. . .." This kind of reasoning may be defined as a *plausible inference* and is applied when conclusions must be drawn despite the absence of total knowledge about a world. These consequences then become a belief that might be modified with subsequent evidence. In a closed world, what is not known to be true must be false. Therefore, one can infer negation if proving the affirmative is not possible. Inferring negation becomes more difficult, of course, in an open world.

A first-order theory implies a monotonic logic; however, a real-world situation is non-monotonic because of gaps or incompleteness in the knowledge base. The default inference can then be used to fill in these gaps, which is very similar to some of Piaget's arguments.

McCarthy (1980) presents an idea that he calls *circumscription*. Circumscription is a rule of conjecture that argues when deriving a conclusion, that the only relevant entities are the facts on hand, and those whose existence follows from these facts. The correctness of the conclusion depends upon all of the relevant facts having been taken into account. Rephrased, if A is a collection of facts, conclusions derived from circumscription are conjectures that A includes all the relevant facts and that the objects whose existence follows from A are all relevant objects.

Reiter (1980), on the other hand, argues for default inferences from a closed-world perspective. Under such an assumption, he asserts that if R is some relation, then one can assume not R (the opposite of R or R does not exist) if assuming not R is consistent to do so. This consistence is based on not being able to prove R from the information on hand. If such a proof cannot be done, then the proof must not be true, or, similarly, if an object cannot be proven to exist in the current world, then the object does not exist.

Looking at the relationship between fuzzy logic and Reiter's form of non-monotonic logic, Reiter asserts that a default inference provides a representation for (almost all) the fuzzy subsets (and with most in terms of defaults). Reiter's assertion is not strictly correct because the inference is either true or not true, whereas a fuzzy grade of membership expresses a degree of belief in the entity.

A fundamental difference between these two theories is that Reiter's theory appears to require a global domain, whereas McCarthy's theory does not. McDermott and Doyle (1980) argue that this may not be a weakness in Reiter's approach. In either case, the intention is to extend a given set of facts (beliefs) by inferring new beliefs from the existing ones. These new beliefs are held until the evidence is introduced to contradict them. When such counterevidence occurs, a reorganization of the belief system is required.

In the discussion of his TMS (Truth Maintenance System) system, Doyle (1979) suggests that such a reorganization may take either of the two forms: *world model reorganization* or *routine revision*. Routine revision requires maintaining a body of facts that are expressed as universally true but may have some exceptions. Such a need usually occurs as a result of inferences, default assumptions, or observations. World model reorganization involves more wholesale restructuring of the model when something goes wrong. The aforementioned world model reorganization is usually quite complex and is typically the result of induction hypothesis, testimony, analogy, and intuition.

Note that these two (monotonic and non-monotonic reasoning) are very similar to Piaget's concepts of assimilation and accommodation. From Doyle's point of view, non-monotonic logic is reasoning with revision and that if a default election is made from a number of possible alternatives based on the alternatives not being believed, then the concept or argument under debate or consideration is not extensible.

1.8 Sets and Logic

1.8.1 Classical Sets

Classical sets are considered crisp because their members satisfy precise properties. For example, for illustration, let H be the set of integer real numbers from 6 to 8. Using set notation, one can express H as:

$$\begin{aligned} H &= \{r \varepsilon R \mid 6 \le r \le 8\} \\ H &= \{6, 7, 8\} \end{aligned} \tag{1.1}$$

One can also define a function $\mu_H(r)$ called a *membership function* to specify the membership of r in the set H,

$$\begin{aligned} \mu_H(r) &= 1 \quad 6 \le r \le 8 \\ &= 0 \quad \text{otherwise} \end{aligned} \tag{1.2}$$

The expression states:

> r is a member of the set H (membership in $H = 1$) if its value is 6, 7, or 8. Otherwise, it is not a member of the set (membership in $H = 0$).

One can also present the same information graphically as in Figure 1.2.

Whichever representation is chosen, it remains clear that every real number, r, is surrounded by crisp boundaries and is either in the set H or not in the set H.

Moreover, because the membership function μ maps the associated *universe of discourse* of every classical set onto the set $\{0, 1\}$, it should be evident that crisp sets correspond to a two-valued logic. An element is either in the set or it is not in the set, and it is either TRUE or it is FALSE.

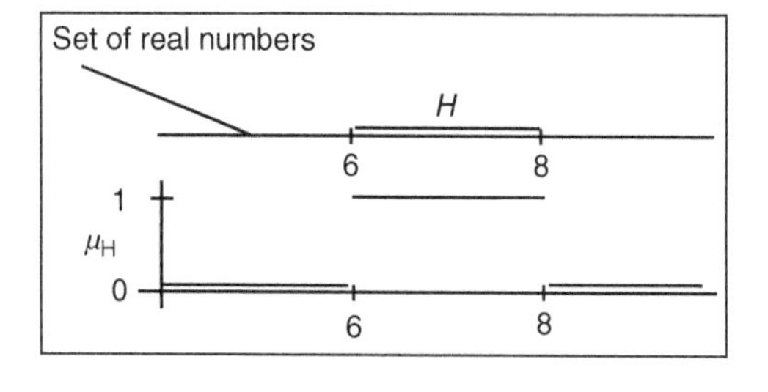

Figure 1.2 Membership in subset H.

1.8.2 Fuzzy Subsets

In relation to crisp sets, as we noted, fuzzy sets are supersets (of crisp sets) whose members are composed of collections of objects that satisfy imprecise properties to varying degrees. As an example, we can write the statement that X is a real number close to 7 as:

F = set of real numbers close to 7

But what do "set" and "close to" mean and how do we represent such a statement in mathematically correct terms?

Zadeh suggests that F is a fuzzy subset of the set of real numbers and proposes that it can be represented by its membership function, mF. The value of mF is the extent or grade of membership of each real number r in the subset of numbers close to 7. With such a construct, it is evident that fuzzy subsets correspond to a continuously valued logic and that any element can have various degrees of membership in the subset.

Let's look at another example. Consider that a car might be traveling on a freeway at a velocity between 20 and 90 mph. In the fuzzy world, we identify or define such a range as the *universe of discourse*. Within that range, we might also say that the range of 50–60 is the *average* velocity.

In the fuzzy context, the term *average* would be classed as a *linguistic* variable. A velocity below 50 or above 60 would not be considered a member of the *average* range. However, values within and equal to the two extrema would be considered members.

1.8.3 Fuzzy Membership Functions

The following paragraphs are partially reused in Chapter 4.

The fuzzy property "close to 7" can be represented in several different ways. Who decides what that representation should be? That task falls upon the person doing the design.

To formulate a membership function for the fuzzy concept "close to 7," one might hypothesize several desirable properties. These might include the following properties:

- *Normality* – It is desirable that the value of the membership function (grade of membership for 7 in the set F) for 7 be equal to 1, that is, $\mu_F(7) = 1$.

  We are working with membership values 0 or 1.

- *Monotonicity* – The membership function should be monotonic. The closer r is to 7, the closer $\mu_F(r)$ should be to 1.0 and vice versa.

  We are working with membership values in the range 0.0–1.0.

- *Symmetry* – The membership function should be one such that numbers equally distant to the left and right of 7 have equal membership values.

  We are working with membership values in the range 0.0–1.0.

It is important that one realize that these criteria are relevant only to the fuzzy property "close to 7" and that other such concepts will have appropriate criteria for designing their membership functions.

Based on the criteria given, graphic expressions of the several possible membership functions may be designed. Three possible alternatives are given in Figure 1.3. Depending upon whether one is working in a crisp or fuzzy domain, the range of grade of membership (vertical) axis should be labeled either {0–1} or {0.0–1.0}.

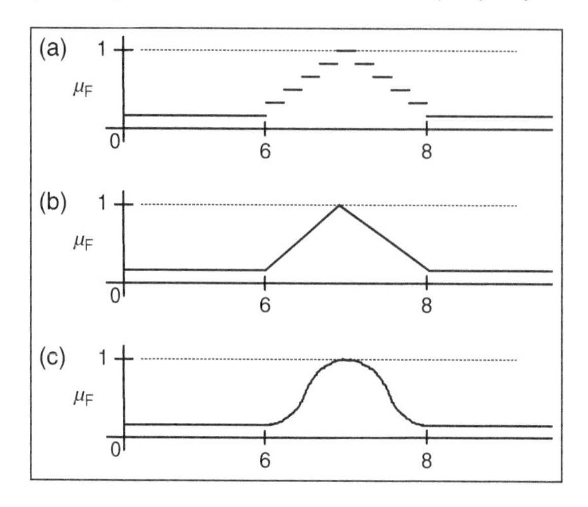

Figure 1.3 Membership functions for "close to 7."

Note that in graph c, every real number has some degree of membership in F although the numbers far from 7 have a much smaller degree. At this point, one might ask if representing the property "close to 7" in such a way makes sense.

Example 1.1
Consider a shopping trip with a friend in Paris who poses the question:

How much does that cost?

To which you answer:

About 7 euros.

which certainly can be represented graphically.

Example 1.2
As a further illustration, let the crisp set H and the fuzzy subset F represent the heights of players on a basketball team. If, for an arbitrary player p, we know that the membership in the set H is given as $\mu_H(p) = 1$, then all we know is that the player's height is somewhere between 6 and 8 ft. On the other hand, if we know that the membership in the set P is given as $\mu_F(p) = 0.85$, we know that the player's height is close to 7 ft. Which information is more useful?

Example 1.3
Consider the phrase:

Etienne is old.

The phrase could also be expressed as:

Etienne is a member of the set of old people.

If Etienne is 75, one could assign a fuzzy truth value of 0.8 to the statement. As a fuzzy set, this would be expressed as:

$$\mu_{\text{old}}\,(\text{Etienne}) = 0.8$$

From what we have seen so far, membership in a fuzzy subset appears to be very much like probability. Both the degree of membership in a fuzzy subset and a probability value have the same numeric range: 0–1. Both have similar values: 0.0 indicating (complete) nonmembership for a fuzzy subset and FALSE for a probability and 1.0 indicating (complete) membership in a fuzzy subset and TRUE for a probability. What, then, is the distinction?

Consider a natural language interpretation of the results of the previous example. If a probabilistic interpretation is taken, the value 0.8 suggests that there is an 80% chance that Etienne is old. Such an interpretation supposes that Etienne is or is not old and that we have an 80% chance of knowing it. On the other hand, if a fuzzy interpretation is taken, the value of 0.8 suggests that Etienne is more or less old (or some other term corresponding to 0.8).

To further emphasize the difference between probability and fuzzy logic, let's look at the following example.

Example 1.4

Let

L = the set of all liquids
P = the set of all potable liquids

Suppose that you have been in the desert for a week with nothing to drink, and you find two bottles (Figure 1.4).

Which bottle do you choose?

Bottle A could contain wash water. Bottle A could not contain sulfuric acid.

> A *membership value* of 0.9 means that the contents of A are very similar to perfectly potable liquids, namely, water.
> A *probability* of 0.9 means that over many experiments that 90% yield B to be perfectly safe and that 10% yield B to be potentially deadly. There is 1 chance in 10 that B contains a deadly liquid.

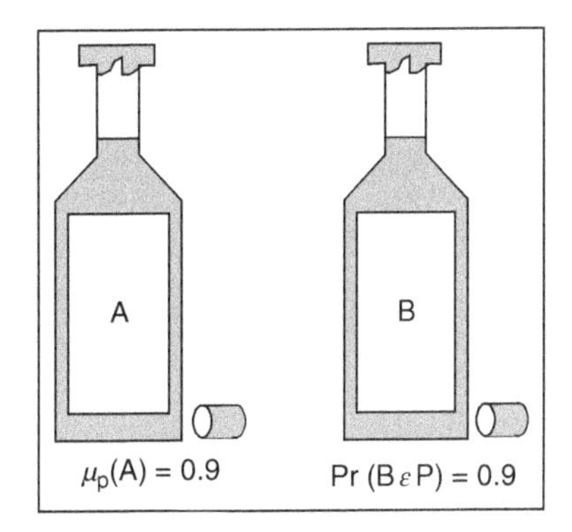

Figure 1.4 Bottles of liquids – front side.

Observation:

What does the information on the back of the bottles reveal?

> After examination, the membership value for A remains unchanged, whereas the probability value for B containing a potable liquid drops from 0.9 to 0.0 (Figure 1.5).

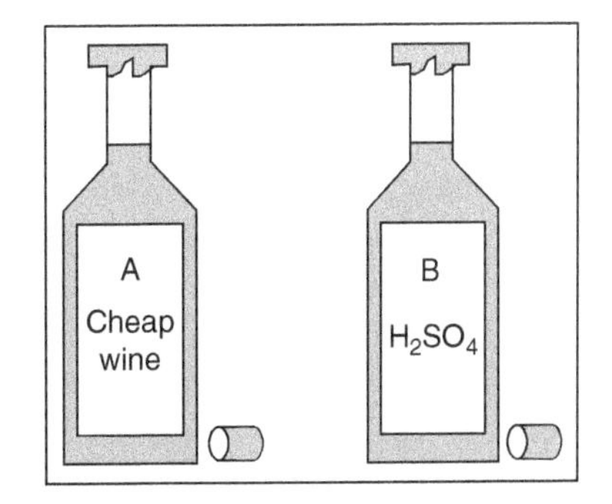

Figure 1.5 Bottles of liquids – back side.

As the previous example illustrates, fuzzy memberships represent similarities to objects that are imprecisely defined, while probabilities convey information about relative frequencies.

1.9 Expert Systems

Expert systems are an outgrowth of postproduction systems augmented with various elements of probability theory and fuzzy logic to emulate human reasoning within constrained domains. The general structure of such systems is a decision-making portion known as an *inference engine* associated with a hierarchical knowledge structure or *knowledge base.*

The knowledge base typically consists of the domain-specific knowledge and at least one level of abstraction. This second level contains the knowledge about knowledge or *meta-knowledge* for the domain. As such, this level of abstraction provides the inference engine with the criteria for making decisions or reasoning within the specific domain of application.

In some cases, these systems have performed with remarkable success. Most notables are *Dendral,* Feigenbaum (1969), *Meta-Dendral*, Buchanan (1978), *Rule-Based Expert Systems: MYCIN Experiments,* Shortliffe (1984), and *Prospector,* Duda (1978). In each case, there have been tens and perhaps hundreds of man-years devoted to tailoring each to a specific task. Nonetheless, those could be integrated, with very little effort, as a brute-force solution to reasoning. These systems have little ability to learn from previous experience and are extremely fragile at the boundaries of their knowledge.

It is clear that human understanding of human knowledge is in its infancy. Sundry schemes have been suggested and tried while attempting to explain and to emulate or simulate the human thought and reasoning processes we so casually take for granted. None has demonstrated, nor even proposed, a universal answer; they merely chip at small corners and suggest that the elephant is a thin line like a rope, round like a cylinder, or flat like a wall.

1.10 Summary

We began this chapter with a look at some of the early works in learning and reasoning. We have seen that vagueness and imprecision are common in everyday life. Very often, the kind of information we encounter may be placed into two major categories: statistical and nonstatistical. The former we model using probabilistic methods, and the latter we model using fuzzy methods.

We introduced and examined the concepts of monotonic, non-monotonic, and approximate reasoning and several models of human reasoning and learning.

We examined the concepts of crisp and fuzzy subsets and crisp and fuzzy subset membership. We learned that the possible degree of membership $\mu_x F()$ (membership of the variable x in the set F) of a variable x in a fuzzy subset spans the range [0.0–1.0] and that when we restrict the membership function so as to admit only two values, 0 and 1, fuzzy subsets reduce to classical sets. We also introduced the membership graph as a tool for expressing membership functions.

Review Questions

1.1 In the chapter opening, we introduced the term learning. Describe what it means.

1.2 Identify the three laws of thought and briefly describe each of them.

1.3 The chapter identified what is termed Feigenbaum's five-phase learning process. Identify and describe each of those phases.

1.4 Based on Feigenbaum's learning process, identify and describe each member of the corresponding six-level taxonomy.

1.5 In the context of this chapter, discuss what the term fuzzy means.

1.6 Identify and describe the difference between a classical set and a fuzzy set.

1.7 What is a fuzzy membership function? What information does such a function give you?

1.8 Formulate and graph a membership function for the concept very tall.

1.9 Formulate and graph membership functions for Examples 1.2 and 1.3.

1.10 What kind of information are we expressing with a crisp membership graph? What kind of information does such a graph give you?

1.11 What kind of information are we expressing with a fuzzy membership graph? What kind of information does such a graph give you?

1.12 Why do you think that membership graphs have different shapes?

1.13 Identify and describe how crisp and fuzzy membership graphs differ.

1.14 The chapter identified two kinds of imprecision. What are they and describe each and the differences.

1.15 To effectively design modern real-world systems, the chapters identified several important things. Identify and describe them.

1.16 Draw and annotate a membership graph for members of the crisp set old people.

1.17 Draw and annotate a membership graph for members of the fuzzy set old people.

2

A Review of Boolean Algebra

THINGS TO LOOK FOR...

- The definition of an algebra or algebraic system.
- Crisp logic and Boolean algebra.
- Fundamental unary and binary relations.
- Huntington postulates.
- Proofs of the Huntington postulates.
- Truth tables and algebraic proof using truth tables.
- Fundamental algebraic theorems.
- The definition and use of minterms and maxterms.
- Using theorems to simplify logic expressions.
- De Morgan's Law and its use.
- Basic logic gates.
- Karnaugh maps or (KM, K-Map, Karnaugh Map) and their use.
- Don't care variables.

2.1 Introduction to Crisp Logic and Boolean Algebra

We have hypothesized how our early ancestors might have evolved a hierarchy from symbols to knowledge with intermediate levels linked by relations. Before moving into the world of fuzzy logic, we will use crisp logic to introduce, express, discuss, and explore such possible relations by using the field of mathematics and logic called Boolean algebra.

The simplest relation involves a single entity and is termed *unary*. In mathematics, a unary operation has only one operand, that is, a single input. A unary operation is in contrast to binary operations, which use two operands. However, single-entity relations really don't make much sense. We have recognized that numbers are relevant only when meaning is assigned and a context specified. However, such an assignment is an essential first step.

The fundamental rules of arithmetic typically specify relations between two or more objects. Therein, we have the basic relationships of add, subtract, multiply, and divide. These relations are simple, and we should be quite familiar with them.

To be able to perform more complex systems of reasoning, we can introduce *logical* relationships. Like fundamental arithmetic, we begin by talking about relations between two things. From such binary relationships, we can go to what are termed *n*-ary relations,

Introduction to Fuzzy Logic, First Edition. James K. Peckol.

that is, relations among n entities. With such relations, we can form a complete system for reasoning. Such a system is also termed a *classic* or *crisp logic*. Relaxing constraints (moving from binary to n-ary) opens the doors to the world of *fuzzy logic*.

2.2 Introduction to Algebra

We will now form what is called algebra. An *algebra* or algebraic system is formally defined as follows:

- A set X together with one or more n-ary operations on X.
  We will define such a set and explore several binary relations.
- Such relations satisfy a specified set of axioms.
  Let's examine these relations and work with Boolean algebra. Boolean algebra was put forth by George Boole in 1854 in *An Investigation into the Laws of Thought*.
  We will initially work with two-valued variables in our set. From such a base, we can easily extend to multiple-valued logics. Our relations will be binary – the logical *OR* and the logical *AND*, which are expressed by the following symbols:
  • meaning AND
  + meaning OR

Now we need some axioms. To that end, we will work with the following axioms or postulates and theorems. An important point here is that a postulate can be untrue and a theorem is always true.

2.2.1 Postulates

The following postulates were presented by Huntington in 1904.
Formally:

Let A be a set of elements.
We define an algebraic system $\{A, \bullet, +, 0, 1\}$.

- The symbols • and + are the operations of AND and OR.
- 0 and 1 are distinguished elements of A

and associated equivalence relation on A such that:

I) For elements a, b, and c in A, the equivalence relations are
   1) Reflexive
      $a = a$ for all a in A
   2) Symmetric
      if $a = b$ then $b = a$ for all a, b in A
   3) Transitive
      if $a = b$ and $b = c$ then $a = c$ for all a, b, and c in A
   4) Substitutive
      if $a = b$ then substituting a for b in any expression will result in an equivalent relation

II) Let operators • and + be defined such that if a and b are elements of set A, then the result of basic computations using a and b and either of the operators will be another number from that set. Such a property is called the *closure* property.

$a \bullet b$ is in A.

$a + b$ is in A.

There exist elements 0 and 1 in A such that

$$a \bullet 1 = a$$

$$a \bullet 0 = 0$$

III) The operators • and + are *commutative* for all elements a and b in A

$$a \bullet b = b \bullet a$$

$$a + b = b + a$$

IV) The operators • and + are *distributive* for all elements a, b, and c in A

$$a + (b \bullet c) = (a + b) \bullet (a + c)$$

$$a \bullet (b + c) = a \bullet b + a \bullet c$$

V) For every element a in A, there exists an element ~a such that

$$a \bullet \sim a = 0$$

$$a + \sim a = 1$$

The symbol ~ indicates *not*, *inverse*, or *complement*.

VI) There are at least two elements a and b in A such that

$$a \neq b$$

Based on these postulates and theorems, the simplest algebra uses:

- The set of elements

$$A = \{0, 1\}$$

- The two operations • and + that are defined as:

For the *and* (•) operator

$$0 \bullet 0 = 0$$

$$0 \bullet 1 = 0$$

$$1 \bullet 0 = 0$$

$$1 \bullet 1 = 1$$

For the *or* (+) operator

$$0+0=0$$
$$0+1=1$$
$$1+0=1$$
$$1+1=1$$

•	0	1
0	0	0
1	0	1

+	0	1
0	0	1
1	1	1

Figure 2.1 Algebraic AND and OR operations.

The operations are also illustrated in the truth tables in Figure 2.1.

For the proposed algebra, we can show that each of the postulates holds.
Using the set A,

I and II are satisfied.
Based on the members in the set {0,1}.
The definitions of • and + given above.
III is satisfied.
Because the elements of the set {0,1} are defined.
The definitions of the operators • and + are specified above.
IV is satisfied.
Based on the definitions of the operators • and +.
V is demonstrated using a truth-table proof.
Consider the example truth table in Figure 2.2.
Observe that the values for the expressions $a+b \bullet c$ and $(a+b) \bullet (a+c)$ on the left- and right-hand sides of the table in Figure 2.2 are the same.

	LHS		RHS		
a b c	b•c	a + b•c	a + b	a + c	(a + b) • (a + c)
0 0 0	0	0	0	0	0
0 0 1	0	0	0	1	0
0 1 0	0	0	1	0	0
0 1 1	1	1	1	1	1
1 0 0	0	1	1	1	1
1 0 1	0	1	1	1	1
1 1 0	0	1	1	1	1
1 1 1	1	1	1	1	1

Figure 2.2 Truth-table proof.

VI is given by exhaustive proof.
The symbol ~ indicates *not*, *inverse*, or *complement*.

$$a \bullet \sim a = 0$$
$$a + \sim a = 1$$

Using the definitions for the operators • and +,

Let $a=0$.

Then for the set A,

$\sim a$ must be 1 since 0 and 1 are the only elements in A.
$a \bullet \sim a = 0 \bullet \sim 0 = 0 \bullet 1 = 0$ from the definition of $\bullet$.
$a + \sim a = 0 + \sim 0 = 0 + 1 = 1$ from the definition of $+$.

Let $a = 1$.

Then for the set A,

$\sim a$ must be 0 since 0 and 1 are the only elements in A.
$a \bullet \sim a = 1 \bullet \sim 1 = 1 \bullet 0 = 0$ from the definition of $\bullet$.
$a + \sim a = 1 + \sim 1 = 1 + 0 = 1$ from the definition of $+$.

VII can be proven similarly.

2.2.2 Theorems

The following fundamental theorems can be now derived from the basic axioms. These theorems are essential building blocks in basic algebra and in crisp logic. Similar components will also be found in fuzzy logic.

Th I	The elements 0 and 1 are unique.
Th II	For every a in A,
	$a \bullet a = a$
	$a + a = a$
Th III	For every a in A,
	$a \bullet 0 = 0$
	$a + 1 = 1$
Th IV	The elements 0 and 1 are distinct and $\sim 0 = 1$ and $\sim 1 = 0$.
Th V	For all a and b in A,
	$a + (a \bullet b) = a$
	$a \bullet (a + b) = a$
Th VI	For all a in A, $\sim a$ is unique.
Th VII	For all a in A, $a = \sim\sim a$.
Th VIII	For all a, b, and c in A,
	$a \bullet ([(a+b)+c]) = a \bullet (a+b) + a \bullet c = a$
Th IX	For all a, b, and c in A,
	$a + (b + c) = (a + b) + c$
	$a \bullet (b \bullet c) = (a \bullet b) \bullet c$
Th X	For all a and b in A,
	$a + \sim a \bullet b = a + b$
	$(a + \sim a) \bullet (a + b)$
	$a \bullet (\sim a + b) = a \bullet b$
	Can be proven using a truth table.

In the domain of set theory, *De Morgan's Laws* relate mathematical statements through their opposites or complements. Here, we apply De Morgan's Laws to the *intersection* and *union* of sets.

Th XI

For all a and b in A,

De Morgan's Law

$$\sim(a \bullet b) = \sim a + \sim b$$
$$\sim(a + b) = \sim a \bullet \sim b$$

Note:

$$\sim(a \bullet b) \neq \sim a \bullet \sim b$$

De Morgan's Law extends to any number of variables.

For three variables, we have:

$$\sim(a \bullet b \bullet c) = \sim a + \sim b + \sim c$$
$$\sim(a + b + c) = \sim a \bullet \sim b \bullet \sim c$$

2.3 Getting Some Practice

Let's now look at working with some of these axioms and theorems. When designing logic circuits, we have several major goals. We might trade these off during the design process. One of our major objectives is simply simplicity. Why is this the case? We have four important criteria:

- Cost
- Lower failure rate
- Easier to build
- Easier to test

As we start, keep in mind an important point. Simplicity is important; however, in the design engineering world where people's lives may be at stake, safety and reliability are the number one and the most important criteria.

To start, we can use the above theorems to simplify logic expressions. Let's try a few examples.

Examples

$$ab + \sim ab \qquad \Rightarrow b$$
$$\sim a \sim b + \sim ab + a \sim b \qquad \Rightarrow \sim a + \sim \mathrm{b}$$
$$\sim a + \sim a \sim b + bc \sim d + b \sim d \qquad \Rightarrow \sim a + b \sim d$$
$$abc + \sim abc + \sim bc \qquad \Rightarrow c$$
$$ab + \sim ac + bc \Rightarrow \sim a \sim bc + \sim abc + ab \sim c + abc \qquad \Rightarrow ab + \sim ac$$
$$a + \sim ab + \sim(a + b)c + \sim(a + b + c)d \qquad \Rightarrow a + b + c + d$$

2.4 Getting to Work

Now that we are armed with the great collection of postulates and theorems and have gotten a little practice with them, let's start putting them to work.

2.4.1 Boolean Algebra

2.4.1.1 Operands

Operands or variables are two valued; that is, they are binary, *true* or *false.*

True	*False*
1	0
True iff not false	False iff not true
	// iff → if and only if

An important point to note is that other algebras such as fuzzy logic, which we'll study later, permit multiple-valued logics.

2.4.1.2 Operators

Boolean algebra supports three basic kinds of operators.

2.4.1.2.1 Unary

Takes single operand

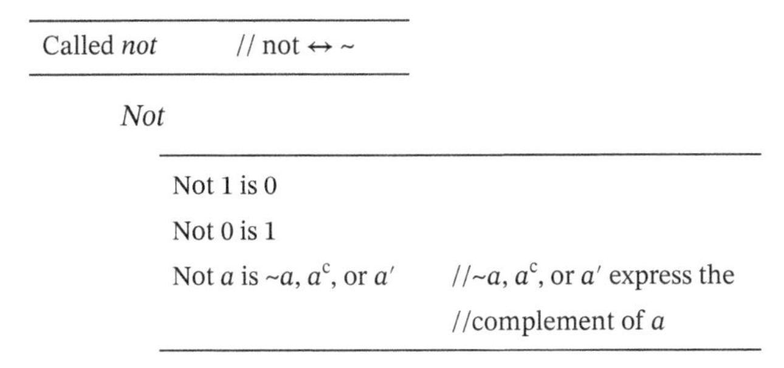

Called *not*	// not ↔ ~

Not

Not 1 is 0	
Not 0 is 1	
Not a is $\sim a$, a^c, or a'	//$\sim a$, a^c, or a' express the //complement of a

2.4.1.2.2 Binary Binary operators take two operands:

Called the AND and the OR
Expressed by the symbols • and +

We've already seen these earlier.

2.4.1.3 Relations

Using the defined operators and operands, we can now begin to express relations between and among things. With Boolean algebra, these relations will be crisp and can be expressed in several ways: in a truth table or an equation.

Truth Table

A *truth table* is a table of combinations that lists all possible values of a set of variables. We can have a single variable, two variables, or n variables.

Single Variable (Figure 2.3)
A (or a) has two values.

0, 1
true, false

In the tabular form.

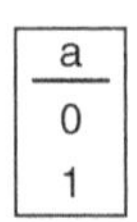

a
0
1

Figure 2.3 Single-variable truth table.

Two Variables (Figure 2.4)
A, B (or a, b).
Each can be either true or false.
Two variables give four combinations.

$$A, B \Rightarrow 0{,}0;\ 0{,}1;\ 1{,}0;\ 1{,}1$$

In the tabular form.

a	b
0	0
0	1
1	0
1	1

Figure 2.4 Two-variable truth table.

n Variables
In general,
Number of combinations

$$N = 2^n$$

n is the number of variables.

We can use a truth table to demonstrate the truth of a proposition. We saw one example earlier. Let's now use another to illustrate the application of De Morgan's Law. Recall that De Morgan's Law states:

> The complement of the *union* (logical *or*) of two sets is equal to the *intersection* (logical *and*) of their complements and the complement of the intersection (logical *and*) of two sets is equal to the *union* of their complements.

	LHS		RHS
a b	a • b	~(a • b)	~a+~b
0 0	0	1	1
0 1	0	1	1
1 0	0	1	1
1 1	1	0	0

Figure 2.5 Verifying De Morgan's Law.

On the left-hand side of the truth table in Figure 2.5, we have the expression $a \bullet b$.

- The value of that expression for each the four combinations of a and b
- The value of the complement of that expression $\sim(a \bullet b)$ for each of the four input combinations

On the right-hand side, we have the expression ~a + ~b.

- The value for each of the four of the combinations of a and b

The results on the left- and right-hand sides agree and thereby confirm De Morgan's Law.

Let's now look at a more complex relationship. Consider serving tea from a vending machine (Figure 2.6).

Tea ⇒ T
Lemon ⇒ L
Milk ⇒ M
Good Tea ⇒ GT
Bad Tea ⇒ BT

	TLM	Good tea	Bad tea
0	0 0 0	0	1
1	0 0 1	0	1
2	0 1 0	0	1
3	0 1 1	0	1
4	1 0 0	1	0
5	1 0 1	1	0
6	1 1 0	1	0
7	1 1 1	0	1

Figure 2.6 De Morgan's tea vending machine.

How many variables and how many combinations can we now extract from the truth table?

From the combinations of tea, lemon, and milk expressed in the table, we have three combinations designated to be good tea and five designated as bad. These lead to the two logic Eqs. (2.1) and (2.2) that follow.

$$\mathrm{GT} = 1 = \mathrm{T} \sim \mathrm{L} \sim \mathrm{M} + \mathrm{TL} \sim \mathrm{M} + \mathrm{T} \sim \mathrm{LM} \tag{2.1}$$

$$\mathrm{BT} = 1 = \sim \mathrm{T} \sim \mathrm{L} \sim \mathrm{M} + \sim \mathrm{T} \sim \mathrm{LM} + \sim \mathrm{TL} \sim \mathrm{M} + \sim \mathrm{TLM} + \mathrm{TLM} \tag{2.2}$$

Minterms

We have two expressions written in the *Sum of products* form. Each product is called a *minterm* and is also known as an *implicant*. They are also written as follows:

$$\mathrm{GT} = \Sigma 4, 5, 6 \qquad (100), (101), (110)$$

$$\mathrm{BT} = \Sigma 0, 1, 2, 3, 7 \quad (000), (001), (010), (011), (111)$$

or

$$\mathrm{GT} = m_4 + m_5 + m_6$$

$$\mathrm{BT} = m_0 + m_1 + m_2 + m_3 + m_7$$

where 4, 5, 6, and so on are binary equivalents of minterm variables. Thus, the digit 4 expresses the binary pattern (1 0 0) corresponding to the binary equivalent of the number 4 and the sequence T ~ L ~ M. We get these by identifying all terms for which the logical expression has a truth value of 1.

Let's now simplify the two expressions:

$$\begin{aligned}
\mathrm{GT} &= \mathrm{T} \sim \mathrm{L} \sim \mathrm{M} + \mathrm{TL} \sim \mathrm{M} + \mathrm{T} \sim \mathrm{LM} && A + A = A \\
&= \mathrm{T} \sim \mathrm{M}(\sim \mathrm{L} + \mathrm{L}) + \mathrm{T} \sim \mathrm{L}(\mathrm{M} + \sim \mathrm{M}) && A + \sim A = 1 \\
&= \mathrm{T} \sim \mathrm{M} + \mathrm{T} \sim \mathrm{L} && A \bullet 1 = A \\
&= \mathrm{T}(\sim \mathrm{M} + \sim \mathrm{L})
\end{aligned}$$

$$\begin{aligned}
BT &= \sim T \sim L \sim M + \sim T \sim LM + \sim TL \sim M + \sim TLM + TLM \\
&= \sim T \sim L(\sim M + M) + \sim TL(\sim M + M) + TLM && A + \sim A = 1 \\
&= \sim T \sim L + \sim TL + TLM && A \bullet 1 = A \\
&= \sim T(\sim L + L) + TLM \\
&= \sim T + TLM && A + \sim A = 1 \\
&= \sim T + LM && A + \sim AB = A + B
\end{aligned}$$

Maxterms

A term called a *maxterm* or *product of sums* expresses the complement of the corresponding *minterm*. Therefore, we write:

$$\text{Maxterm } M_i = \sim m_i$$

We get a different form of expression if we cover 0s, that is, terms for which the value is False.

We now write the expressions for the alternative tea selections in the *maxterm* format, which can also be written as:

$$\begin{aligned}
GT &= \Pi\, 0, 1, 2, 3, 7 \\
BT &= \Pi\, 4, 5, 6
\end{aligned}$$

where the symbol Π indicates product or:

$$\begin{aligned}
GT &= M_0 \bullet M_1 \bullet M_2 \bullet M_3 \bullet M_7 \\
BT &= M_4 \bullet M_5 \bullet M_6
\end{aligned}$$

These give a listing of the terms for which the logical expression is 0.

Using De Morgan's Law,

$$\begin{aligned}
GT &= 0 \\
&= (T + L + M)(T + L + \sim M)(T + \sim L + M)(T + \sim L + \sim M)(\sim T + \sim L + \sim M)
\end{aligned}$$

$$\begin{aligned}
BT &= 0 \\
&= (\sim T + L + M)(\sim T + \sim L + M)(\sim T + L + \sim M)
\end{aligned}$$

2.5 Implementation

Let's now go into building these logical functions. Looking at the algebraic operations, we see three functions are necessary.

AND
OR
NOT

The physical hardware used to implement such relations is called gates or logic gates. Such a term was probably derived from the concept of gating a signal or allowing a signal to pass. We will learn to implement such logic devices using threshold logic, fuzzy logic, and perceptron logic in the future chapters.

We have used three logic symbols in Figure 2.7 to represent such logic gates and the corresponding logical functions. Note that two devices are illustrated with two inputs each; the physical devices can be implemented or drawn with any number of inputs. Practical values are typically 2, 3, or 4.

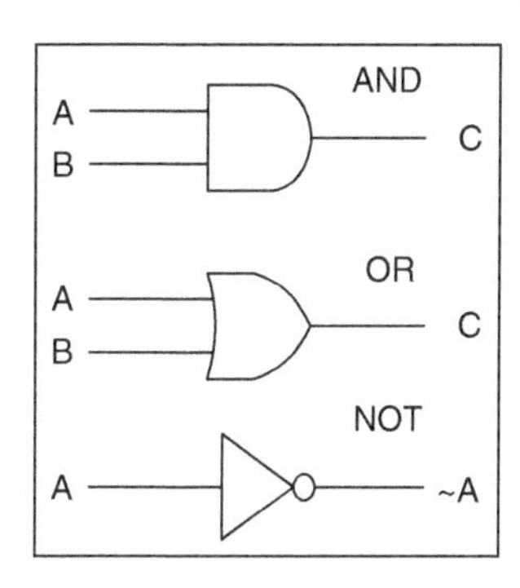

Figure 2.7 Basic logic devices.

Accompanying such logical functions, we may use additional pieces of supporting hardware, which are termed resistors, capacitors, and transistors.

The following examples illustrate how logical operations are physically implemented.

Example 2.1

Consider the logical function in Figure 2.8

$$F = A \bullet B + C \bullet D \tag{2.3}$$

The function can be decomposed as follows:

$$F = S + T$$
$$S = A \bullet B$$
$$T = C \bullet D$$

First, we implement the OR part: $S + T$.

Then, the two AND parts: $A \bullet B$, $C \bullet D$.

Finally, we bring them together as shown in Figure 2.9. The outputs of the two AND gates serve as inputs to the OR gate to produce the output F.

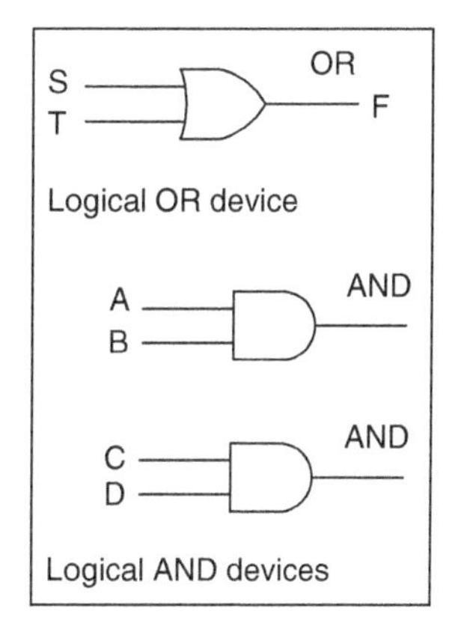

Figure 2.8 Logic gates.

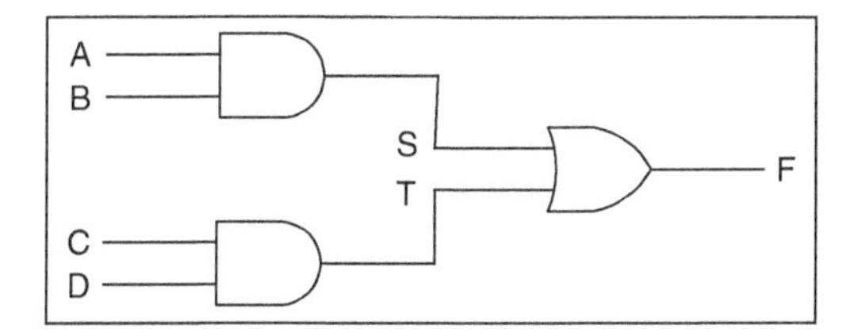

Figure 2.9 Logic circuit.

Example 2.2

Now, as shown in Figure 2.10, let's try:

$$F = A \sim B$$

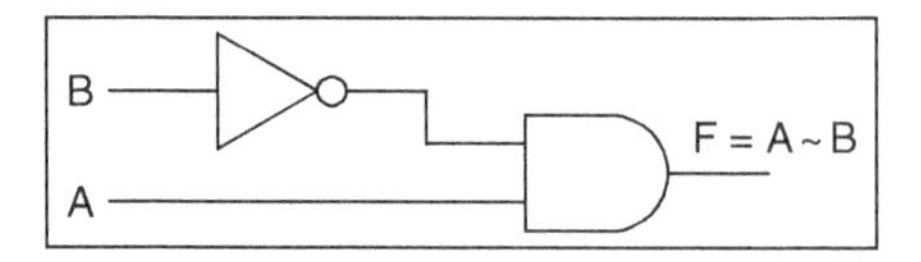

Figure 2.10 Logic circuit.

Example 2.3

Let's now return to the tea problem.

The output is implemented as:

$$\mathrm{T}(\sim \mathrm{M} + \sim \mathrm{L}) \rightarrow (\mathrm{T} \sim \mathrm{M}) + (\mathrm{T} \sim \mathrm{L})$$

Observe in Figure 2.11 that the output could have been written as:

$$\mathrm{T}(\sim \mathrm{M} + \sim \mathrm{L}) \rightarrow \mathrm{T} \sim (\mathrm{ML})$$

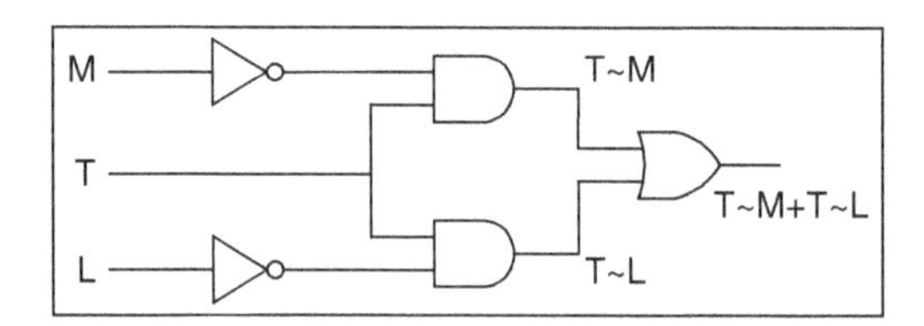

Figure 2.11 Tea circuit.

2.6 Logic Minimization

As we mentioned earlier, when we design something, we would like the circuit to be as simple as possible for a number of reasons: cost, power, reliability, weight, and size with safety and reliability considered as prime concerns.

2.6.1 Algebraic Means

We use the theorems and postulates discussed earlier to aid in minimizing the circuit logic. Let's look at some examples.

Example 2.4

$$F = \Sigma 1, 3, 7$$

When we write this expression, how many variables do we need? We have three real numbers that can be expressed as minterms and written as:

$$F = \sim A \sim BC + \sim ABC + ABC \qquad // \ 001 + 011 + 111$$

Observe that the choice of specific variables is arbitrary in this case.
Simplifying gives:

$$\begin{aligned} F &= \sim AC(\sim B + B) + BC(\sim A + A) \\ &= \sim AC + BC \\ &= C(\sim A + B) \end{aligned}$$

2.6.2 Karnaugh Maps

A *Karnaugh maps,* also known as a *K-map,* K-Map, or KM, give us a graphical or pictorial method for grouping algebraic expressions with common factors together. We can use such a view to identify, visually apply the rules of Boolean algebra, and thus potentially eliminate unwanted variables.

We can also view the *Karnaugh map* as a special arrangement of a truth table. The tool was developed by Maurice Karnaugh and introduced in 1953. The K-Map can be used for manipulating both crisp and fuzzy logic expressions.

2.6.2.1 Applying the K-Map

As a first step, let's now look at the following equation. How can it be reduced?

$$F = A \sim B + AB = 2 + 3$$

How about these three?

$$\begin{aligned} G &= AB \sim C + ABC = 6 + 7 \\ H &= A \sim BC + ABC = 5 + 7 \\ I &= \sim ABC + ABC = 3 + 7 \end{aligned}$$

Let's also recall a couple of truth tables, one with a slight reordering yet keeping the same terms. Looking at Figure 2.12, what do we see between adjacent rows with new ordering? Looking at the ordering of the terms in the two new orderings, we see that the sequences are using a *Gray code,* that is, a code with a single bit (or digit) change between adjacent rows.

	A	B
0	0	0
1	0	1
2	1	0
3	1	1

	A	B
0	0	0
1	0	1
3	1	1
2	1	0

	A	B	C
0	0	0	0
1	0	0	1
2	0	1	0
3	0	1	1
4	1	0	0
5	1	0	1
6	1	1	0
7	1	1	1

	A	B	C
0	0	0	0
1	0	0	1
3	0	1	1
2	0	1	0
6	1	1	0
7	1	1	1
5	1	0	1
4	1	0	0

Figure 2.12 Basic truth tables.

Observe

If we combine minterms in any two adjacent rows in the reordered truth tables, one variable drops out.

Case I

$$\text{Rows } 0+1 \rightarrow 00+01 =\sim A$$
$$\text{Rows } 1+3 \rightarrow 01+11 = B$$
$$\text{Rows } 3+2 \rightarrow 11+10 = A$$

Tricky one, the top and bottom rows

$$\text{Rows } 0+2 \rightarrow 00+10 =\sim B$$

Case II

$$\text{Rows } 0+1 \rightarrow 000+001 =\sim A \sim B$$
$$\text{Rows } 1+3 \rightarrow 001+011 =\sim A \bullet C$$

etc.

2.6.2.2 Two-Variable K-Maps

A	0	1	B
0	0	1	
1	2	3	

Let's put this to work. We begin with two variables and create a matrix-like map. Observe that each cell has two adjacent cells and those cells differ by a single variable. The cells are numbered as minterms. Note that this is a Gray sequence.

A	0	1	B
0	1	1	
1	0	0	

Let's go back to Case I. Let minterms 0 and 1 be True and 2 and 3 be False. We enter this information into the map as seen in the adjacent diagram.

Now the simplification of the expression

$$F = \sim A \sim B + \sim AB =\sim A$$

is the same as looking at the map and combining the two true minterms.

A	0	1	B
0	0	0	
1	1	1	

Now, let's reverse the two rows. Now, the simplification of the expression

$$F = A \sim B + AB = A$$

once again is the same as looking at the map and combining two true minterms.

A	0	1	B
0	0	1	
1	1	0	

A	0	1	B
0	1	0	
1	0	1	

These two maps contain minterms 1 and 2 and 0 and 3.

$$F = \sim AB + A \sim B$$

$$F = \sim A \sim B + AB$$

Now, observe that no logic simplification is possible. In general, to be able to combine the terms, they must be adjacent, that is, differ by single variable change.

2.6.2.3 Three-Variable K-Maps

We now extend to three variables. The map is basically the same. Observe that once again, we use a Gray code pattern as illustrated in the adjacent figure. Let's now look at Case II from Figure 2.13 again using three-variable maps.

A	B	0	1	C
0	0	0	1	
0	1	2	3	
1	1	6	7	
1	0	4	5	

Figure 2.13 Basic truth table.

A	B	0	1	C
0	0	1	1	
0	1	0	0	
1	1	0	0	
1	0	0	0	

$$\sim A \sim B \sim C + \sim A \sim BC$$

$$000 + 001 = \sim A \sim B$$

A	B	0	1	C
0	0	0	1	
0	1	0	1	
1	1	0	0	
1	0	0	0	

$$\sim A \sim BC + \sim ABC$$

$$001 + 011 = \sim AC$$

As we see from the K-Maps, these two reduce to ~*A*~*B* and ~*AC*.
Note that we are combining or grouping adjacent terms.

A	B	0	1	C
0	0	0	0	
0	1	0	0	
1	1	0	1	
1	0	1	1	

Let's now try a more complex pattern. From Figure 2.13, what does this give?

$$\begin{aligned} H &= \sum 4,5,7 \\ &= A \sim B \sim C + A \sim BC + ABC \\ &= 100 + 101 + 111 = A \sim B + AC \end{aligned}$$

A	B	0	1	C
0	0			
0	1	1		
1	1	1		
1	0	1	1	

In practice, we typically do not write the 0 terms. Let's now try:

$$\begin{aligned} J &= \sum 2, 4, 5, 6 \\ &= \sim AB \sim C + A \sim B \sim C + A \sim BC + AB \sim C \\ &= 010 + 100 + 101 + 110 = A \sim B + B \sim C \end{aligned}$$

A	B	0	1	C
0	0			
0	1	1		
1	1		1	
1	0	1	1	

Note that we have several ways of combining. Now we try:

$$\begin{aligned} K &= \sum 2, 4, 5, 7 \\ &= \sim AB \sim C + A \sim B \sim C + A \sim BC + ABC \\ &= 010 + 100 + 101 + 111 = \sim AB \sim C + A \sim B + AC \end{aligned}$$

Now observe that the top and bottom rows are adjacent, and we now have four terms that are adjacent. The resulting function is:

$$N = \Sigma 0, 1, 4, 5$$
$$N = \sim A \sim B \sim C + \sim A \sim BC + A \sim B \sim C + A \sim BC$$
$$= 000 + 001 + 100 + 101$$

A	B	0	1	C
0	0	1	1	
0	1			
1	1			
1	0	1	1	

$$N = \sim B(\sim A \sim C + \sim AC + A \sim C + AC)$$
$$N = \sim B(00 + 01 + 10 + 11)$$
$$N = \sim B(0 + 1)$$
$$= \sim B$$

2.6.2.4 Four-Variable K-Maps

Adding one more variable is not too much more complex, and certainly it adds more adjacencies. Exactly, where are new adjacencies? Now we have in Figure 2.14:

Corners

Reflecting minterms 0, 2, 8, and 10

Edges

Top and bottom
Left and right

A B	0 0	0 1	1 1	1 0	C D
0 0	0	1	3	2	
0 1	4	5	7	6	
1 1	12	13	15	14	
1 0	8	9	11	10	

Figure 2.14 Four-variable K-Map.

As with our earlier maps, the numerics are expressing the 16 possible patterns for a four-variable minterm. Let's examine several possible four-variable maps.

Let's start with the accompanying map in Figure 2.15. What is the best way to solve this? The K-Map is suggesting two groups of four minterms. The group in the upper left-hand corner reduces to *~A~C*, and the group in the lower right reduces to *AC*, which gives the final relation:

$$P = \sim A \sim C + AC$$

A B	0 0	0 1	1 1	1 0	C D
0 0	1	1			
0 1	1	1			
1 1			1	1	
1 0			1	1	

Figure 2.15 Four-variable K-Map.

The next two maps seem to offer more of a challenge. A closer examination reveals that they are not too difficult. Starting with the first map in Figure 2.16, we have

A B	0 0	0 1	1 1	1 0	C D
0 0	1			1	
0 1		1	1		
1 1		1	1		
1 0	1			1	

Figure 2.16 Four-variable K-Map.

a group of four adjacent variables in the center and a second group of four occupying the four corners. The center group reduces to *BD*, and the corners combine to *~B~D*, giving the final relation:

$$R = BD + \sim B \sim D$$

A B	0 0	0 1	1 1	1 0	C D
0 0	1			1	
0 1		1			
1 1			1		
1 0	1			1	

Figure 2.17 Four-variable K-Map.

In the third map in Figure 2.17, once again the four corners yield *~B~D*; however, the remaining two variables are not adjacent. Consequently, they cannot be combined with any of the other variables to yield a simpler expression. Thus, in the final relation, they appear as four-variable minterms:

$$S = \sim B \sim D + \sim AB \sim CD + ABCD$$

2.6.2.5 Going Backward

We have seen how to take a (reduced) function from the map. Let's see how to put one on and then remove it in simpler form.

Two Variables

Let's start by placing the following two-variable function onto the K-Map.

$$F = A + \sim AB + AB$$

A	0	1	B
0	0	1	
1	1	1	

Placing the variables onto the map as follows:

A is bottom row.
AB is minterm 3 // Placed in the lower right hand corner.
~AB is minterm 1 // Placed in the upper right hand corner.

Let's now take a reduced expression of the map.

$M = \prod(1, 2, 3)$ // The $\prod$ symbol indicates the logical OR.
Grouping the two terms in the bottom row gives us *A*.
Grouping the two terms in the right-hand column gives us *B*.

Grouping and extracting, we have the reduced expression:

$$F = A + B$$

Three Variables

A	B	0	1	C
0	0			
0	1			
1	1	1	1	
1	0	1	1	

Working with more variables is more interesting. Let's work with the following expression:

$$F = AB + A(\sim B + C)$$

First, placing the expression onto the K-Map.

The term *AB* is entered into the third row of the K-Map.
Expanding $A(\sim B + C)$ gives $A{\sim}B + AC$.
Entering these two expressions onto the K-Map places $A{\sim}B$ into row 4 and *AC* into the last two spots in the right-hand column.

A	B	0	1	C
0	0		1	
0	1			
1	1	1	1	
1	0		1	

Next extracting the reduced function, a group of four variables, from the map gives us:

$$F = A$$

Let's now try:

$$G = AB \sim C + AC + \sim BC$$

First, placing the expression onto the K-Map

The term $AB{\sim}C$ is entered into row 3 in column 1 of the K-Map.
The term *AC* is entered into rows 3 and 4 in column 2.
The term ${\sim}BC$ is entered into rows 1 and 4 in column 2.

Next extracting the reduced function from the map, which comprises both terms from row 3 and the terms from rows 1 and 4 in column 2, gives us:

$$G = AB + \sim BC$$

Four Variables

Moving to four variables, we'll start with the following function:

A	B	0	0	0	1	1	1	1	0	C	D
0	0							1			
0	1										
1	1	1		1				1			
1	0	1		1				1			

$$H = A \sim C + AB \sim D + \sim BCD$$

Entering the function onto a K-Map, we have:

The reduced function now becomes:

$$H = A \sim C + A \sim D + \sim BCD$$

A	B	0	0	0	1	1	1	1	0	C	D
0	0	1		1				1			
0	1			1		1					
1	1					1					
1	0										

Trying another function, we'll work with:

$$J = \sim A \sim B \sim C + \sim A \sim C \sim D + BCD + \sim A \sim BC \sim D$$

The reduced function now becomes:

$$J = \sim A \sim B \sim D + \sim A \sim C D + BCD$$

2.6.2.6 Don't Care Variables

As we have seen, three binary variables give us eight combinations.

If we have a system with three inputs, those inputs potentially can take on up to eight values (Figure 2.18). For one reason or another, some of those values may never occur. Under those circumstances, we don't care what values they take on. That is, we can give them values of 0 or 1 in a K-Map to help in simplification because we don't care what value such terms take on. Thus, we call them *don't cares*.

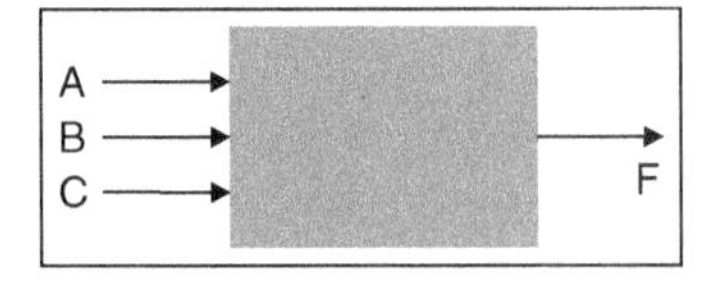

Figure 2.18 Three-input system.

A	B	0	1	C
0	0	1	1	
0	1			
1	1			
1	0		1	

Consider the following function:

$$F = \Sigma 0,1,5$$

We enter these minterms as 1s as usual.
As the map stands, we get:

$$F = \sim A \sim B + \sim BC$$

Now, assume that combinations given by minterms 4 and 6 never occur, or if they do, they are irrelevant. Thus, we have the don't care terms (dct):

A	B	0	1	C
0	0	1	1	
0	1			
1	1	X		
1	0	X	1	

$$\text{dct} = \Sigma 4, 6$$

We enter these on the K-Map as *x*'s.
Now, if we assign values of

0 for don't care minterm 6
1 for don't care minterm 4

With don't care terms, the function *F* reduces to:

$$F = \sim B$$

Let's now look at a more extensive example. We want to design a code translator that translates from BCD (binary coded decimal) to Excess 3 code. The values for the BCD to Excess 3 codes are given in the table in Figure 2.19.

Decimal number	BCD inputs				Excess 3 output			
	w	x	y	z	f4	f3	f2	f1
0	0	0	0	0	0	0	1	1
1	0	0	0	1	0	1	0	0
2	0	0	1	0	0	1	0	1
3	0	0	1	1	0	1	1	0
4	0	1	0	0	0	1	1	1
5	0	1	0	1	1	0	0	0
6	0	1	1	0	1	0	0	1
7	0	1	1	1	1	0	1	0
8	1	0	0	0	1	0	1	1
9	1	0	0	1	1	1	0	0

Figure 2.19 BCD to Excess 3 decoder.

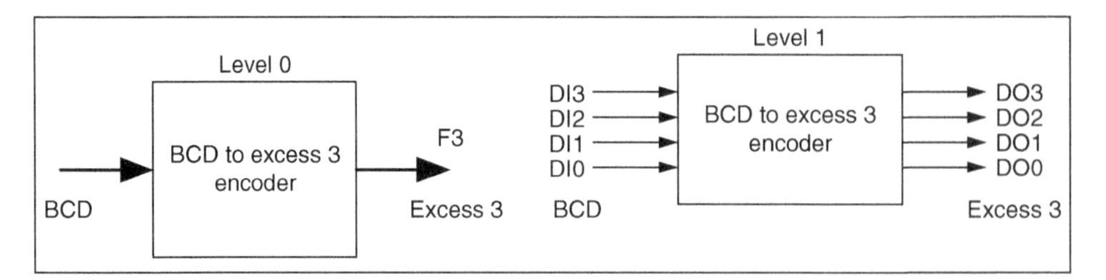

Figure 2.20 BCD to Excess 3 encoder block diagrams.

The diagrams in Figure 2.20 express a block diagram in two levels of detail.

Let's look at the logic for the Level 0 diagram, which has one output line that we'll specify as *F*3. We'll choose the output line from Figure 2.20. Note that the input minterms 10. . .15 are don't cares since these values do not occur in a *BCD* code.

We enter the function for the output term Excess 3, including the don't care terms in the following K-Map.

From the K-Map, we have:

$$F3 = B \sim C \sim D + \sim BD + \sim BC$$

A B	0 0	0 1	1 1	1 0	C D
0 0		1	1	1	
0 1	1				
1 1	x	x	x	x	
1 0		1	x	x	

2.7 Summary

In this chapter, beginning with the underlying concepts of crisp logic, we opened with the definition of an algebra or algebraic system and reviewed the fundamentals of Boolean algebra. We followed with the Huntington postulates and worked through their proofs.

We then discussed the concept of a truth table and demonstrated algebraic proof using such tables. We then learned that the entries in such a table are called minterms and that a minterm is a binary aggregate of logical 0s and 1s that sets the logical value, true or false, of single cell entries in truth tables.

Moving from the basic truth table, we introduced and reviewed the Karnaugh Map, or K-Map. The K-Map is a pictorial tool for grouping logical expressions with shared or common factors. Such sharing enables the elimination of unwanted variables, thereby simplifying a logical expression. As a further addition to the K-Map, the concept of *don't care* variables was introduced. Such variables are determined to never occur in an input pattern and thus can be entered into a K-Map and used to help simplify the logic yet never affect the state of the system's logical output.

The material that we have covered here lays the groundwork for relaxing the precision of classic logic and introduces concepts and tools similar to those that we'll apply and work with in the fuzzy world.

In our next chapter, we will explore and learn to work with sets and set operations.

Review Questions

2.1 According to the chapter, what is an algebra or algebraic system?

2.2 Identify and explain each of the Huntington postulates.

2.3 From the list of fundamental theorems, choose six and explain what each means.

2.4 When designing logic circuits, one of our major objectives is simply simplicity. Why is this the case, and what does it mean?

2.5 What are other important criteria when executing a design?

2.6 How many kinds of basic operators does Boolean algebra have? What are these? How many values do operands have? What are these?

2.7 What is a truth table? A truth-table proof? For what might these be used?

2.8 How many variables can a truth table have?

2.9 What are minterms and maxterms? Explain their use.

2.10 What are the three basic logic devices or types of gates? Using truth tables, illustrate the behavior of each device.

2.11 What is a Karnaugh Map, and what is it used for?

2.12 Explain how a Karnaugh Map can be used to simplify a logic expression.

2.13 In a Karnaugh Map, what is a *don't care* variable?

3

Crisp Sets and Sets and More Sets

THINGS TO LOOK FOR...

- The fundamental concept of sets.
- Sets and set membership.
- Universe of discourse.
- Concept of classic or crisp sets and set membership.
- Classic or crisp set membership and characteristic or membership functions.
- The foundations of classic or crisp logic.
- Fundamental terminology.
- The properties of and logical operations on crisp sets.
- Basic crisp membership applications.

3.1 Introducing the Basics

In the last chapter we reviewed Boolean algebra. We now raise this question: What is Boolean algebra, and what is its purpose? As engineers, our main goals are to recognize, understand, and solve problems. To aid in our quest, we seek tools that will facilitate that process. Boolean algebra brings us some of those tools in the form of variables and equations. Variables give us a tool to identify and to quantify real-world entities. Equations give us a means to specify and work with relations among those variables as we pursue our goal and the challenge of accurately solving problems.

The immediate question is how much information does a variable give us. Let's hypothesize the problem of measuring the circumference of Australia. Working with a tool that is 500 km long, how accurate is the measurement going to be? How accurate will the measurement be if that tool is replaced with one that is 5 m long or one that is 500 km long with marks every 1 m or 1 cm? The resolution of data values is an important difference between classic logic and fuzzy logic.

Classic logic has been a very powerful tool for millennia and remains so. As technology evolves, however, the tools needed to solve problems must also evolve. A major difference between classic and fuzzy logic is that a classic logic variable is either true or false, whereas a fuzzy variable can have degrees of truth or falseness. This difference will become more significant as we move into sets and set membership.

Introduction to Fuzzy Logic, First Edition. James K. Peckol.

A common example used for illustrating the difference between crisp and fuzzy logic is an individual's tallness. In the classic world, the person is tall or not tall. In the fuzzy world, the person can be not tall, not too tall, somewhat tall, tall, very tall, or very very tall, etc.

With that brief introduction, in this chapter and the next, we introduce and examine classical and fuzzy logic, sets, and the concept of set membership for both classical and fuzzy sets and subsets. More formally, we then take the first step on the road to getting acquainted with the logical operations on and with fuzzy sets and subsets. We should all feel comfortable talking about the concept of sets in general. Remember that sets play an important role in the field of mathematics; however, they are not limited to that world.

Let's begin by introducing a few familiar sets. . .

- I have a set of colored crystal wine glasses.
- Oops, I dropped them and broke the red subset.
- She has a set of silver eating utensils.
- He has a set of Dicken's works.
- This set of tires is perfect for my car.
- This set of numbers is the combination to his bicycle lock.

Typically, we see a set simply as a collection of similar objects of one sort or another. However, do we understand or can we explain or characterize the concept of sets? Do we understand when something can or cannot be included in or as a member of a set? Is that decision important? Can a wine glass be included in the set of Dicken's works or can one of Dicken's works be used as eating utensils? Let's explore.

As we explore, we enter the portion of the mathematical world of logic known as *set theory*. The modern study of this field was initiated in the 1870s by Richard Dedekind and Georg Cantor. As we may have surmised, any comparable or similar type of objects such as those described above can be collected into a set. However, most often we find *set theory* linked with the field of mathematics.

We say that an object is or is not a *member* of a specific set. In formal terminology, we have a *binary relationship* between a set and an object. An essential key point is that for any such specific collection, the properties of *all* of the members are well defined and constrained to meet specified properties or characteristics within that collection.

Further, in formal terms, when such a constrained collection is referred to or discussed, it is designated the *domain* or *universe of discourse*. Such a domain comprises the total range of ideas, entities, attributes, events, relations, etc. that are expressed, assumed, implied, or hypothesized with respect to that collection during the course of the discussion.

Consider the following three collections or sets in Figure 3.1. Looking at the individual sets, what can we say about the members in each? We can see that each of these sets

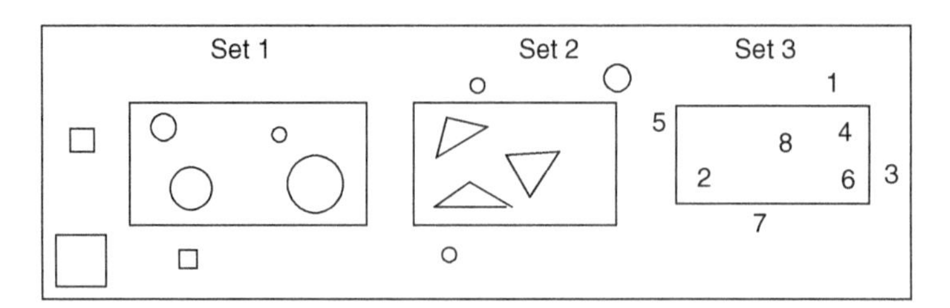

Figure 3.1 Three sets.

contains various sizes and shapes of different geometric figures and that each set contains a single type of figure. The surrounding rectangles merely delimit the sets; they are not part of the sets.

Let's first examine the members of Sets 1 and 2. The figures (members or entities) within each of these two specific sets or *universes of discourse* are all of the same type, either circles or triangles but no other types. They are also of different sizes and different orientations; nevertheless, each is an instance of a specific type of geometric figure.

Such types of requirements seem important, and for our work, they will be. A further apparent restriction is that a circle, or any other figure, is not a member of the triangle set or universe, and a triangle, or any other figure, is not a member of the circle set or universe.

In the region around the two sets, we also find various sizes and shapes of different geometric figures. Based on the evidence we see, squares cannot be a member of any of the three sets, whereas the various circles could be a member of the circle set. Why is there a difference with respect to which figure is accepted into a specific set, and what establishes that difference?

The members of Set 3, also delimited by a rectangle, are integer numbers, all of which are even numbers, i.e. divisible by 2. Surrounding Set 3, but not within the rectangle, we also find a collection of integer numbers. This external collection, also a possible set, contains only odd numbers. Thus, based on the graphic evidence, its members, apparently, cannot be members of any of the other sets.

Looking at our three sets under discussion, empirically we can associate a *degree* or *grade of membership* descriptor or qualifier for the objects within a specific set and a similar descriptor or qualifier for the objects outside of that set that don't meet the constraints or criteria of the designated set. We're seeing a pattern.

Formally, for *classic* or *crisp* sets, the grade of membership qualifier for an object that meets a particular set's restrictions or requirements is set to "1" or [1], which thereby indicates permissible membership in that set. For those that do not meet a particular set's qualifications, the grade of membership qualifier for that object is set to "0" or [0], which thereby indicates nonmembership in that particular set. Nonmembership for a particular set, however, does not preclude the possibility of membership in a different set.

Once again, if we empirically consider each of the sets in Figure 3.1.

Set 1

Contains four circles of different sizes. Apparently, to be a member of Set 1, an object or entity must be a circle; size apparently doesn't matter. Since the objects in the Set 1 universe are all circles, the designated *grade of membership* for each contained object must be 1 and that for all other entities, three squares and four odd integers, the grade of membership in the Set 1 universe must be 0. What about the three circles outside of Set 1? Could they be a member of Set 1? We do not consider members of Set 2 or Set 3.

Set 2

Contains three triangles of different sizes. Apparently, to be a member of Set 2, an object must be a triangle. Since the objects in the Set 2 universe are all triangles, independent of type, the grade of membership for each contained object is [1] and that for all other entities, three squares, three circles, and four odd integers, membership in Set 2 is [0]. We do not consider members of Set 1 or Set 3.

Set 3

Contains four different even integers. Apparently, to be a member of Set 3, an object must be an even integer. Since the objects in the Set 3 universe are all even integers, the grade of membership for each contained object is [1] and that for all other entities: three squares, three circles, and four odd integers, the membership in Set 3 is [0]. We do not consider members of Set 1 or Set 2.

3.2 Introduction to Classic Sets and Set Membership

Now that we've begun to explore classic sets and set membership from a high-level perspective, let's now venture into the formal details.

3.2.1 Classic Sets

Let's examine the concept of sets and set membership for classical or crisp sets more formally. We then examine fuzzy sets and set membership in the next chapter.

3.2.2 Set Membership

We begin with a review of some of the basic definitions and concepts of the theory of classical or crisp sets by defining set membership for classical sets.

In the opening paragraphs of this chapter, we identified a set or subset as a collection of objects and asserted that a particular object may or may not be a member of a certain set or subset. In the classical world, the degree or extent of set membership is defined as binary, *1* or *0*, *true* or *false*. That is, there are only two possibilities, an element *does* or *does not* meet the requirements to be a member of a particular set; there is no middle ground.

Formally:

Let

S be an arbitrary set of elements
T be any subset of S

A subset T of S is formally defined as follows:

Definition 3.3.1:

A subset T of S is the set of ordered pairs, (x_{1i}, x_{2i}), such that x_{1i} is an element of S and x_{2i} is an element of the set $\{0,1\}$.

The element x_{2i} shall have the value 1 if x_{1i} is in T and 0 otherwise.

Accordingly, the subset T will contain, at most, one and only one ordered pair (x_{1i}, x_{2i}) for each element of S. Such a relationship defines a subset T of S as a mapping from S into the set $\{0,1\}$. As a result, if S is the universe of discourse, the truth or falsity of any statement $x1_i$ in T with respect to S is determined by the ordered pair having the statement $x1_i$ as its first element and $x2_i$ as the second element. The statement is *TRUE* if the second element $x2_i$ is 1 and *FALSE* otherwise.

Let us now examine the concept of set membership for crisp subsets more formally. We begin with a brief review of some of the basic definitions and concepts of the theory of classical or crisp sets and then extend these to the theory of fuzzy subsets in Chapter 4.

A classical or crisp set is a set with well-defined boundaries that contains a collection of entities that have common properties or characteristics that distinguish them from other entities outside of the set that lack those properties. Such entities are defined as *members* of the set. They can potentially be further collectively defined as a *subset* of a *superset* that is also called a *universe of discourse*. The term universe of discourse refers to a collection of entities that are currently being discussed, analyzed, examined, etc. as illustrated in Figure 3.2.

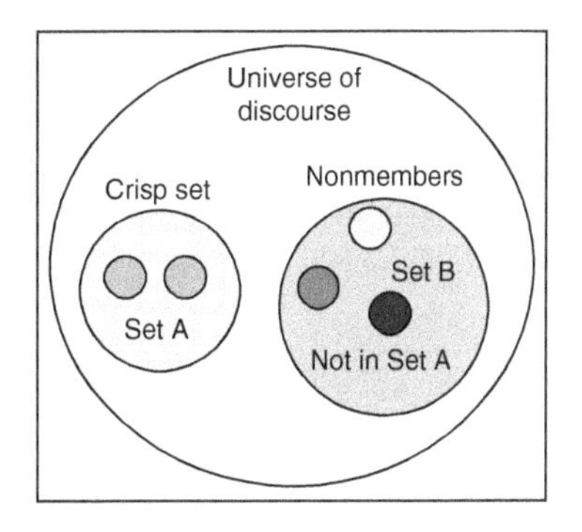

Figure 3.2 Crisp set and universe of discourse.

As we see in Figure 3.2, at the fundamental level, a *universe of discourse* with respect to a set in this classical world is split into two segments: one that contains members of the set (Set A) and one that contains nonmembers (Set B). A universe of discourse, however, is not constrained to containing only a single crisp set. Consider a discussion about the best after-dinner drink: beer, wine, tea, or coffee. Here, we have four different crisp sets. We further note that a particular entity within a universe of discourse may or may not be a member of a particular subset or set.

In the classical world, we have learned that the degree or extent of set membership is defined as binary, *true* or *false*. That is, there are only two possibilities, an element *is* or *is not* a member of a particular set; there is no middle ground. Plato was right. The *degree of membership* of an entity is specified as 0 if it is not in the set or 1 if it is in the set. Classic or crisp set membership can be expressed in several different ways:

1) *Mathematically*, where the expression $\mu_A(x)$ is a membership function for the set A.

$\mu_A(x) = 0$	if $x \notin A$	//x is not a member of A
$\mu_A(x) = 1$	if $x \in A$	//x is a member of A

2) *Graphically*, as illustrated in Figure 3.3.

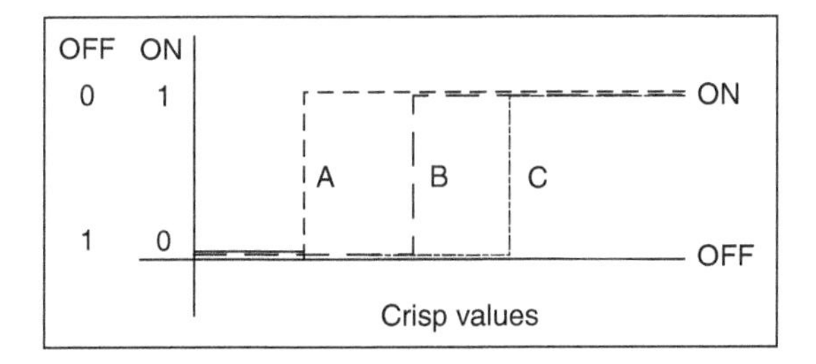

Figure 3.3 Two sets of light bulb states.

Now consider that we have a room with three light bulbs as in Figure 3.4. Let S be our *universe of discourse* with two crisp sets, P and Q. The state of the light bulbs will be our discriminating factor for set membership. The members in set P will be the light bulbs that are OFF and the members in set Q the bulbs that are ON. It is important to note once again that when working with classic or crisp sets, set membership values must be either 0 or 1. Intermediate values are not permitted.

When we first enter the room, all the lights are OFF. Thus, we have the set, P, with three members (a, b, c), those lights that are OFF, and the set Q with no members, no lights that are ON. If the lights are successively turned on, we move through the following sequence of set memberships for P and Q:

	P (a, b, c)	Q (a, b, c)
1.	OFF {1, 1, 1},	ON{0, 0, 0}
2.	OFF {0, 1, 1},	ON{1, 0, 0}
3.	OFF {0, 0, 1},	ON{1, 1, 0}
4.	OFF {0, 0, 0},	ON{1, 1, 1}

We start with the *universe of discourse* S, which is composed of the elements "s_i." The subscript i simply tags or identifies each of the elements in S.

Then, we define P and Q as crisp subsets of S. The term *support* or *sup* identifies all elements "s_i" whose value meets the criteria for acceptance into a crisp subset of S. Membership in the designated subset is supported.

$S = \{A, B, C\}, s = \{0 \text{ or } 1\}$

$P = \Sigma_{sup} S_i\, \mu P(S_i)/s$ // P will comprise the accepted elements S_i

$Q = \Sigma_{sup} S_i\, \mu Q(S_i)/s$ // Q will comprise the accepted elements S_i

$\sup(S_i)P = \{S_i \in S/\mu P(S_i) = 1\}$ // if $s_i \in S$ and the light is OFF

$\sup(S_i)Q = \{S_i \in S/\mu Q(S_i) = 1\}$ // if $s_i \in S$ and the light is ON

The expressions $\mu P(s_i)$ and $\mu Q(s_i)$ are membership functions for sets P and Q, respectively. They stipulate the degree of membership of the specific element (s_i) in the designated set. Once again, for crisp sets, the membership degree will be either 0 or 1.

Using the set membership functions $\mu_P(s)$ and $\mu_Q(s)$, where s is a working variable referring to one of the members of the universe of discourse S, we write:

1) $\mu_P(s) = \{(a,1), (b,1), (c,1)\}$ // 3 members, 3 lights OFF
   $\mu_Q(s) = \{(a,0), (b,0), (c,0)\}$ // 0 members, 0 lights ON
2) $\mu_P(s) = \{(a,0), (b,1), (c,1)\}$ // 2 members, 2 lights OFF
   $\mu_Q(s) = \{(a,1), (b,0), (c,0)\}$ // 1 member, 1 light ON
3) $\mu_P(s) = \{(a,0), (b,0), (c,1)\}$ // 1 member, 1 light OFF
   $\mu_Q(s) = \{(a,1), (b,1), (c,0)\}$ // 2 members, 2 lights ON
4) $\mu_P(s) = \{(a,0), (b,0), (c,0)\}$ // 0 members, 0 lights OFF
   $\mu_Q(s) = \{(a,1), (b,1), (c,1)\}$ // 3 members, 3 lights ON

We now look at some of the operations that we can apply to crisp sets. We'll begin by defining two sets, X{1, 2, 3, 4, 5} and Y{2, 4, 7, 8}. We note that there are four basic operations that are fundamental to both classical and fuzzy sets, which are given as follows:

1) Union
   The result of the *union* of two sets (X) and (Y) includes all elements in set X *or* Y *or* both.
   The union operation is designated: $X \cup Y$. The union of the two sets will be {1, 2, 3, 4, 5, 7, 8}. The union is a logical *OR* operation.
2) Intersection
   The result of the *intersection* of two sets (X) and (Y) includes all elements that are in both sets X *and* Y.
   The intersection operation is designated: $X \cap Y$. The intersection of the two sets will be {2, 4}. The intersection is a logical *AND* operation.
3) Complement
   The result of the *complement* of a set (X) includes all elements not in the designated set (X).
   The complement operation is designated: X^c or $\sim X$. Assume that we are dealing with the two sets (X) and (Y) and that we are only dealing with 8 numbers. The complement X^c of (X) will be {6, 7, 8} and Y^c of (Y) will be {1, 3, 5, 6}.
4) Difference
   The result of the *difference* between two sets (X) and (Y) includes all elements in set X and not in Y or vice versa.
   The difference between (X) and (Y) is {1, 3, 5} and between (Y) and (X) is {7, 8}.

The four basic operations are illustrated graphically using what are called Venn diagrams in Figure 3.4.

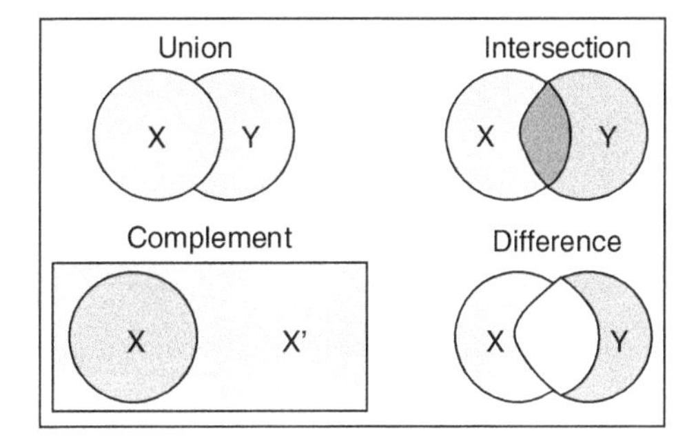

Figure 3.4 Four basic logical operations.

3.2.3 Set Operations

Let's now put what we've just learned to work. Consider, once again, our room with three light bulbs, and we define two possible sets based on the state of the light bulbs. One set contains those that are OFF, and a second set contains those that are ON. Classic or crisp set membership can be expressed graphically as illustrated in Figure 3.3.

When we first enter the room, all the lights are OFF. Thus, we have one set with 3 members (A, B, C), indicating lights that are OFF, and one set with 0 members, indicating lights that are ON. As the lights are successively turned on, we move through the following sequence of set memberships:

1) OFF {1, 1, 1} , ON{0,0,0}
2) OFF {0, 1, 1} , ON{1,0,0}
3) OFF {0, 0, 1} , ON{1,1,0}
4) OFF {0, 0, 0} , ON{1,1,1}

The same three binary bits can be organized into two sets and used for specifying and controlling a motor speed {stop, slow, medium, fast} and turning a direction

{clockwise, counterclockwise}.

The two sets {clockwise, counterclockwise} and members (0.7) are given as:

Clockwise		*Counterclockwise*		
0 00	(0)	1 00	(4)	Stop
0 01	(1)	1 01	(5)	Slow
0 10	(2)	1 10	(6)	Medium
0 11	(3)	1 11	(7)	Fast

The universe of discourse and two subsets are given in Figure 3.5. Encircled entities from the universe of discourse indicated by the dashed line and tagged with a 0 cannot be entered

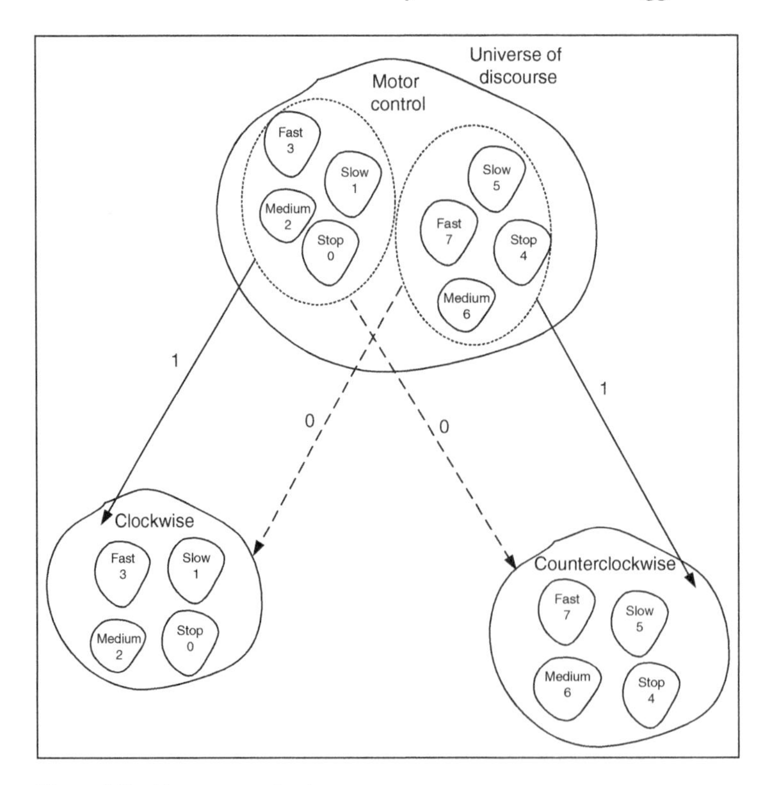

Figure 3.5 Motor control subsets.

into the target subset. Those indicated by the solid line and tagged with the 1 can be entered into the target subset.

It is important to stress once again that when working with classic or crisp sets, the degree or grade of membership must be either [0] or [1]. Intermediate values are not permitted. We now look at some of the operations that apply to classic or crisp sets.

3.2.4 Exploring Sets and Set Membership

Building on the following concepts:

- The three laws of thought: the *Law of Identity*, the *Law of Excluded Middle*, and the *Law of Contradiction*
- The early work of Georg Cantor and Richard Dedekind on set theory in the latter quarter of the 1800s
- The later work of George Boole

we move into the mathematical domain known by the names *Boolean, classical,* or *crisp* logic. Such a logic is built upon the basic concept of a *set*. We shall begin by defining set membership for classical sets and then extend the idea to fuzzy subsets in the next chapter. Classical sets are considered crisp because their members satisfy precise properties.

3.2.5 Fundamental Terminology

Each component, element, or entity in or of a set is called a *member*. That entity is said to have *membership* in the named set. The concepts of *empty* and *size*, the relationships of *equality* and *containment*, and the operations of *union*, *intersection*, and *negation* can be defined on a set or on a collection of sets.

The following basic operators and symbols in Figure 3.6 are traditionally used in specifying relations and operations on classic or crisp sets and set members.

3.2.6 Elementary Vocabulary

Let's continue in the crisp world with some of the elementary vocabulary and operations that we find in basic/underlying mathematics. As we wander through that world, we encounter various collections of numbers, words, sounds, observations, measurements, and physical objects. We often refer to such collections as *information* or *data*.

In our very early years, our forebears discovered useful ways to manipulate such data to understand and gain great varieties of knowledge to solve contemporary problems. Today, we have what we call equations, procedures, or operations that we also apply to or on such data that we call *operands* to gain knowledge and, once again, to solve problems. The simple function is a good starting example.

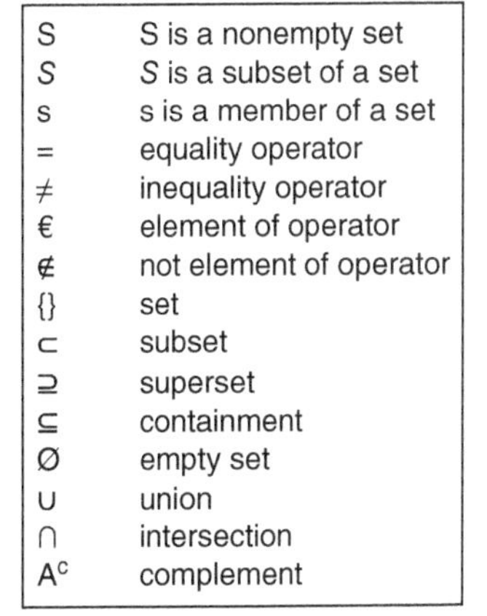

S	S is a nonempty set
S	*S* is a subset of a set
s	s is a member of a set
=	equality operator
≠	inequality operator
∈	element of operator
∉	not element of operator
{}	set
⊂	subset
⊇	superset
⊆	containment
Ø	empty set
∪	union
∩	intersection
A^c	complement

Figure 3.6 Basic crisp set symbols.

We write such a function as follows.

$$f(x) = y$$

f is a mathematical operation being performed.
x is the *domain* of the function, that is, the entity or entities to which we are applying the function.
y is the *range* of the function, the entity or entities that result from applying the function.

Let the basic integer addition operation (indicated by the operator +) be our function and let the set of integers $X = (1, 2, 3, 4, 5, 6)$ be the elements of the domain to which we apply the function.

Thus, we have our function and the domain to which the function is repeatedly applied:

$$(1+2) \rightarrow (3)$$

$$(3+4) \rightarrow (7)$$

$$(5+6) \rightarrow (11)$$

to yield the range:

$$(3, 7, 11).$$

The operations of subtraction, multiplication, and division follow similarly.

Observe that our function operated on two entities or integers at a time. Such operations or relations are termed *binary*.

We now form the set A_0 based on the entities X_i and label the individual members, $i = \{1 \ldots 6\}$.

$$A_0 = \{x_1, x_2, x_3, x_4, x_5, x_6\}$$

In the crisp or classic world, we now define a *characteristic function*, $\mu_{A0}(x_i)$, to designate a member of the set A_0. We read: x_i *is a member of the set* A_0. In the fuzzy world, the term *characteristic function* is typically called the *membership function* and the associated sets called *fuzzy sets*.

If the designated member is not in the set, the *characteristic or membership function* returns a 0; otherwise, it returns a 1. For the set A_0, selecting any of the six members will return a 1. Thus, as a result, for each term that is a member of the set, we can now write:

$$\mu_{A0}(x_1) = 1, \mu_{A0}(x_2) = 1, \mu_{A0}(x_3) = 1, \mu_{A0}(x_4) = 1, \mu_{A0}(x_5) = 1, \mu_{A0}(x_6) = 1$$

Let's now define set A_1 as $= \{x_1, x_3, x_5\}$.

For set A_1, entities x_2, x_4, x_6 are not members of the set and selecting any of them should return a 0, while selecting entities x_1, x_3, x_5, which are members, should return a 1. For each member of the set A_1, we can now write:

$$A_1 = \{x_1, x_3, x_5\}$$

$$\mu_{A1}(x_1) = 1, \mu_{A1}(x_2) = 0, \mu_{A1}(x_3) = 1, \mu_{A1}(x_4) = 0, \mu_{A1}(x_5) = 1, \mu_{A1}(x_6) = 0$$

Using the *element of* : € and *not element of* : ∉ operators, we can also express set membership.

x_1 € A1, x_2 ∉ A1, x_3 € A1, x_4 ∉ A1, x_5 € A1, x_6 ∉ A1

We can also pair each of the members in A_1 with their grade of membership in a classic set and express the members of the set and their grade as follows:

$A_1 = (x_1, 1), (x_2, 0), (x_3, 1), (x_4,0), (x_5, 1), (x_6, 0)$

3.3 Classical Set Theory and Operations

3.3.1 Classical Set Logic

Such a logic is built upon the basic concept of a *set*, *membership* within the set, and the *operations* thereon. We begin by reiterating the basic concepts of a *set* and *set membership.* As we stated in the introduction, we define a *set* as a collection of (not necessarily ordered) well-defined objects sharing common properties. Associated with the set is a *sharp*, *crisp*, and *unambiguous* distinction between objects that are members of the set and those that are nonmembers of the set. Such a membership relationship is also termed *binary,* which indicates that an object either fully meets the membership conditions or it does not. The following illustrates and formalizes such a relationship.

Let

S be an arbitrary set of elements
T be any subset of S

A subset T of S is defined as follows:

Definition 3.1 A subset T of S comprises the set of all ordered pairs:

$\left(x_{1i}, x_{2i}\right)$.

The first member of the pair x_{1i} is an element of T.
The second member of the pair x_{2i} is the degree of membership of x_{1i} in the subset T.

Since we are working with classic/crisp sets, the second member of the pair, x_{2i}, can only be an element of the set {0,1}. The element x_{2i} will have the value 1 if x_{1i} is in T, i.e. *true,* and 0, i.e. *false* otherwise.

Accordingly, the subset T can contain one and only one ordered pair for each element of S. Such a relationship defines a subset T of S as a mapping from S into the set {0,1}.

If S is the universe of discourse, the truth or falsity of any statement "*X is in T with respect to S*" is determined by the ordered pair having *X* as its first element, (x_{1i}). The statement is TRUE if the second element (x_{2i}) is 1 and FALSE otherwise, i.e. 0.

We now review some of the basic terminology, definitions, and concepts comprising the theory of ordinary or classical sets.

3.3.2 Basic Classic Crisp Set Properties

We begin with a review some of the basic properties, definitions, and concepts of the theory of ordinary or classical sets and extend these to the theory of fuzzy subsets in the next chapter. The concepts or properties of *empty* and *size*, the relationships of *equality* and *containment*, and the operations of *union*, *intersection*, and *negation* may be defined on a set or a collection of sets.

To illustrate the concepts, properties, and definitions, let S (the universe of discourse) be the set of five small integers as shown in Figure 3.7, and let A, B, and C be subsets of S containing 3, 4, and 4, small integers, respectively.

S = {1, 2, 3, 4, 5}

The subsets A, B, and C are written with their value and degree of membership shown: (x_{1i}, x_{2i}) from Definition 3.1.

A = {(1,1), (2,1), (3,1), (4,0), (5,0)}

B = {(1,1), (2,1), (3,1), (4,1), (5,0)}

C = {(1,1), (2,1), (3,1), (4,0), (5,1)}

The subsets A, B, and C may also be written as:

A = {1, 2, 3}

B = {1, 2, 3, 4}

C = {1, 2, 3, 5}

with degree of membership implied.

Figure 3.7 Basic universe of discourse.

Property 3.1 Empty

A subset P is *empty* (P = ∅) iff (if and only if) for all ordered pairs (x_{1i}, x_{2i}) in P the element $x_{2i} = 0$.

Accordingly, none of the subsets A, B, or C is empty.

Property 3.2 Size

The *size* of a subset P is the number of ordered pairs (x_{1i}, x_{2i}) in P for which the element $x_{2i} = 1$.

From Figure 3.4, the sizes of subsets A, B, and C are:

$$A = \{(1,1), (2,1), (3,1)\}$$
$$B = \{(1,1), (2,1), (3,1), (4,1)\}$$
$$C = \{(1,1), (2,1), (3,1),(5,1)\}$$

which gives 3, 4, and 4, respectively.

Property 3.3 Equal

Two subsets P and R are *equal* (P = R) iff for each element (x_{1i}, x_{2i}) in P there is a corresponding element (y_{1j}, y_{2j}) in R such that $x_{1i} = y_{1j}$ and $x_{2i} = y_{2j}$, and for each element (y_{1j}, y_{2j}) in R there is a corresponding element (x_{1i}, x_{2i}) in P such that $y_{1j} = x_{1i}$ and $y_{2j} = x_{2i}$.

Here, the subsets A, B, and C are not equal.

A = {1, 2, 3}
B = {1, 2, 3, 4}
C = {1, 2, 3, 5}
A≠B, A≠C, B≠C

Here, the subsets D and E are equal.

D = {(1,1), (2,1), (5,1)}
E = {(1,1), (2,1), (5,1)}

Thus, subset D = subset E.

Property 3.4 Containment
A subset P is contained in a subset R (P⊆R) iff for each element (x_{1i}, x_{2i}) in P there is an element (y_{1j}, y_{2j}) in R such that $x_{1i} = y_{1j}$ and $x_{2i} = y_{2j}$.

The definition of *containment* is looser than that of equality, but if the relationship holds for all elements in P and all elements in R, then containment reduces to equality and P = R.

For subsets A, B, and C given in Figure 3.7,

A = {1, 2, 3}
B = {1, 2, 3, 4}
C = {1, 2, 3, 5}

The subset A is contained in the subset B:

A⊆B //the subset {1, 2, 3} is contained in the subset{1, 2, 3, 4}
{(1,1), (2,1), (3,1), (4,0), (5,0)}⊆{(1,1), (2,1), (3,1), (4,1), (5,0)}

The subset A is contained in the subset C.

A⊆C//the subset {1, 2, 3} is contained in the subset{1, 2, 3, 5}
{(1,1), (2,1), (3,1), (4,0), (5,0)}⊆{(1,1), (2,1), (3,1), (4,0), (5,1)}

Property 3.5 Union
The *union* of two sets is defined as a set containing all the elements that are members of either individual set alone or both sets. Here that set, the union, will be Z.

For the two subsets P and R, the union of (P∪R) is the subset of ordered pairs (z_{1k}, z_{2k}) formed from all ordered pairs (x_{1i}, x_{2i}) in P and (y_{1j}, y_{2j}) in R such that $x_{1i} = y_{1j} = z_{1k}$ and $z_{2k} = \max(x_{2i}, y_{2j})$ or equivalently $\max(\mu_P(x_{1i}), \mu_R(y_{1i}))$.

Examining the union, we see that $z_{2k} = \max(x_{2i}, y_{2j})$. Observe that with the expression: $z_{2k} = \max(x_{2i}, y_{2j})$, x_{2i} or y_{2i} or both (*can*, not *must*) be a binary 1 value so that if either term or both are a binary 1, the term(s) with a membership of 1 will appear in the set z.

With the union, we are dealing with an *OR*, not an *AND* operation.

Here, the union of A and B from Figure 3.7 is given as:

$$\begin{aligned} A \cup B &= \{(1, \max(1,1)), (2, \max(1,1)), (3, \max(1,1)), (4, \max(0,1)), (5, \max(0,0))\} \\ &= \{(1,1), (2,1), (3,1), (4,1), (5,0)\} \\ A \cup B &= \{1, 2, 3, 4\} \end{aligned}$$

One or both of the subsets, A and B, have the objects 1, 2, 3, or 4.

The membership for term 4, max(0,1) = 1, appears in the subset B but not the subset A; therefore, it can appear in the union. However, the term 5 does not appear in either the subset A or subset B, max(0,0) = 0; thus, it will not appear in the union.

Note that we are performing a logical OR operation.

Property 3.6 Intersection

The *intersection* of two subsets P and R, $(P \cap R)$, is the subset of ordered pairs (z_{1k}, z_{2k}) formed from all ordered pairs (x_{1i}, x_{2i}) in P *and* (y_{1j}, y_{2j}) in R such that $x_{1i} = y_{1j} = z_{1k}$ and $z_{2k} = \min(x_{2i}, y_{2j})$ or equivalently $\min(\mu_P(x_{1i}), \mu_R(y_{1i}))$.

Observe that with the expression: $z_{2k} = \min(x_{2i}, y_{2j})$, if either x_{2i} or y_{2i} or both are a binary 0 value, that term cannot appear in the subset of ordered pairs (z_{1k}, z_{2k}).

With the intersection, we are dealing with an AND not an OR operation. A term must appear in both subsets to be included in the intersection of the subsets that comprise the intersection operation.

Here, the intersection of A and B is given as:

$$\begin{aligned} A \cap B &= \{(1, \min(1,1)), (2, \min(1,1)), (3, \min(1,1)), (4, \min(0,1)), \\ &\quad (5, \min(0,0))\} \\ &= \{(1,1), (2,1), (3,1), (4,0), (5,0)\} \\ A \cap B &= \{1,2,3\} \end{aligned}$$

Here, the term 5 does not appear in either the set A or the set B and the term 4 does not appear in the subset A; thus, neither will not appear in the union. Note that we are performing a logical AND operation.

Property 3.7 Negation

The *negation* or *complement* of a subset $P(P')$ or $P(P^c)$ is the set of all ordered pairs (z_{1k}, z_{2k}) such that for all ordered pairs (x_{1i}, x_{2i}) in P, the element $z_{1k} = x_{1i}$ and $z_{2k} = 1 - x_{2i}$ or equivalently, $1 - \mu_P(x_{1i})$.

$$\begin{aligned} A &= \{1(1-1), 2(1-1), 3(1-1), 4(1-0), 5(1-0)\} \\ B &= \{1(1-1), 2(1-1), 3(1-1), 4(1-1), 5(1-0)\} \\ C &= \{1(1-1), 2(1-1), 3(1-1), 4(1-0), 5(1-1)\} \end{aligned}$$

Here, the negations or complements of subsets A, B, and C are given as:

$$\begin{aligned} A^c &= \{1(0), 2(0), 3(0), 4(1), 5(1)\} \\ &= \{4,5\} \end{aligned}$$

$$\begin{aligned}B^c &= \{1(0), 2(0), 3(0), 4(0), 5(1)\}\\ &= \{5\}\\ C^c &= \{1(0), 2(0), 3(0), 4(1), 5(0)\}\\ &= \{4\}\end{aligned}$$

Property 3.8 Commutation
The commutation or interchange of two subsets is given as follows.

Commutation of a Union of Two Subsets

$$\begin{aligned}A &= \{(1,1), (2,1), (3,1), (4,0), (5,0)\}\\ B &= \{(1,1), (2,1), (3,1), (4,1), (5,0)\}\end{aligned}$$

For two subsets A and B, the *commutation* of the *union* is given as:

$$\begin{aligned}A \cup B &= B \cup A\\ A \cup B &= \{(1, \max(1,1)), (2, \max(1,1)), (3, \max(1,1)), (4, \max(0,1)), (5, \max(0,0))\}\\ &= \{(1, 1), (2, 1), (3, 1), (4, 1), (5, 0)\}\\ A \cup B &= \{1, 2, 3, 4\}\\ B \cup A &= \{(1, \max(1,1)), (2, \max(1,1)), (3, \max(1,1)), (4, \max(0,1)), (5, \max(0,0))\}\\ &= \{(1, 1), (2, 1), (3, 1), (4, 1), (5, 0)\}\\ B \cup A &= \{1, 2, 3, 4\}\\ A \cup B &= B \cup A\\ \{1, 2, 3, 4\} &= \{1, 2, 3, 4\}\end{aligned}$$

Commutation of an Intersection of Two Subsets

$$\begin{aligned}A &= \{(1,1), (2,1), (3,1), (4,0), (5,0)\}\\ B &= \{(1,1), (2,1), (3,1), (4,1), (5,0)\}\end{aligned}$$

For two subsets A and B, the *commutation* of the *intersection* is given as:

$$\begin{aligned}A \cap B &= B \cap A\\ A \cap B &= \{(1, \min(1,1)), (2, \min(1,1)), (3, \min(1,1)), (4, \min(0,1)), (5, \min(0,0))\}\\ &= \{(1, 1), (2, 1), (3, 1), (4, 0), (5, 0)\}\\ A \cap B &= \{1, 2, 3\}\\ B \cap A &= \{(1, \min(1,1)), (2, \min(1,1)), (3, \min(1,1)), (4, \min(1,0)), (5, \min(0,0))\}\\ &= \{(1, 1), (2, 1), (3, 1), (4, 0), (5, 0)\}\\ B \cap A &= \{1, 2, 3\}\\ A \cap B &= B \cap A\\ \{1, 2, 3\} &= \{1, 2, 3\}\end{aligned}$$

Property 3.9 Associativity
The *associativity* property applies to the *union* and *intersection* of subsets. We start with subsets A, B, and C.

$$\begin{aligned}A &= \{1, 2, 3\}\\ B &= \{1, 2, 3, 4\}\\ C &= \{1, 2, 3, 5\}\end{aligned}$$

Here, the *associativity* of subsets A, B, and C is given as:

Union of Subsets

$$(A \cup B) \cup C = A \cup (B \cup C)$$
$$(A \cup B) \cup C$$
$$(\{1, 2, 3\} \cup \{1, 2, 3, 4\}) \cup \{1, 2, 3, 5\} = \{1, 2, 3, 4,5\}$$
$$A \cup (B \cup C)$$
$$(\{1, 2, 3\} \cup (\{1, 2, 3, 4\} \cup \{1, 2, 3, 5\})) = \{1, 2, 3, 4,5\}$$

Intersection of Subsets

$$(A \cap B) \cap C = A \cap (B \cap C)$$
$$(A \cap B) \cap C = (\{1, 2, 3\} \cap \{1, 2, 3, 4\}) \cap \{1, 2, 3,5\} = \{1, 2, 3\}$$
$$A \cap (B \cap C) = \{1, 2, 3\} \cap (\{1, 2, 3, 4\}) \cap \{1, 2, 3, 5\}) = \{1, 2, 3\}$$

Property 3.10 Distributivity

The *distributivity* property applies to the union and intersection of subsets.

We start with subsets A, B, and C.

For subsets A, B, and C

$$A = \{1, 2, 3\}$$
$$B = \{1, 2, 3, 4\}$$
$$C = \{1, 2, 3, 5\}$$

Here, the *distributivity* of subsets A, B, and C is given as:

Union of Subsets

$$A \cup (B \cap C) = (A \cup B) \cap (A \cup C)$$

$$\{1, 2, 3\} \cup (\{1, 2, 3, 4\} \cap \{1, 2, 3, 5\}) = \{1, 2, 3\} \cup \{1, 2, 3\}$$
$$= \{1, 2, 3\}$$
$$(\{1, 2, 3\} \cup (\{1, 2, 3, 4\}) \cap (\{1, 2, 3\} \cup \{1, 2, 3, 5\})$$
$$= \{1, 2, 3, 4\} \cap \{1, 2, 3, 5\}$$
$$= \{1, 2, 3\}$$

Intersection of Subsets

$$A \cap (B \cup C) = (A \cap B) \cup (A \cap C)$$

$$\{1, 2, 3\} \cap (\{1, 2, 3, 4\} \cup \{1, 2, 3, 5\}) = \{1, 2, 3\} \cap \{1, 2, 3, 4, 5\}$$
$$= \{1, 2, 3\}$$
$$(\{1, 2, 3\} \cap \{1, 2, 3, 4\}) \cup \{1, 2, 3\} \cap \{1, 2, 3, 4, 5\})$$
$$=\{1, 2, 3\} \cup (\{1, 2, 3\}$$
$$= \{1, 2, 3\}$$

Property 3.11 Idempotence

Idempotence is the property of a mathematical operation in which that operation can be applied multiple times to an operand without changing the result following the first application. Consider the trivial case of multiplication by 0. Following the initial multiplication, the result will always be 0.

Examples of two operations that can be applied multiple times on a subset are given as follows.

Working with subset A from the set S

$$A = \{(1,1), (2,1), (3,1), (4,0), (5,0)\}$$
$$A = \{1, 2, 3\}$$

The idempotent property on the subset A of S is given as:

$$A \cup A = A$$
$$A \cap A = A$$

Union Operation

$$A \cup A = \{(1,1), (2,1), (3,1)\} \cup \{(1,1), (2,1), (3,1)\}$$
$$= \{(1,1), (2,1), (3,1)\}$$
$$\{A \cup A\} \cup A = \{(1,1), (2,1), (3,1)\} \cup \{(1,1), (2,1), (3,1)\}$$
$$= \{(1,1), (2,1), (3,1)\}$$

Intersection Operation

$$A \cap A = \{(1,1), (2,1), (3,1)\} \cap \{(1,1), (2,1), (3,1)\}$$
$$= \{(1,1), (2,1), (3,1)\}$$
$$\{A \cap A\} \cap A = \{(1,1), (2,1), (3,1)\} \cap \{(1,1), (2,1), (3,1)\}$$
$$= \{(1,1), (2,1), (3,1)\}$$

Property 3.12 Transitivity

Transitivity is a relation between three elements such that if the relation holds between the first and second and it also holds between the second and third, it then holds between the first and third. Containment is a good example. If A is contained in B and B is contained in C, A must therefore contain C.

$$\text{If } A \subseteq B \subseteq C, \text{ then } A \subseteq C.$$

Property 3.13 Excluded Middle

The *excluded middle* property for a subset and its complement within set, S, is given as follows.

The *excluded middle* property of an operation on a subset A and its complement is given as:

$$A \cup A^c = S$$
$$A \cup A^c = \{(1,1), (2,1), (3,1), (4,0), (5,0)\} \cup \{1(0), 2(0), 3(0), 4(1), 5(1)\}$$
$$= \{1, 2, 3,\} \cup \{4, 5\} = \{1, 2, 3, 4, 5\}$$

Property 3.14 Contradiction

The *contradiction* property for a subset and its complement within a set, S, is given as follows.

The *contradiction* property of an operation on a subset A and its complement is given as:

$$A \cap A^c = \varnothing$$

$$\{1,2,3,\} \cap \{4,5\} = \{\varnothing\}$$

Property 3.15 De Morgan
De Morgan's Law simply states that the complement of the union of two sets is the same as the intersection of their complements and that the complement of the intersection of two sets is the same as the union of their complements.

For subsets A and B,

$$A = \{1, 2, 3\}$$
$$B = \{1, 2, 3, 4\}$$

The application of De Morgan's Law to the union of two subsets, A and B, is given as:
Complement of the Union Operation

$$(A \cup B)^c = A^c \cap B^c$$
$$(\{1, 2, 3,\} \cup \{1, 2, 3,4\})^c = \{1, 2, 3,4\}^c \cap \{1, 2, 3,4\}^c$$
$$(\{1, 2, 3, 4\})^c = \{5\} \cap \{5\}$$
$$\{5\} = \{5\}$$

Complement of the Intersection Operation

$$(A \cap B)^c = A^c \cup B^c$$
$$(\{1, 2, 3\} \cap \{1, 2, 3, 4\})^c = \{1, 2, 3\}^c \cup \{1, 2, 3,4\}^c$$
$$(\{1, 2, 3\})^c = \{4, 5\} \cup \{5\}$$
$$\{4, 5\} = \{4, 5\}$$

3.4 Basic Crisp Applications – A First Step

We have looked at the basic properties of crisp sets. Let's now take a first look at how we might begin to apply those tools.

Example 3.1 A Crisp Activity
Let the activity be two friends, John and Mary, going to a movie. From a crisp perspective, each person can either like the movie or not like the movie, no in-between. Thus, each person's opinion of the movie can be assigned a value of either 0 or 1 in the set *like*, that is, either *not in* the set *like* or *in* the set *like*.

Combining the two possible opinions from John and Mary gives a crisp set, *Like*, with four possible combinations or four possible crisp subsets.

Here, the intersection of John and Mary's opinions of the movie gives:

$$\text{John} \cap \text{Mary} = \{(0, 0), (0, 1), (1, 0), (1, 1)\}$$

The scenario is expressed in Figure 3.8 graphically and as a truth table.

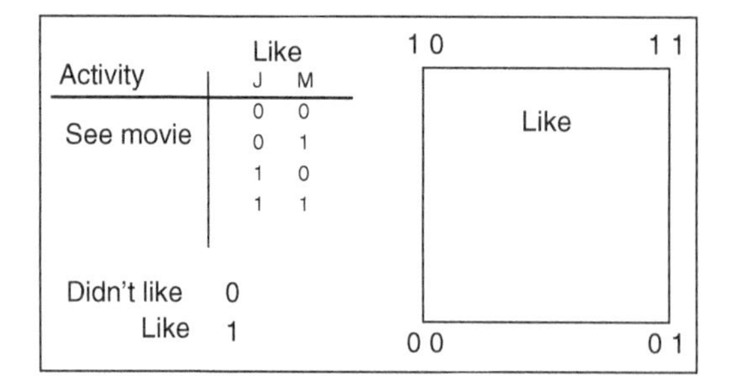

Figure 3.8 Viewing a movie.

In the diagram, we have the classic or crisp set *Like*, L, that has four crisp subsets. Each crisp subset has two members expressing membership in the set *Like*: *not a member* – 0 and a member, *like* –1, reflecting the possible opinions of the movie. Each subset occupies one of the four corners of the diagram labeled *Like*.

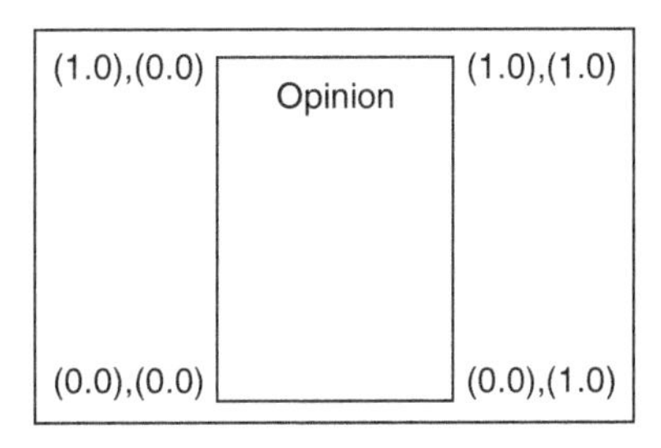

Figure 3.9 Expressing an opinion-0.

The membership functions in the four subsets, expressing possible opinions of the attendees, are given as:

$$\mu L(J) = 0, J \notin L \qquad \mu L(M) = 0, \quad M \notin L$$
$$\mu L(J) = 1, J \in L \qquad \mu L(M) = 1, \quad M \in L$$

Note that the model limits or restricts the attendees to two extreme choices, either like or dislike the movie, and restricts the subsets to the four corners of the block. Let's now anticipate where we are going and express the opinions as floating-point variables as shown in Figure 3.9.

3.5 Summary

In this chapter, we introduced the fundamental concept of sets focusing on what are known as classical sets. We began with an introduction of some of the elementary terminology and vocabulary. We then reviewed some of the principle definitions and concepts of the theory of ordinary or classical sets. We then introduced and examined the concept of classic or crisp sets and then explored set membership in such subsets. Set membership led to the concept of membership functions and the basic crisp set symbols.

We learned that four basic operations are fundamental to both classical and fuzzy sets. We then took a first step in applying those operations as we demonstrated each of them for classical sets. With the fundamentals of sets and set membership established, we examined the basic concepts of classic or crisp logic. We then moved to the details of the properties and logical operations using crisp sets and developing crisp membership applications.

Review Questions

3.1 Following our review of Boolean algebra, what are the primary things that it brings us?

3.2 In the design process, how much information does a *classic variable* or a *fuzzy variable* give us?

3.3 Explain the expressions domain and universe of discourse.

3.4 Explain the terms *degree* or *grade* of *membership* for classic sets and fuzzy sets. What is the difference? What are the constraints on set membership for classic sets? What are the constraints on set membership for fuzzy sets?

3.5 Identify and explain the four basic operations that are fundamental to both crisp and fuzzy sets.

3.6 Explain the terms *range* and *domain* for a function with an example other than what is in the chapter.

3.7 Explain the term subset of a set and illustrate with an example.

3.8 The chapter identifies 15 properties of crisp sets. Identify and illustrate 8 of them with examples that are different from those in the chapter.

3.9 Two groups of five friends each go to a rock concert. Within each group or set, some like the music and some do not. For each group or set, compute the member subsets and then compute union and intersection of the subsets.

3.10 Working with Question 3.9, compute the complement of the subsets within each group or set and then compute union and intersection of the subsets.

4

Fuzzy Sets and Sets and More Sets

THINGS TO LOOK FOR...OF THE RESULTING ENTRIES IN THE UNION
• What is fuzzy logic? • The foundations of fuzzy logic. • A review of the fundamental concept of sets. • The concept of fuzzy sets and set membership. • More sets and set membership. • Fuzzy set imprecision, membership, and membership functions. • Comparing fuzzy subset membership and probability values. • Basic fuzzy terminology, membership properties, and applications. • The properties and logical operations of fuzzy sets and subsets. • Graphical presentation of membership functions and degree of membership.

4.1 Introducing Fuzzy

In routine conversation or discussion, the word fuzzy will typically mean *indistinct, uncertain, vague, blurred*, or *not focused or sharp*. In academic, scientific, or technical worlds, the word fuzz or fuzzy is used as a qualifying adjective in an attempt to describe a sense of *imprecision, potential uncertainty, ambiguity,* or *vagueness,* and is often associated with *human concepts*. For many years, such imprecision was frowned upon. That view has gradually changed. Today, it's important to recognize and to note that the term fuzzy does not automatically mean errors, false statements, or confusion.

In daily life, we find that there are two kinds of imprecision: statistical and nonstatistical. Statistical or random imprecision is that which arises from events such as the outcome of a coin toss or card game. Nonstatistical or nonrandom imprecision, on the other hand, is that which we find in driving instructions such as "Begin to apply the brakes soon" or "Start slowing down" or in winter "Be careful, that walk is very icy." What do the terms "soon," "slow down," or "very icy" really mean, or do they have a precise meaning? This latter type of imprecision is what we call fuzzy; generally, in the appropriate context, we do understand what is meant by such expressions.

In our studies of classical set theory, we have learned that set membership is stated in binary terms. An element either belongs to the set (or subset) or it does not. In contrast,

Introduction to Fuzzy Logic, First Edition. James K. Peckol.

in fuzzy set theory, set membership is stated in more degreed terms, occupying the domain between and including the two classical extremes. In both cases, set membership is specified with the support of a *membership function*.

To effectively design modern, real-world hardware, software, or a mix of both types of systems, it is important that one be able to recognize, represent, manipulate, and interpret statistical and nonstatistical uncertainties. One should use *fuzzy* models to capture and to quantify nonrandom imprecision and *statistical* models to capture and quantify random imprecision.

4.2 Early Mathematics

Fuzzy logic, with roots in early Greek philosophy, finds a wide variety of contemporary applications ranging from self-driving vehicles, the manufacture of cement, to the control of high-speed trains or auto-focus cameras. Yet, early mathematics began by emphasizing precision, i.e. crisp logic and frowning on uncertainty. The central theme in the philosophy of Aristotle and many others was the search for perfect numbers or golden ratios. Pythagoras and his followers kept the discovery of irrational numbers a secret. Their mere existence was counter to many fundamental religious teachings.

Later mathematicians continued the search for precision and were driven toward the goal of developing a concise theory of mathematics. One such effort was *The Laws of Thought* published by Korner in 1967 in the Encyclopedia of Philosophy. Korner's work included a contemporary version of the *Law of Excluded Middle* that stated that every proposition could only be TRUE or FALSE . . . there could be no in between. An early version of this law, proposed by Parmenides in approximately 400 BC, met with immediate and strong objections. Heraclitus, a fellow philosopher, countered that propositions could simultaneously be both TRUE and NOT TRUE. Clearly, he had learned from Plato.

4.3 Foundations of Fuzzy Logic

Classic or crisp logic has been the foundation upon which the fields of digital and embedded systems have been built and grown for many years. Over the years, technology has advanced with demands for decreasing system sizes, increasing speeds, and for higher and higher signal resolution. Note that in the previous list of demands, several are fuzzy in nature.

Signals in classic digital logic have traditionally been limited to two states expressing a logic 0 or a logic 1. For many years, a logic 0 has been 0 V and a logic 1 has been 5.0 V. Advances in semiconductor design have been able to successfully bring the logic 1 to lower voltage levels. However, the logical decisions remained constrained at true or false. Variable signal levels had to be relegated to the analog world.

Translating and implementing expressions such as the following definitely presented a challenge:

- Change the state when the sensor signal gets *close to* 4 V.
- Turn the heat off when the temperature gets *too high*.
- Reduce the sample rate if the signal frequency falls *too far below normal*.

The world of fuzzy logic and the concept of *hedges* evolved to begin to address such problems. As challenges advance, our engineers and our technology often advance to address them.

In addition, the real-world AI-related fields of neural networks, expert systems, and machine learning have entered the picture coupled with the need to interact with human thinking and reasoning. Such thinking and reasoning involve inherently fuzzy and fragmented information that typically can be well understood by humans but can be difficult for machines to comprehend and process.

The need for the capabilities of fuzzy logic has continued to increase. Today, we want our systems to be able to grasp and manage imprecise, incomplete, potentially unreliable, or contradictory information and give expert opinions and develop expert solutions to contemporary, real-world problems.

We are in the quest of designs that can perform as captured graphically in Figure 4.1. It is important to be aware, and to remember, that it is not magic that is coming out of the right-hand side of the following figure but thoughtful and careful design management and control of typically electrical signals to the implementing devices to achieve the desired goal.

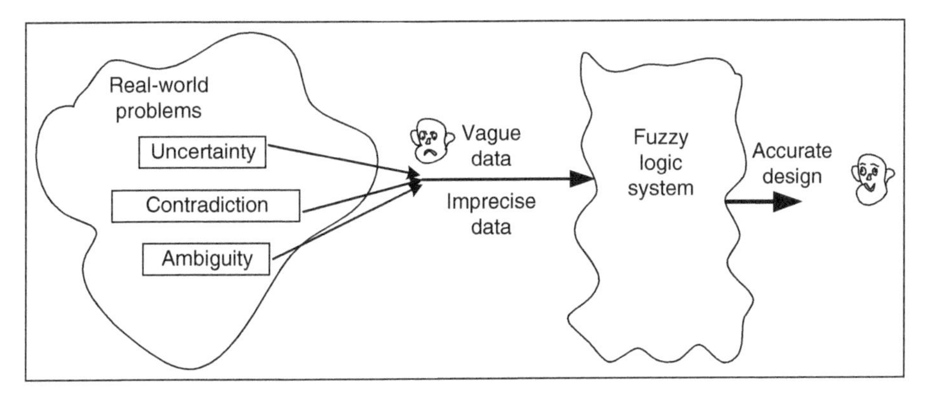

Figure 4.1 Real-world problems fuzzy logic solution.

Plato, a student of Socrates, was among the first to attempt to quantify a possible alternative state of existence. Countering his teacher, as some students are wont to do, in the province of logic and reasoning, he proposed the existence of a third region, beyond TRUE and FALSE, in which "*opposites tumbled about.*" Many modern philosophers, including Hegle, Heraclitus, and Engels, have supported and further developed Plato's early ideas.

The contemporary first formal steps, away from classical logic, were taken by the Polish mathematician, Jan Lukasiewicz (also the developer of *Reverse Polish Notation*, RPN). He proposed a three-valued logic in which the third value, called *possible*, was to be assigned a numeric value somewhere between TRUE and FALSE. Lukasiewicz also developed an entire set of notation and an axiomatic system for his logic. His intention was to derive modern mathematics.

In later work, he also explored four- and five-valued logics before declaring that there was nothing to prevent the development of infinite-valued logics. Donald Knuth proposed a similar three-valued logic and suggested using the values of −1, 0, 1. The idea never received much support.

The birth of modern fuzzy logic is usually traced to the seminal paper *Fuzzy Sets* published in 1965 by Lotfi A. Zadeh (1965). In his paper, Zadeh described the mathematics of fuzzy subsets and, by extension, the mathematics of fuzzy logic. He suggested that membership of an element in a set need not be restricted to the values 0 and 1 (corresponding to FALSE and TRUE) but could be extended to include all real numbers in the range 0.0–1.0. In addition, he proposed the collection of operations for his new logic.

Zadeh introduced his ideas as a new way of representing the vagueness common in everyday language and human interaction. His fuzzy sets are a natural generalization of classic sets, which are one of the basic structures underlying contemporary mathematics.

4.4 Introducing the Basics

We have explored many of the details of classic sets, set membership, and a bit of the history of fuzzy logic. Now, we'll venture into the world of fuzzy sets. As we explore, once again we also enter the portion of the world of mathematical logic known as *set theory*. The modern study of this field was initiated in the 1870s by Richard Dedekind and Georg Cantor. As we may have surmised, any type of object, such as those described earlier, can be collected into a set. However, most often we find *set theory* closely linked with entities closely related to mathematics.

In formal terminology, as we have seen, fundamentally, we have a *binary relationship* between a set and an object. We say that the object is or is not a *member* of a specific set. A key point for a specific collection is that the properties of *all* of the members of the collection are well defined and constrained within that collection. Further, as identified in Chapter 3, when such a constrained collection is referred to or discussed, it is designated as the *domain* or *universe of discourse*. Let's look at the diagram in Figure 4.2.

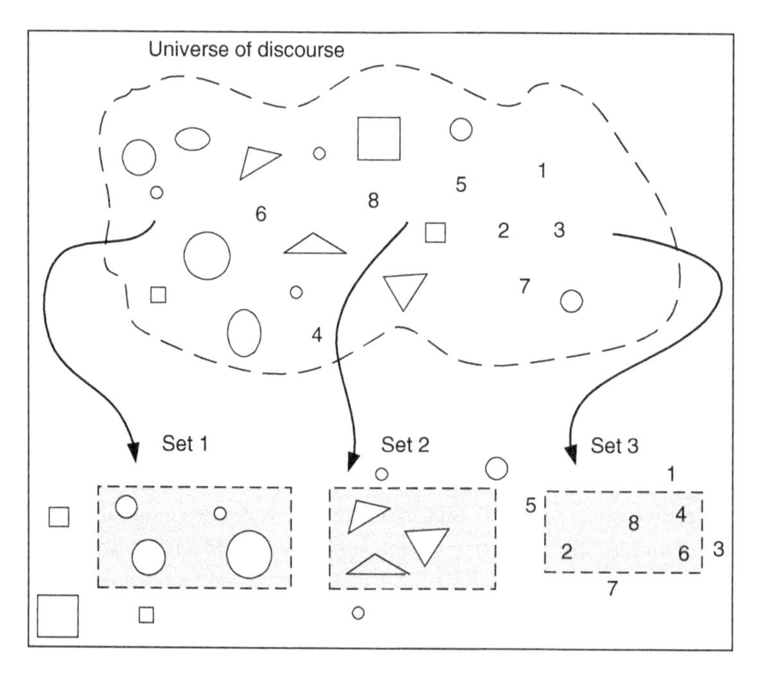

Figure 4.2 Universe of discourse and three subsets.

We have the *universe of discourse* comprising a collection of numbers and a collection of various geometric figures that we can examine and analyze and from which we can define three collections or sets. Looking at the individual sets, what can we say about the members in each? Each of these sets contains various sizes and shapes of different geometric figures. The surrounding rectangles merely delimit the sets; they are not part of the sets.

Let's first examine the members of Sets 1 and 2. The figures (members) within each of those two specific sets are all of the same type, that is, either circles or triangles but no other types. The members are of different sizes and different orientations; nonetheless each is an instance of a specific type of geometric figure. A further restriction is that a circle, or any other figure, cannot be a member of the triangle set, and a triangle, or any other figure, cannot be a member of the circle set.

In the region around two of the sets, we again find different sizes of different geometric figures. The squares cannot be a member of any of the sets, whereas the circles could be a member of the circle set.

The members of Set 3, also delimited by a rectangle, are integer numbers, all of which are even numbers, i.e. divisible by 2. Surrounding Set 3, but not within the rectangle, we also find a collection of integer numbers. However, this external collection, also a set, contains only odd numbers, and thus its members cannot be a member of any of the other sets.

Looking at our three sets under discussion, we can associate a degree or *grade membership* descriptor or qualifier for the objects within a specific set and a similar descriptor or qualifier for the objects outside of that set that don't meet the constraints of the designated set. Let the grade of membership qualifier for an object within a set that meets the restrictions of that set be "1" and for those outside the set that don't meet the set's qualifications be "0." Consider. . .

Set 1

Four circles of different sizes. To be a member of Set 1, an object must be a circle. Since the objects in Set 1 are all circles, the designated *grade of membership* for each contained object is 1. For the three squares and four odd integers, the grade of membership in Set 1 is 0. What about the three circles outside of Set 1? Could they be a member of Set 1?

Set 2

Three triangles of different sizes. To be a member of Set 2, an object must be a triangle. Since the objects in Set 2 are all triangles, the grade of membership for each contained object is 1 and that for all other entities three squares, three circles, and four odd integers, grade of membership in Set 2 is 0.

Set 3

Four different even integers. To be a member of Set 3, an object must be an even integer. Since the objects in Set 3 are all even integers, the grade of membership for each contained object is 1 and that for all other entities three squares, three circles, and four odd integers, the grade of membership in Set 3 is 0.

Up to this point, we've been dealing with crisp sets. However, in the universe of discourse, there are two objects that look somewhat like circles, but they do not meet the standard geometric criteria for a circle so they cannot be members of Set 1. They may be a diamond in the rough, let's keep an eye on them.

4.5 Introduction to Fuzzy Sets and Set Membership

We have explored many of the details of classic sets and set membership. Now, we'll venture into the world of fuzzy sets and examine details there. According to what we have learned for a crisp set S, a subset T of S will contain one and only one ordered pair (x_{1i}, x_{2i}) for each element or member of S. Is that the same for fuzzy sets? Let's wait and see.

Fuzzy Sets

Most of the familiar and traditional tools that we use in formal logic, i.e. reasoning, modeling, or computing, are crisp, precise, and deterministic. When we studied crisp logic, we learned about sets, some of their properties, and that sets contained objects. With respect to an algorithm or problem solution, when we refer to objects in a set, we are generally referring to the steps and their sequence that comprise the algorithm or path to a solution.

We have also learned that those objects are either in the set or they are not. In crisp logic, an algorithm or the steps in the algorithm are either correct or they aren't; they work or they don't. Their degree or *grade of membership* in the set is either [0], FALSE, not in the set, or [1], TRUE, in the set. The algorithm's steps are not partially in the set or partially out. In a design, a solution is either feasible or it isn't.

Building upon the basic concepts that we studied for crisp logic, we now move to the branch of mathematics called *fuzzy logic*. We understand the term logic; however, what is fuzzy and what does it have to do with logic? That whole concept seems a bit fuzzy. Let's see if we can get it sorted out.

4.5.1 Fuzzy Subsets and Fuzzy Logic

Earlier, we examined set membership for crisp sets and the set operations for classical sets. Because fuzzy set theory is a basic extension and a generalization of the basic theory of crisp sets, the set operations from classical set theory can now be extended to fuzzy set theory by relaxing the restriction on the values for the degree of membership in the set or subset. In such a case, the concepts of *empty* and *size*, the relationships of *equality* and *containment*, and the operations of *union, intersection* and *negation* for fuzzy set theory are valid and remain identical to those for classical set theory. Set membership for fuzzy subsets follows analogously.

Fuzzy logic is derived from fuzzy set theory and relaxes the restriction on the entities' values for the degree of membership in a subset by allowing any membership value (and gradual transition) between [0.0, 1.0]. In addition, support is added for the linguistic form of variables such as *very*, *slightly*, *quickly*, *slowly*. Such modifications thereby enable support for approximate reasoning and the ability to handle imprecise, uncertain, or vague concepts, which are discussed in Chapter 5.

In contrast to crisp sets, fuzzy subsets are subsets (of crisp sets) whose members are composed of collections of objects that have imprecise properties to varying degrees. As an example, we can write the statement: *X is a real number close to 7* as:

F = set of real numbers close to 7

But what do "set" and "close to" mean? How close is close, and how do we represent such a statement mathematically?

Zadeh suggests that F is a fuzzy subset of the set of real numbers and proposes that it can be represented by its membership function. Note, he does not say a set of integer numbers. The value of $\mu_F(r) \rightarrow f(r \in F)$ is the *extent* or *grade* of membership of each real number "*r*" in the subset of numbers (F) close to 7. With such a construct, it is evident that fuzzy subsets correspond to a continuously valued logic and that any element can have a degree of membership between 0.0 and 1.0 inclusive of the endpoints as illustrated in the graphic in Figure 4.3.

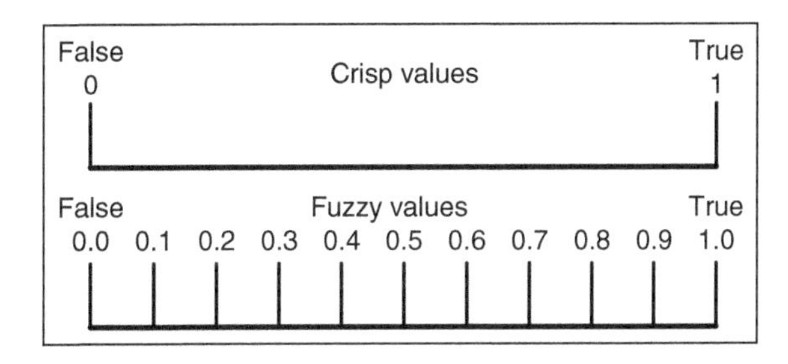

Figure 4.3 Crisp and fuzzy membership values.

Fuzzy Sets

Moving now from the crisp world into the fuzzy, in this chapter we introduce and examine fuzzy sets and the concept of set membership for fuzzy subsets more formally. As we do so, we will take the first step to get acquainted with the logical operations on and with fuzzy sets. At this stage, we should all feel comfortable talking about the concept of sets, subsets, and grade of membership in general. As we begin, we repeat some of the introductory material from the previous chapter.

Set Membership

Let us now examine the concept of sets and set membership for fuzzy subsets. The protocol for managing set membership for fuzzy subsets generally follows that for crisp set membership. There will be times, however, when the set membership rules become a bit more detailed.

Formally:

Let

S be an arbitrary set of elements

F be a (fuzzy) subset of S

A fuzzy subset F of S is defined as follows:

Definition 4.1 Here we have the first major difference with crisp logic. With crisp logic, the grade of membership in a set is limited to the two values {0, 1}. With fuzzy logic, that expands to include the range {0.0 . . . 1.0}.

Similar to classic set membership, the subset F may contain, at most, one and only one ordered pair for each element of S. Like crisp logic, a fuzzy subset F of S is the set of all ordered pairs, (x_{1i}, x_{2i}), such that x_{1i} is an element of S, and the value of element x_{2i} indicates the degree of membership of the element x_{1i} in the subset F. In contrast to crisp logic, x_{2i} is a real number in the range [0.0, 1.0]. A value x_{2i} of 0.0 shall indicate complete nonmembership, a value of 1.0 shall indicate complete or full membership, and intermediate values shall indicate various or fractional degrees of membership. For example, we have the following vector (an ordered set of numbers) indicating degrees of membership of seven entities in a fuzzy subset.

$$\{0.1, 0.2, 0.3, 0.4, 0.7, 0.8, 0.9\}$$

Such a relationship defines a subset F of S as a mapping from the set S into the interval [0.0, 1.0]. Note that the definition of a fuzzy subset reduces to that for a classical set if values for x_{2i} are restricted to the endpoints [0.0, 1.0] of the interval.

From *Definition* 4.1, let S be the *universe of discourse* and T be a subset of S. The degree of truth of any statement: X in S with respect to T is determined or specified by the ordered pair having X as its first element. The second element is the degree of truth of the statement X in T. If the second element is 0.0, the statement is totally FALSE, and if it is 1.0, the statement is totally TRUE. Intermediate values are permitted and specify varying degrees of truth. For example, if the value of the second element is 0.9, the statement X is almost TRUE and if it is 0.1, the statement is almost FALSE.

Ah Ha, now we know. Those two ellipse-shaped objects in the universe of discourse in Figure 4.2 were not perfect circles; however, they almost were. Let's say that they were 0.8 circle. They could have been members of a fuzzy subset. Note that the definition of a fuzzy subset reduces to that for a classical set if values for x_{2i} are restricted to the endpoints of the interval {0.0, 1.0}.

4.6 Fuzzy Membership Functions

From Chapter 1, consider the fuzzy property "*close to 7*." Such a property can be represented in several different ways. Who decides what that representation should be? That task falls upon the person doing the design.

To formulate a membership function for the fuzzy concept "*close to 7*," one might hypothesize several desirable properties. These might include the following properties:

Normality – A membership function that has a grade of "1" is called normal. Thus, it is desirable that the value of the membership function for 7 in the set F be equal to 1.0, that is, $\mu_F(7) = 1.0$. See Figure 4.4b.

Monotonicity – The membership function should be monotonic. The closer an object x is to 7, the closer the grade of membership $\mu_F(x)$ should be to 1.0 as the slope of the graph increases and vice versa as it decreases. See Figure 4.4b and c.

Symmetry – The membership function should be one such that numbers equally distant to the left and right of 7 are symmetric, that is, they should have equal membership values. See Figure 4.4b and c.

It is important that one realize that these criteria are relevant only to the fuzzy property "*close to 7*" and that other such concepts will have appropriate criteria for designating their membership functions. Consider the numeric range 6–8.

Based on the criteria given, several possible graphic membership functions may be designed. Three possible alternatives are given in Figure 4.4.

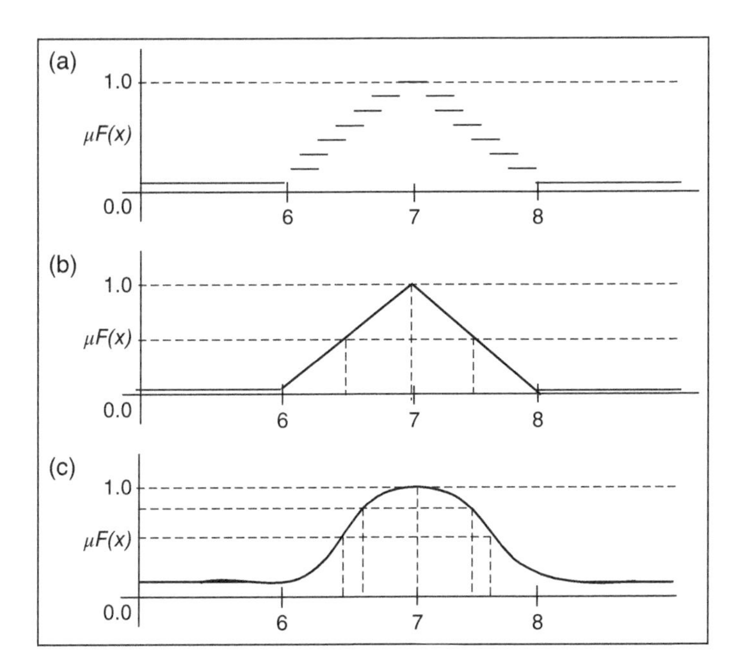

Figure 4.4 Membership functions for "*close to 7*."

Note that in graph c, every real number has some degree of membership in fuzzy set F, $\mu_F(\,)$, although the numbers far from 7 have very little degree of membership. Observe in graph b that the dashed line indicates that the number 6.5 has a grade of membership of 0.5 in the set of numbers close to 7.

At this point, one might ask if representing the property "*close to 7*" in such a way makes sense. If one poses the question

How much does it cost?

The answer

About 7 euros

can certainly be represented appropriately and seems to make sense in a routine response or conversation.

As a further illustration, let the crisp subset H and the fuzzy subset F represent the heights, 0–8 feet, of players on a basketball team. If, for an arbitrary player p, we know that $\mu H(p) = 1$, then all we know is that the player's height is probably somewhere between 6 and 8 ft. On the other hand, if we know that $\mu F(p) = 0.85$, we know that the player's height is close to 7 ft.

Why are we saying that the height of a player whose membership in subset H is 1 is probably somewhere between 6 and 8 ft tall rather than exactly 8 ft? Which information, the crisp or the fuzzy, is more useful? Ultimately, that depends upon the designer and the problem being solved.

Example 4.1 Comparing Fuzzy Subset Membership and Probability Values – Step 1
Consider the phrase:

Etienne is old.

The phrase could also be expressed as:

Etienne is a member of the set of old people.

If Etienne is 75, one could assign a truth value of 0.8 to the statement. As a fuzzy set, this would be expressed as: $\mu_{\text{old}}(\text{Etienne}) = 0.8$.

That is, Etienne's membership in the set of old people is 80%. From what we have seen so far, membership in a fuzzy subset appears to be very much like probability. Both the degree of membership in a fuzzy subset and a probability value have the same numeric range: 0.0–1.0. Both have similar values: 0.0 indicating (complete) nonmembership for a fuzzy subset and FALSE or 0 for a probability and 1.0 indicating (complete) membership in a fuzzy subset and TRUE or 1 for a probability. What, then, is the distinction?

Consider a natural language interpretation of the results of the previous example. If a probabilistic interpretation is taken, the value 0.8 suggests that there is an 80% chance that Etienne is old. Such an interpretation supposes that Etienne is or is not old and that we have an 80% chance of knowing his age or guessing it correctly. On the other hand, if a fuzzy interpretation is taken, the value of 0.8 suggests that Etienne is more or less old (or some other term corresponding to a membership of 0.8 in the set *old*).

To further emphasize the difference between probability and fuzzy logic, let's look once again at the following example from Chapter 1.

Example 4.2 Comparing Fuzzy Subset Membership and Probability Values in Figure 4.5 – Step 2

Figure 4.6 illustrates the backs of the bottles

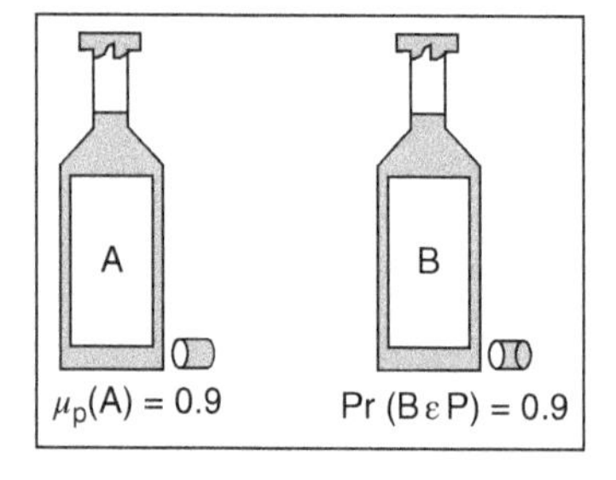

Figure 4.5 Fuzzy set membership vs. probability.

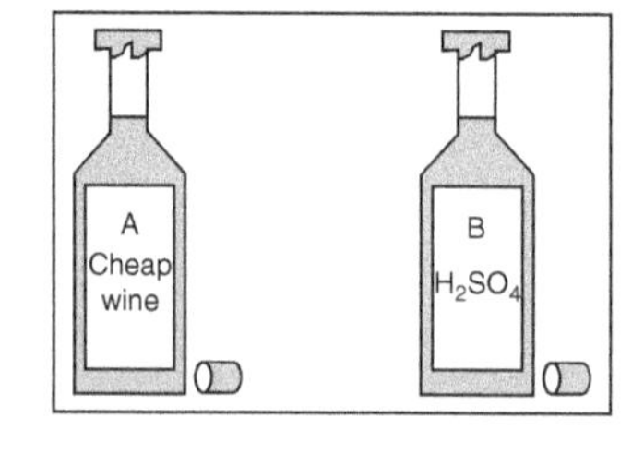

Figure 4.6 Information from the backs of the bottles.

Let

L be the set of all liquids

P be the set of all potable liquids

Suppose that you have been in the desert for a week with nothing to drink and you find two bottles. Which bottle do you choose?

Bottle A could contain wash water; however, Bottle A could not contain sulfuric acid. A fuzzy set membership value of 0.9 in the set of potable liquids means that the content of A is very similar to perfectly potable liquids, potentially water.

On the other hand, a probability of 0.9 means that over many experiments 90% yield B to be perfectly safe and that 10% yield B to be potentially unsafe. There is 1 chance in 10 that B contains a deadly liquid.

Observation:

What does the information on the back of the bottles reveal?

After examination, the fuzzy set membership value for A remains unchanged, whereas the probability value that B contains a potable liquid drops from 0.9 to 0.0. As the example illustrates, fuzzy memberships represent similarities to imprecisely defined objects, while probabilities convey information about relative frequencies.

4.7 Fuzzy Set Theory and Operations

We began with a review of some of the basic terminology, definitions, and concepts comprising the theory of ordinary or classical sets. As we now move more into to fuzzy sets, we will find a lot of similarity.

4.7.1 Fundamental Terminology

As with the crisp sets, each component or element of a fuzzy set is called a *member*. The concepts of *empty* and *size*, the relationships of *equality* and *containment*, and the operations of *union*, *intersection,* and *negation* may be defined on a fuzzy set or a collection of fuzzy sets.

We find that the following set of basic fuzzy set operators and symbols in Figure 4.7 are also traditionally used in specifying relations and operations on crisp sets and set members. In Figure 4.7, we are working with two sets and a set member:

S a nonempty set
F a fuzzy subset of set S
s a (fuzzy) set member of *F*

4.7.2 Basic Fuzzy Set Properties and Operations

As noted earlier, four basic operations are fundamental to both classical and fuzzy sets. Repeating from Chapter 3, those operations are given as follows. We'll begin by defining two crisp sets, X{1, 2, 3, 4, 5} and Y{2, 4, 7, 8}.

S	S is a nonempty set
F	*F* is a fuzzy subset of a set S
μF: S	membership function for F in S
s	s is a member of a fuzzy set
$\mu F(s)$	degree of membership of s in *F*
sup(F)	support for a subset of S whose members have nonzero membership in F
=	equality operator
≠	inequality operator
∈	element of operator
∉	not element of operator
{}	set
⊂	subset
⊆ or ≤	containment
⊇	superset
Ø	empty set
∪	union
∩	intersection
F^c	complement of a fuzzy subset

Figure 4.7 Basic fuzzy set symbols.

1) Union
   The result of the *union* of two sets (X) and (Y) includes all elements in set X *or* Y *or* both. The *or* is key here.
   The union operation is designated: X ∪ Y. The union of the two sets will be {1, 2, 3, 4, 5, 7, 8}. The union is a logical *OR* operation.
2) Intersection
   The result of the *intersection* of two sets (X) and (Y) includes all elements that are in both sets X *and* Y. The *and* is key here.
   The intersection operation is designated: X ∩ Y. The intersection of the two sets will be {2, 4}. The intersection is a logical *AND* operation.
3) Complement
   The result of the *complement* of a set (X) includes all elements in the domain of discourse, not in the designated set (X).
   The complement operation is designated: X^c or ~X. Assume that we are dealing with the two sets (X) and (Y) and that we are only dealing with eight numbers. The complement X^c of (X) will be {6, 7, 8} and Y^c of (Y) will be {1, 3, 5, 6}.
4) Difference
   The result of the *difference* between two sets (X) and (Y) includes all elements in set X and not in Y or vice versa.
   The difference between X and Y is {1, 3, 5} and between Y and X is {7, 8}.

The four basic operations are illustrated graphically using what are called Venn diagrams in Figure 4.8.

Figure 4.8 Four basic logical operations.

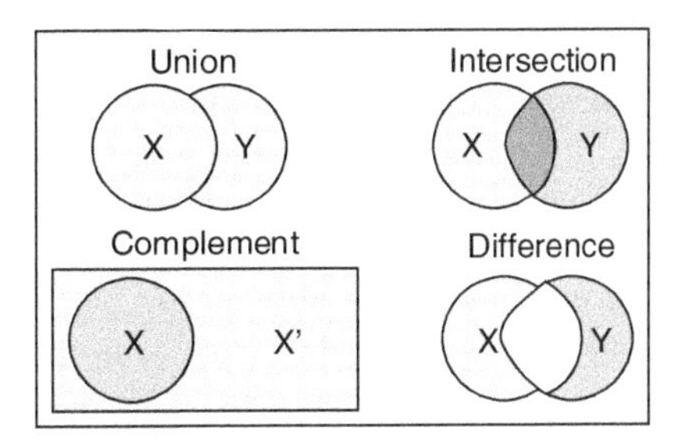

Moving now to fuzzy sets, we illustrate the concepts and definitions of *empty* and *size*, the relationships of *equality* and *containment*, and the operations of *union*, *intersection*, and *negation* in the fuzzy world. Let S (the universe of discourse) be the finite set of five arbitrary people. Now, let T and B be fuzzy subsets of S containing the tall people (T) and the blond (B) people, respectively, of the set S given in Figure 4.9.

S = {Jean, Marie, Pierre, David, Stephanie}
T = {(Jean, 0.9), (Marie, 0.7), (Pierre, 0.8), (David, 0.6), (Stephanie, 0.6)}
B = {(Jean, 0.5), (Marie, 0.9), (Pierre, 0.7), (David, 0.4), (Stephanie, 0.0)}

Figure 4.9 Basic universe of discourse with two subsets.

A fuzzy subset P, for example, may also be written using the notation expressed in Eq. (4.1).

$$P = \sum_{\sup x} \mu_P(x)/x \tag{4.1}$$

$$\sup(x) = \{x \in X | \mu P(x) \neq 0\} \tag{4.2}$$

Here, P is the fuzzy subset of X (the universe of discourse) composed of the elements x of X. The term *support* or *sup* identifies all elements x whose value meets the criteria for acceptance into a fuzzy subset of X. We read Eq. (4.2) as: the support of x is the set of all elements x in X such that the membership of x in P is not 0.0.

Using Eq. (4.1) and the data in Figure 4.9, we can express the fuzzy subset Tall as:

$$\mathrm{T} = 0.9/\mathrm{Jean} + 0.7/\mathrm{Marie} + 0.8/\mathrm{Pierre} + 0.6/\mathrm{David} + 0.6/\mathrm{Stephanie}$$

and the fuzzy subset Blond as:

$$\mathrm{B} = 0.5/\mathrm{Jean} + 0.9/\mathrm{Marie} + 0.7/\mathrm{Pierre} + 0.4/\mathrm{David} + 0.0/\mathrm{Stephanie}$$

From the subset B, Stephanie does not have blonde hair. Perhaps she has brown hair or she has colored it red, blue, or green.

Property 4.1 Empty

A subset P is empty (P = ∅) iff (if and only if) for all ordered pairs (x_{1i}, x_{2i}) in P, the element x_{2i} = 0.0 or, equivalently, $\mu_P(x_{2i})$ = 0.0.

Accordingly, neither of the subsets T or B is empty.

Property 4.2 Size

The size of a subset P is the number of ordered pairs (x_{1i}, x_{2i}) in P for which the element $x_{2i} > 0.0$ or, equivalently, $\mu_P(x_{2i}) > 0.0$.

Thus, the size of subset T is 5, that of subset B is 4.

Property 4.3 Equal

The current use of the term *equal* implies that the objects being compared are of the same type. We are not comparing pigs and trees.

Two fuzzy subsets P and R are equal (P = R) iff (if and only if)

For each element (x_{1i}, x_{2i}) in P there is an element (y_{1j}, y_{2j}) in R such that $x_{1i} = y_{1j}$ and $x_{2i} = y_{2j}$ and

for each element (y_{1j}, y_{2j}) in R there is an element (x_{1i}, x_{2i}) in P such that $y_{1j} = x_{1i}$ and $y_{2j} = x_{2i}$.

$$T = 0.9 / \text{Jean} + 0.7 / \text{Marie} + 0.8 / \text{Pierre} + 0.6 / \text{David} + 0.6 / \text{Stephanie}$$

$$B = 0.5 / \text{Jean} + 0.9 / \text{Marie} + 0.7 / \text{Pierre} + 0.4 / \text{David} + 0.1 / \text{Stephanie}$$

In the two subsets, for example, we see that Jean and Marie are members of both subsets, but Jean is taller than Marie and that Marie is more blond than Jean. There are similar differences with the remaining members; thus, the subsets T and B are not equal for several reasons.

$$T \neq B$$

Property 4.4 Containment

A fuzzy subset P is contained in a fuzzy subset R $(P \subseteq R)$ iff for each element (x_{1i}, x_{2i}) in P there is an element (y_{1j}, y_{2j}) in R such that $x_{1i} = y_{1j}$ and $x_{2i} \subseteq y_{2j}$ or, equivalently, $\mu_P(x_{1i}) \leq \mu_R(y_{1j})$.

As we saw with classical sets, the definition of containment is looser than the definition of equality, but if the relationship holds for all elements in P and all elements in R, then containment reduces to equality and P = R. Working with our two subsets:

$$T = 0.9 / \text{Jean} + 0.7 / \text{Marie} + 0.8 / \text{Pierre} + 0.6 / \text{David} + 0.6 / \text{Stephanie}$$

$$B = 0.5 / \text{Jean} + 0.9 / \text{Marie} + 0.7 / \text{Pierre} + 0.4 / \text{David} + 0.0 / \text{Stephanie}$$

Here, the fuzzy subset T is not contained in B and vice versa.

Reversing Marie's degree of membership in the two subsets:

$$T = 0.9 / \text{Jean} + 0.9 / \text{Marie} + 0.8 / \text{Pierre} + 0.6 / \text{David} + 0.6 / \text{Stephanie}$$

$$B = 0.5 / \text{Jean} + 0.7 / \text{Marie} + 0.7 / \text{Pierre} + 0.4 / \text{David} + 0.0 / \text{Stephanie}$$

Now the fuzzy subset B is contained in T.

Property 4.5 Union

The union of two fuzzy sets is defined as the set containing all the elements that are members of either individual set alone or members of both sets. Consider the two sets, T and B; see Figure 4.9. The union is expressed as T ∪ B from which we have:

$$T \cup B = \{x \mid x \in T \text{ or } x \in B\}$$

The union operation can be extended for any number of sets.

The union of two fuzzy subsets P and R ($P \cup R$) is defined as the fuzzy subset of ordered pairs (z_{1k}, z_{2k}) formed from all ordered pairs (x_{1i}, x_{2i}) in P and (y_{1j}, y_{2j}) in R such that $x_{1i} = y_{1j} = z_{1k}$ and $z_{2k} = \max(x_{2i}, y_{2j})$ or, equivalently, $\max(\mu_P(x_{1i}), \mu_R(y_{1j}))$.

Why do we apply the max operator to the degree of membership pair?

The interpretation of the members of an ordered pair is as follows. The first member is an instance of the entity type within the fuzzy subset, and the second member is the strength or degree of membership of an instance of that entity within the fuzzy subset.

When the union of two (or more) pairs is performed, the result is also an ordered pair. The first member of the resulting pair is the type of the comprising entities within the fuzzy subset, and the second member, using the max operator, is the largest grade of membership of the two (or more) of the comprising entities.

Does such a definition give results that are consistent with our expectations? Also, refer back to the crisp implementation of a union.

Now, working with the fuzzy subsets of T, the tall people, and B, the blond people, in the universe of discourse S given in Figure 4.9, we have:

$$T = 0.9/\text{Jean} + 0.7/\text{Marie} + 0.8/\text{Pierre} + 0.6/\text{David} + 0.6/\text{Stephanie}$$

$$B = 0.5/\text{Jean} + 0.9/\text{Marie} + 0.7/\text{Pierre} + 0.4/\text{David} + 0.0/\text{Stephanie}$$

Here, the union of T and B is given as:

$$T \cup B = \{ (\text{Jean}, \max(0.9, 0.5)), (\text{Marie}, \max(0.7, 0.9)), (\text{Pierre}, \max(0.8, 0.7)), (\text{David}, \max(0.6, 0.4)), (\text{Stephanie}, \max(0.6, 0.0)) \}$$

We now select the max grade of membership for each of the resulting entries in the union. Does this make sense?

$$T \cup B = \{ (\text{Jean}, 0.9), (\text{Marie}, 0.9), (\text{Pierre}, 0.8), (\text{David}, 0.6), (\text{Stephanie}, 0.6) \}$$
$$= 0.9/\text{Jean} + 0.9/\text{Marie} + 0.8/\text{Pierre} + 0.6/\text{David} + 0.6/\text{Stephanie}$$

When expressed in a natural language, the union of these two fuzzy subsets describes the subset of people who are either tall or blond.

> If Jean's membership in the fuzzy subset Tall is 0.9, then we may write Jean is a tall person.
> If Jean's membership in the fuzzy subset Blond is 0.5, then we may write Jean is more or less blond.

It seems natural, therefore, that Jean's tallness rather than his blondness should determine the strength of his membership in the composite subset. Why? Because that is the larger or stronger grade of membership.

The application of the max operator to the membership functions of the composite subsets does, in fact, yield a result that is consistent our intuitive ideas. As we mentioned in the discussion of crisp sets, the union operation is an OR type function.

If one takes a probabilistic interpretation of the union of two fuzzy subsets, one reaches a somewhat different result. For independent events, which are certainly tallness and blondness:

$$P(T \cup B) = P(T) + P(B) - P(T)P(B)$$

If membership functions and probabilities were equal, then for Jean:

$$T \cup B = (\text{Jean}, \max(0.9, 0.5) \rightarrow 0.9) \quad //\text{grade of membership}$$

$$\begin{aligned} P(T \cup B) &= 0.9 + 0.5 - 0.9 \cdot 0.5 \\ &= 1.4 - 0.45 \\ &= 0.95 \quad //\ \text{probability} \end{aligned}$$

which is certainly different from 0.9.

Property 4.6 Intersection

The intersection of two sets is the set containing all the elements that are members of both sets. Consider the two sets T and B in Figure 4.9. The intersection is expressed as T ∩ B from which we have:

$$T \cap B = \{x | (x \in T \text{ and } x \in B)\}$$

That is, for all elements "x," each must be a member of both sets. The consequence of such a requirement leads to:

$$T \cap T^c = 0$$

since a set and its complement cannot have elements in common.

That said, working once again with the fuzzy subsets T tall people and B blond people in the universe of discourse S given in Figure 4.9, we have:

$$T = 0.9/\text{Jean} + 0.7/\text{Marie} + 0.8/\text{Pierre} + 0.6/\text{David} + 0.6/\text{Stephanie}$$

$$B = 0.5/\text{Jean} + 0.9/\text{Marie} + 0.7/\text{Pierre} + 0.4/\text{David} + 0.0/\text{Stephanie}$$

In the following analysis, for each ordered pair, the terms x_{1i} and y_{1i} refer to the names of the individual people and x_{2i} and y_{21} refer to the grade of membership for each person in the two sets. The intersection of two fuzzy subsets P and R written as (P ∩ R) is the fuzzy subset of ordered pairs (z_{1k}, z_{2k}) formed from all ordered pairs (x_{1i}, x_{2i}) in P and (y_{1j}, y_{2j}) in R such that the names and grades of membership meet the following criteria: $x_{1i} = y_{1j} = z_{1k}$ and $z_{2k} = \min(x_{2i}, y_{2j})$ or, equivalently, $\min(\mu_P(x_{1i}), \mu_R(y_{1j}))$.

Here, the intersection of T and B is given as:

$$\begin{aligned} T \cap B = \{ & \\ & (\text{Jean}, \min(0.9, 0.5)), (\text{Marie}, \min(0.7, 0.9)), \\ & (\text{Pierre}, \min(0.8, 0.7)), (\text{David}, \min(0.6, 0.4)), \\ & (\text{Stephanie}, \min(0.6, 0.0)) \\ \} & \end{aligned}$$

from which we have:

$$\begin{aligned} T \cap B = \{ & \\ & (\text{Jean}, 0.5), (\text{Marie}, 0.7), (\text{Pierre}, 0.7), (\text{David}, 0.4), \\ & (\text{Stephanie}, 0.0) \\ \} & \\ = & \; 0.5/\text{Jean} + 0.7/\text{Marie} + 0.7/\text{Pierre} + 0.4/\text{David} + \\ & \; 0.0/\text{Stephanie} \end{aligned}$$

When expressed in a natural language, the intersection of these two fuzzy subsets describes the subset of people who are both tall and blond. Also, refer back to the crisp implementation of an intersection.

Following the same line of reasoning we used to justify the definition of union, it should now be Jean's blondness rather than his tallness that determines the strength of his membership in the composite subset. Why is this?

As we mentioned in the discussion of crisp sets, the intersection operation is an AND type function.

Once again, if we take a probabilistic interpretation of the intersection of two fuzzy subsets, the joint probability of two independent events is given by:

$$P(T \cap B) = P(T)P(B)$$

If membership functions and probabilities were equal, then for Jean:

$$\begin{aligned} T \cap B &= 0.9 \bullet 0.5 \\ &= 0.45 \quad // \text{ which is different from } 0.5. \end{aligned}$$

Property 4.7 Negation

The *negation* or *complement* of a fuzzy subset P (P′ or P^c) is the set of all ordered pairs (z_{1k}, z_{2k}) such that for all ordered pairs (x_{1i}, x_{2i}) in P, the element $z_{1k} = x_{1i}$and $z_{2k} = 1 - x_{2i}$ (the complement of z_{2k}) or, equivalently, $1 - \mu_P(x_{1i})$. See Figure 4.9.

$$T = 0.9/\text{Jean} + 0.7/\text{Marie} + 0.8/\text{Pierre} + 0.6/\text{David} + 0.6/\text{Stephanie}$$

$$B = 0.5/\text{Jean} + 0.9/\text{Marie} + 0.7/\text{Pierre} + 0.4/\text{David} + 0.0/\text{Stephanie}$$

Here, the negation or complement of the subsets T and B is given as:

$$T^c = \{ (\text{Jean}, 1-0.9), (\text{Marie}, 1-0.7), (\text{Pierre}, 1-0.8), (\text{David}, 1-0.6), (\text{Stephanie}, 1-0.6) \}$$

$$= 0.1/\text{Jean} + 0.3/\text{Marie} + 0.2/\text{Pierre} + 0.4/\text{David} + 0.4/\text{Stephanie}$$

$$B^c = \{ (\text{Jean}, 1-0.5), (\text{Marie}, 1-0.9), (\text{Pierre}, 1-0.7), (\text{David}, 1-0.4), (\text{Stephanie}, 1-0.0) \}$$

$$= 0.5/\text{Jean} + 0.1/\text{Marie} + 0.3/\text{Pierre} + 0.6/\text{David} + 1.0/\text{Stephanie}$$

Property 4.8 Commutation
The commutation (interchange) of two subsets is given as follows. Here, the commutation of subsets T and B is given as:

Union Operation

$$T \cup B = B \cup T$$

$$T \cup B = \{(\text{Jean}, \max(0.9, 0.5)), (\text{Marie}, \max(0.7, 0.9)), (\text{Pierre}, \max(0.8, 0.7)), (\text{David}, \max(0.6, 0.4)), (\text{Stephanie}, \max(0.6, 0.0))\}$$

$$B \cup T = \{(\text{Jean}, \max(0.5, 0.9)), (\text{Marie}, \max(0.9, 0.7)), (\text{Pierre}, \max(0.7, 0.8)), (\text{David}, \max(0.4, 0.6)), (\text{Stephanie}, \max(0.0, 0.6))\}$$

which gives:

$$\{(\text{Jean}, 0.9), (\text{Marie}, 0.9), (\text{Pierre}, 0.8), (\text{David}, 0.6), (\text{Stephanie}, 0.6)\}$$

$$0.9/\text{Jean} + 0.9/\text{Marie} + 0.8/\text{Pierre} + 0.6/\text{David} + 0.6/\text{Stephanie}$$

Intersection Operation

$$T \cap B = B \cap T$$

$$T \cap B = \{(\text{Jean}, \min(0.9, 0.5)), (\text{Marie}, \min(0.7, 0.9)), (\text{Pierre}, \min(0.8, 0.7)), (\text{David}, \min(0.6, 0.4)), (\text{Stephanie}, \min(0.6, 0.0))\}$$

$$B \cap T = \{(\text{Jean}, \min(0.5, 0.9)), (\text{Marie}, \min(0.9, 0.7)), (\text{Pierre}, \min(0.7, 0.8)), (\text{David}, \min(0.4, 0.6)), (\text{Stephanie}, \min(0.0, 0.6))\}$$

from which we have:

$$\{(\text{Jean}, 0.5), (\text{Marie}, 0.7), (\text{Pierre}, 0.7), (\text{David}, 0.4), (\text{Stephanie}, 0.0)\}$$

$$0.5 / \text{Jean} + 0.7 / \text{Marie} + 0.7 / \text{Pierre} + 0.4 / \text{David} + 0.0 / \text{Stephanie}$$

Property 4.9 Associativity
The associativity property states that an operation on a group of subsets is independent of how the subsets are grouped with respect to the operation.

Consider the following three subsets and the union operation:

$$A = \{(\text{Jean}, 0.9), (\text{Marie}, 0.6), (\text{Pierre}, 0.3), (\text{David}, 0.6)\}$$

$$B = \{(\text{Jean}, 0.5), (\text{Marie}, 0.7), (\text{Pierre}, 0.9), (\text{David}, 0.6)\}$$

$$C = \{(\text{Jean}, 0.4), (\text{Marie}, 0.8), (\text{Pierre}, 0.6), (\text{David}, 0.5)\}$$

Here, the associativity of subsets A, B, and C under the union and intersection operations is given as:

Union Operation

$$(A \cup B) \cup C = A \cup (B \cup C)$$

$$(A \cup B) = \{(\text{Jean}, \max(0.9, 0.5)), (\text{Marie}, \max(0.6, 0.6)), (\text{Pierre}, \max(0.3, 0.9)), (\text{David}, \max(0.6, 0.6))\} \cup$$

$$C = \{(\text{Jean}, 0.4), (\text{Marie}, 0.8), (\text{Pierre}, 0.6), (\text{David}, 0.5)\}$$

$$(A \cup B) \cup C = \{(\text{Jean}, \max(0.9, 0.4)), (\text{Marie}, \max(0.6, 0.8)), (\text{Pierre}, \max(0.9, 0.6)), (\text{David}, \max(0.6, 0.5))\}$$
$$= \{(\text{Jean}, 0.9), (\text{Marie}, 0.8), (\text{Pierre}, 0.9), (\text{David}, 0.6)\}$$

$$A = \{(\text{Jean}, 0.9), (\text{Marie}, 0.6), (\text{Pierre}, 0.3), (\text{David}, 0.6)\}$$

$$(B \cup C) = \{(\text{Jean}, \max(0.5, 0.4)), (\text{Marie}, \max(0.7, 0.8)), (\text{Pierre}, \max(0.9, 0.6)), (\text{David}, \max(0.6, 0.5))\}$$

$$A \cup (B \cup C) = \{(\text{Jean}, 0.9), (\text{Marie}, 0.8), (\text{Pierre}, 0.9), (\text{David}, 0.6)\}$$

Property 4.10 Distributivity

The distributivity of two subsets is given as follows.

Consider the following three subsets and the union operation:

$$A=\{(\text{Jean}, 0.9), (\text{Marie}, 0.6), (\text{Pierre}, 0.3), (\text{David}, 0.6)\}$$

$$B=\{(\text{Jean}, 0.5), (\text{Marie}, 0.7), (\text{Pierre}, 0.9), (\text{David}, 0.6)\}$$

$$C=\{(\text{Jean}, 0.4), (\text{Marie}, 0.8), (\text{Pierre}, 0.6), (\text{David}, 0.5)\}$$

Here, the distributivity of subsets A, B, and C under the union operation is given as:

Union Operation

$$A\cup(B\cap C)=(A\cup B)\cap(A\cup C)$$

$$A = \{(\text{Jean}, 0.9), (\text{Marie}, 0.6), (\text{Pierre}, 0.3), (\text{David}, 0.6)\}$$

$$\cup$$

$$(B\cap C)= \{(\text{Jean}, \min(0.5, 0.4)), \text{Marie}, \min((0.7, 0.8)), (\text{Pierre}, \min(0.9, 0.6)), (\text{David}, \min(0.6\ 0.4)\}$$

$$A\cup(B\cap C)=\{(\text{Jean}, \max(0.9, 0.4)), (\text{Marie}, \max(0.6, 0.7)), (\text{Pierre}, \max(0.3\ 0.6)), (\text{David}, \max(0.6, 0.4)$$

$$=\{(\text{Jean}, 0.9), (\text{Marie}, 0.7), (\text{Pierre}, 0.6), (\text{David}, 0.6)\}$$

$$(A\cup B)=\{(\text{Jean}, \max(0.9, 0.5), \text{Marie}, \max(0.6, 0.7), (\text{Pierre}, \max(0.3, 0.9), (\text{David}, \max(0.6, 0.6)\}$$

$$\cap$$

$$(A\cup C)=\{(\text{Jean}, \max(0.9, 0.4), \text{Marie}, \max(0.6, 0.8), (\text{Pierre}, \max(0.3, 0.6), (\text{David}, \max(0.6, 0.5)\}$$

$$(A\cup B)\cap(A\cup C)=\{(\text{Jean}, \min(0.9, 0.9)), \text{Marie}, \min(0.7, 0.8)), (\text{Pierre}, \min(0.9, 0.6)), (\text{David}, \min(0.6\ 0.6)\}$$

$$= \{(\text{Jean}, 0.9), (\text{Marie}, 0.7), (\text{Pierre}, 0.6), (\text{David}, 0.6)\}$$

Property 4.11 Idempotence

The idempotence property of an operation on a subset is given as follows.

The idempotent property of an operation means that the operation can be repeatedly applied without changing the result after the first application.

The idempotent property of an operation on a subset A is given under the union operation as:

Union Operation

$$A \cup A \cup A = A$$

Consider the fuzzy subset A.

$$A = \{(\text{Jean}, 0.9), (\text{Marie}, 0.6), (\text{Pierre}, 0.3), (\text{David}, 0.6)\}$$

$$\begin{aligned} A \cup A = \{&(\text{Jean}, \min(0.9, 0.9)), \text{Marie}, \min(0.6, 0.6)), \\ &(\text{Pierre}, \min(0.3, 0.3)), (\text{David}, \min(0.6\ 0.6)\} \\ = \{&(\text{Jean}, 0.9), (\text{Marie}, 0.6), (\text{Pierre}, 0.3), (\text{David}, 0.6) \end{aligned}$$

$$\begin{aligned} A \cup A \cup A = \{&(\text{Jean}, \min(0.9, 0.9)), \text{Marie}, \min(0.6, 0.6)), \\ &(\text{Pierre}, \min(0.3, 0.3)), (\text{David}, \min(0.6\ 0.6)\} \\ = \{&(\text{Jean}, 0.9), (\text{Marie}, 0.6), (\text{Pierre}, 0.3), (\text{David}, 0.6)\} \end{aligned}$$

Property 4.12 Excluded Middle
The excluded middle property does not apply for a fuzzy subset.
Why would that be?

Property 4.13 Contradiction
The contradiction property does not apply for a fuzzy subset.
Why would that be?

Property 4.14 Complement
The complement of a fuzzy subset *F* is given as:

$$\mu F(s)^c = 1 - \mu F(s)$$

The complement may also be written as: $\sim\mu F(s)$.

The complement of the membership of s in the fuzzy subset *F* is 1 minus the membership of s in *F*.

$$A = \{(\text{Jean}, 0.9), (\text{Marie}, 0.7), (\text{Pierre}, 0.8), (\text{David}, 0.6), (\text{Stephanie}, 0.6)\}$$

$$A^c = \{(\text{Jean}, 0.1), (\text{Marie}, 0.3), (\text{Pierre}, 0.2), (\text{David}, 0.4), (\text{Stephanie}, 0.4)\}$$

Property 4.15 De Morgan
The application of De Morgan's Law on a pair of subsets is given as follows.

$$A = \{(\text{Jean}, 0.9), (\text{Marie}, 0.7), (\text{Pierre}, 0.8), (\text{David}, 0.6), (\text{Stephanie}, 0.6)$$

$$B = \{(\text{Jean}, 0.5), (\text{Marie}, 0.9), (\text{Pierre}, 0.7), (\text{David}, 0.4), (\text{Stephanie}, 0.1)\}$$

$$(A \cup B)^c = A^c \cap B^c$$

$$(A \cap B)^c = A^c \cup B^c$$

$$\begin{aligned} A \cup B = \{&(\text{Jean}, \max(0.9, 0.5)), (\text{Marie}, \max(0.7, 0.9)), \\ &(\text{Pierre}, \max(0.8, 0.7)), (\text{David}, \max(0.6, 0.4)), \\ &(\text{Stephanie}, \max(0.6, 0.1))\} \end{aligned}$$

$$\begin{aligned} (A \cup B)^c = \{&(\text{Jean}, 0.1), (\text{Marie}, 0.1), (\text{Pierre}, 0.2), \\ &(\text{David}, 0.4), (\text{Stephanie}, 0.4)\} \end{aligned}$$

$$\begin{aligned} A^c \cap B^c = \{&(\text{Jean}, \min(0.1, 0.5), (\text{Marie}, \min(0.3, 0.1), \\ &(\text{Pierre}, \min(0.2, 0.3), (\text{David}, \min(0.4, 0.6), \\ &(\text{Stephanie}, \min(0.4.0.9)\} \\ = \{&(\text{Jean}, 0.1)), (\text{Marie}, 0.1), (\text{Pierre}, 0.2), \\ &(\text{David}, 0.4), (\text{Stephanie}, 0.4)\} \end{aligned}$$

Property 4.16 Difference

The difference of two fuzzy sets A and B is a new fuzzy set A-B, which is defined as:

$$A \quad B = (A \cap B^c)$$

$$A = \{(\text{Jean}, 0.9), (\text{Marie}, 0.7), (\text{Pierre}, 0.8), (\text{David}, 0.6), (\text{Stephanie}, 0.6)\}$$

$$B = \{(\text{Jean}, 0.5), (\text{Marie}, 0.9), (\text{Pierre}, 0.7), (\text{David}, 0.4), (\text{Stephanie}, 0.1)\}$$

$$\begin{aligned} B^c &= 1 - B \\ &= \{(\text{Jean}, 0.5), (\text{Marie}, 0.1), (\text{Pierre}, 0.3), (\text{David}, 0.6), (\text{Stephanie}, 0.9)\} \end{aligned}$$

$$(A \cap B^c) =$$

$$\begin{aligned} A = \{&(\text{Jean}, 0.9), (\text{Marie}, 0.7), (\text{Pierre}, 0.8), (\text{David}, 0.6), \\ &(\text{Stephanie}, 0.6) \\ &\cap \end{aligned}$$

$$\begin{aligned} B^c = \{&(\text{Jean},0.5)),(\text{Marie},0.1),(\text{Pierre},0.3),(\text{David},0.6), \\ &(\text{Stephanie},0.9)\} \\ = \{&(\text{Jean},\min(0.9,0.5)),(\text{Marie},\min(0.7,0.1)), \\ &(\text{Pierre},\min(0.8,0.3)),(\text{David},\min(0.6,0.6)) \\ &(\text{Stephanie},\min(0.6,0.9))\} \\ = \{&(\text{Jean}(0.5)),(\text{Marie}(0.1)), \\ &(\text{Pierre}(0.3)),(\text{David}(0.6)), \\ &(\text{Stephanie}(0.6,))\} \end{aligned}$$

4.8 Basic Fuzzy Applications – A First Step

We have looked at the basic properties of fuzzy sets and the application of crisp tools. Let's now take a first look at how we might begin to apply those tools in the fuzzy world. We'll start with a model that is subject to the membership constraints of the crisp world.

4.8.1 A Crisp Activity Revisited

In Chapter 3, we proposed an activity to be two friends, John and Mary, going to a movie. From a crisp perspective, each person's view is constrained to either liking the movie or not liking the movie. Thus, each person's opinion of the movie can be assigned a membership value in the set *like* of either 0 or 1, that is, either *not* in the set *like* or in the set *like*. Combining the two possible opinions from John and Mary gives a crisp set, *Like*, with four possible combinations or four possible crisp subsets.

The scenario, from the previous chapter, is expressed in Figure 4.10 graphically and as a truth table.

Figure 4.10 Viewing a movie.

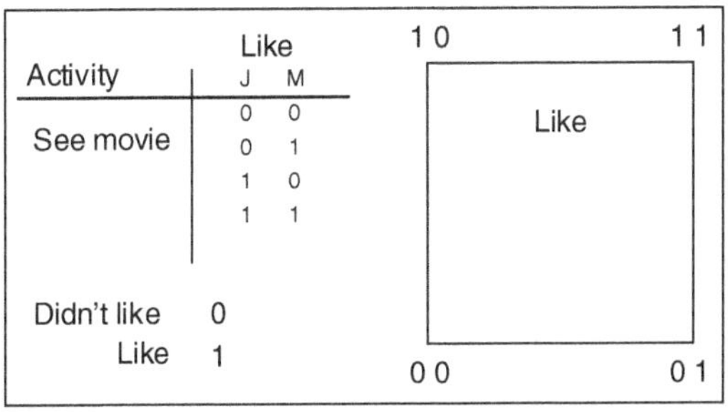

Activity	Like J	Like M
See movie	0	0
	0	1
	1	0
	1	1

Didn't like 0
Like 1

1 0 | 1 1
Like
0 0 | 0 1

In the diagram, we have the classic or crisp set *Like*, L, that has four crisp subsets. Each crisp subset has two members expressing membership in the set *Like*: *not a member* – (0) and a member, *like* – (1), reflecting John's and Mary's possible opinions of the movie. Each two-member subset occupies one of the four corners of the diagram labeled *Like*.

The membership functions in the four subsets, expressing possible opinions of the attendees, are given as:

$$\mu_L(J)=0, J \notin L \quad \mu L(M)=0, M \notin L$$

$$\mu_L(J)=1, J \in L \quad \mu L(M)=1, M \in L$$

Here, the intersection of John's and Mary's opinions of the movie gives:

$$\text{John} \cap \text{Mary} = \{(0,0),(0,1),(1,0),(1,1)\}$$

Example 4.3 A Fuzzy Activity – A First Step

That said, let's look at this model from a slightly different perspective. As we noted in our initial view of the model, the possible opinions of the movie were limited to those expressed at the four corners of the block, which leaves the *Like* block internally void.

If we look closer at the block's four edges, we note that each of the sides is one unit in length. The one-unit length has no inherent meaning, however. We are simply using the endpoints as a tool to reflect or indicate membership in a subset. Let's now increase the resolution of an opinion to include maybe. An opinion can now be 0.0, 0.5, 1.0.

Let's examine the bottom edge of the block in greater detail. From the two extremes of that edge, we observe that at one endpoint, we have the pair (0 0) expressing the opinions {0.0, 0.0}, indicating that neither person's opinion was a *liked* member in the subset. That is, neither person liked the movie. On the opposite endpoint, we have (0 1) expressing the subset {0.0, 1.0}, indicating that one person's opinion was a *liked* member in the subset and the other person's was not. Using our model, what might a subset or pair of values in the middle of that edge reflect?

Following our model, a subset at the midpoint of the edge could have one member with a value of 0.0 and a second member with a value of 0.5. What would this mean? Is 0.5 a crisp value?

Using the same reasoning that we used with the two extreme opinions of the film, we can use the values in such a subset to represent the fact that one person still did not like the movie, i.e. had a 0.0 membership in the *liked* set. That is, they still disliked the film, and the other person had a membership of 0.5 in the *liked* set. We can interpret such a membership as a somewhat mixed opinion or membership in the *liked* set.

Our set membership expressing John's and Mary's opinions now becomes:

$$\text{John} = \{(0.0),(0.5),(1.0)\}$$
$$\text{Mary} = \{(0.0),(0.0),(0.0)\}$$

If we allow Mary's opinion of the movie, that is, permit membership in the *liked* set, to range between 0 and 1, we have taken the first step into the fuzzy world. Mary's opinion can now range from total *dislike* {0.0}, through degrees of like {0.0 to 1.0}, and ultimately to a total *like* of {1.0}.

$$\text{John} = \{(0.0),(0.5),(1.0)\}$$
$$\text{Mary} = \{(0.0),(0.5),(1.0)\}$$

As a next step, we can allow one or the other (but not both) person's membership in the *liked* set to take values along an edge while the others are constrained to 0.0 or 1.0. Such alternatives are reflected in Figure 4.11a and b.

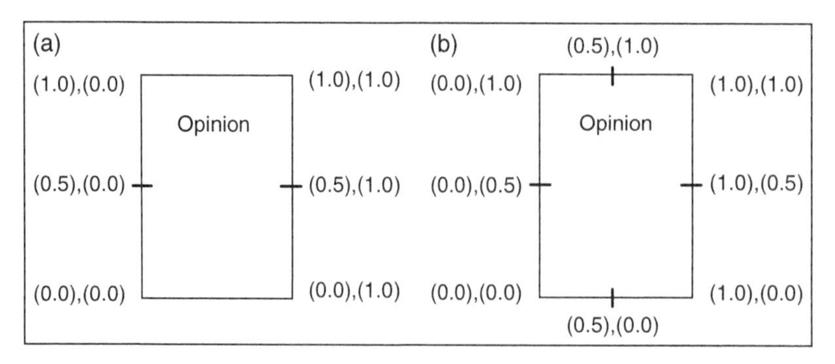

Figure 4.11 Expressing an opinion-1, 2.

Staying on the four-edge periphery, we can relax our constraints to permit one person to voice an opinion as represented by any position on one of two opposing edges and the other to express a view as represented on either of the two remaining opposing edges as illustrated in Figure 4.12.

Finally, moving to any position on either of the two edges as shown in Figure 4.13, we could have subsets in which both memberships could have (different) values ranging from 0.0 to 1.0.

Staying with our line of reasoning, we could use such subsets to represent the possibility that both people have somewhat mixed opinions.

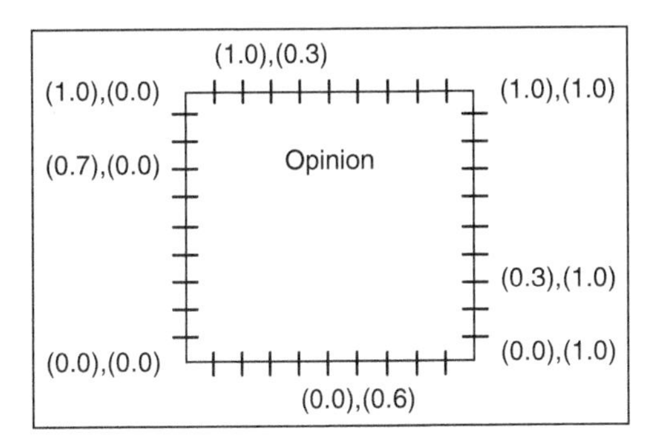

Figure 4.12 Expressing an opinion-3.

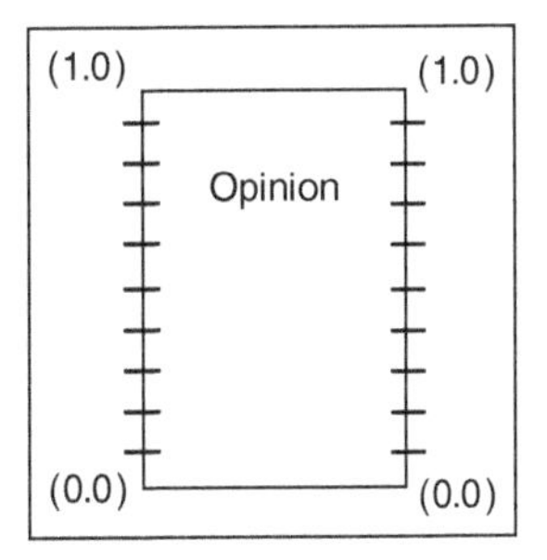

Figure 4.13 Expressing an opinion-4.

Example 4.4 Fuzzy Activity – A Second Step

We now expand the activity from a simple activity to a set of many activities. We define:

Fun be a fuzzy subset of enjoyable activities.

Inexpensive be the fuzzy subset of inexpensive activities.

Table 4.1 lists activities and their grades of membership in the two fuzzy subsets.

Table 4.1 Activities and grades of membership.

Activity	Fun	Inexpensive
Walking	0.5	0.9
Film	0.3	0.8
Theatre	0.8	0.7
Dining	0.7	0.6
Hot air balloon	0.2	0.5
Travel	0.9	0.4

The fuzzy subset of all activities that are either fun or inexpensive is given as:

$$
\begin{aligned}
F \cup I = {} & 0.9/\text{Walking} + 0.8/\text{Film} + 0.8/\text{Theatre} + 0.7/\text{Dining} \\
& + 0.5/\text{Hot Air Balloon} + 0.9/\text{Travel}
\end{aligned}
$$

We have the union of two sets. (MAX)

The fuzzy subset of all activities that are both fun and inexpensive is given as:

$$F \cap I = 0.5/\text{Walking} + 0.3/\text{Film} + 0.7/\text{Theatre} + 0.6/\text{Dining}$$
$$+ 0.2/\text{Hot Air Balloon} + 0.4/\text{Travel}$$

We have the intersection of two sets. (MIN)

4.9 Fuzzy Imprecision And Membership Functions

We have been working with grades of membership that express an entity's level or grade of membership within a set or subset as a numeric digit in the range [0.0, 1.0]. Earlier we used an expression of the form $\mu_S(x)$ as the membership function to map entities (x) in a universe of discourse (S) onto membership values or grades of membership in the range [0.0, 1.0] in a designated subset. Let's now take a first look at a graphical approach to affect the designated mapping.

The fuzzy logic membership graph follows the traditional X, Y graph format. The vertical Y axis presents possible grades of membership as a set of single fractional decimal digits spanning the range [0.0–1.0]. The horizontal X axis presents the *domain* or the range of values of the set of tentative members of a fuzzy subset.

The graphical membership function is expressed as a subjective or arbitrary curve defined to suit the requirements of the problem or design at hand. Examples of such membership functions include graphs formulated from straight or curved lines, triangular or rectangular shapes, or quadratic or polynomial curves. The various shapes and constraints within the graphs will enable us to specify and control how our (potentially imprecise) signals or data will affect the behavior of the system under design and ultimately operation. Let's now look at some introductory examples of working with graphical membership functions.

Example 4.5 Crisp vs Fuzzy Using a Graphic Membership Function

Starting with a crisp graphic, in the United States the typical age of attaining legal adulthood is 18. The crisp and fuzzy membership graphs for the set of people under age 18 or over 18 are given in Figure 4.14.

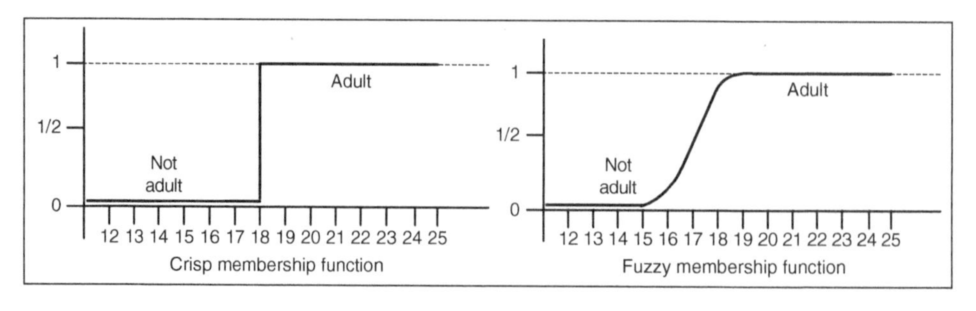

Figure 4.14 Simple crisp and fuzzy adult membership functions.

4.9.1 Linear Membership Functions

The straight-line graph is generally among the simplest and easiest fuzzy set or subset to approximate unknown concepts. We begin our study of graphic membership functions with the basic linear model and then examine the curved version.

Example 4.6 Basic Linear Membership Function
We start with the two fundamental straight-line graphs presented in Figure 4.15. At this stage, no specific meaning other than increasing or decreasing has been associated with the graphs. The basic straight-line graph supports two possible versions: one with states increasing and one with states decreasing. The former starts at a domain value expressing [0.0] membership and increases to full membership [1.0] on the left-hand side. The graph on the right starts with full membership [1.0] and decreases to [0.0].

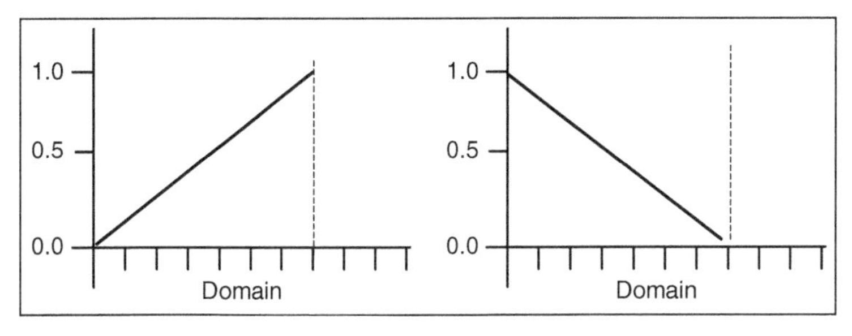

Figure 4.15 Linear increasing and decreasing fuzzy membership functions.

Possible membership values are indicated on the vertical or Y axis. Values or degrees of membership range in 0.1 increments from [0.0] to [1.0]. Data on the horizontal or X axis indicates the values of the fuzzy subset variables that are intended as members of the fuzzy set or subset under analysis. The range of values on the horizontal axis, the problem portion of the design, and the full design under analysis are determined by the designer.

Figure 4.16 illustrates a linear graph with the degree of membership in a fuzzy subset being examined. The degrees of membership in the fuzzy subset for entities with the values 11 and 14 reflect the degrees of membership of 0.25 and 0.75, respectively. The graph shows

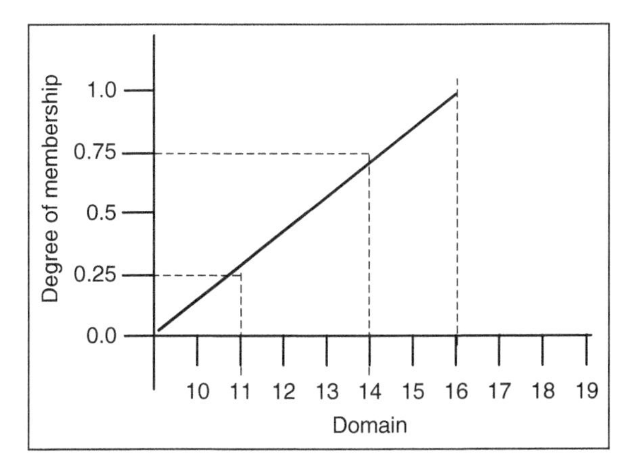

Figure 4.16 Degree of membership.

that the entity with a value of 14 has a much stronger degree of membership in the specific subset, which can be very good or very bad depending upon the particular design.

Example 4.7 Basic Linear Membership Function Linear Up and Linear Down
The following straight-line graphs are similar to those presented in Figure 4.15. Each of the graphs in Figure 4.17 specifies a range of values supported by the fuzzy subset denoted a *span* and a possible change in state, increasing or decreasing, starting from [0.0] or [1.0]. The figures also specify a portion of each span over which there is no change in the membership value.

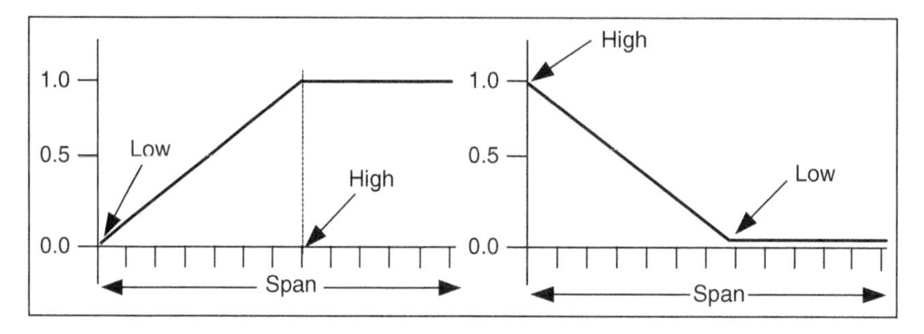

Figure 4.17 Linear increasing and decreasing fuzzy membership functions.

Example 4.8 Basic Linear Membership Function Linear Up and Down with Offset
The straight-line graphs presented in Figure 4.18 are similar to those presented in Figure 4.17. The graphs in Figure 4.18 also illustrate two possible state changes, increasing or decreasing, starting from [0.0] or [1.0] and also specify an offset, a portion of the span that specifies a duration before activity takes place.

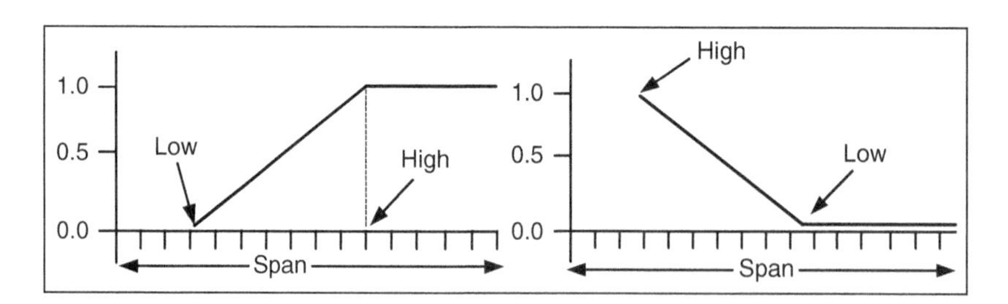

Figure 4.18 Increasing and decreasing fuzzy membership functions.

Example 4.9 Basic Linear and Bell Shaped Membership Functions – Around Graph
The straight-line graph on the right in Figure 4.19 representing a fuzzy number illustrates an example of the concept of a number, in this case 5, which has a grade of membership in a fuzzy set of 1. Surrounding the number 5 are numbers that are *approximately equal* to, *close* to, *near* to, or *around* 5 but have a lower grade of membership than 5 in the subset. In the figure, the subset on the left is first modeled as a bell curve and on the right as a basic linear triangle. Laying the two graphics on top of one another will show how close the two

are. In the diagram, once again, we can see that each of the graphs is centered around the number 5, hence the name *Around* graph.

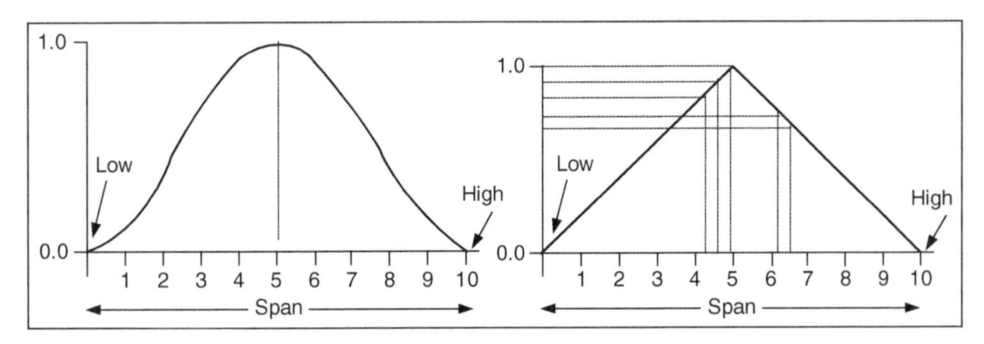

Figure 4.19 Basic linear and bell-shaped fuzzy membership functions around 5.

Example 4.10 Basic Linear Membership Functions Around and Restricted Domains
The straight-line graph representing a fuzzy number in Figure 4.19 illustrated an example of values around a number, in that case 5. The graphs in Figure 4.20 illustrate an *around* graph for numbers around 50 with a domain 20–80 labeled X_0–X_1. The same basic graph is then illustrated for a case with the subrange of the domain elements labeled X_a–X_b restricted.

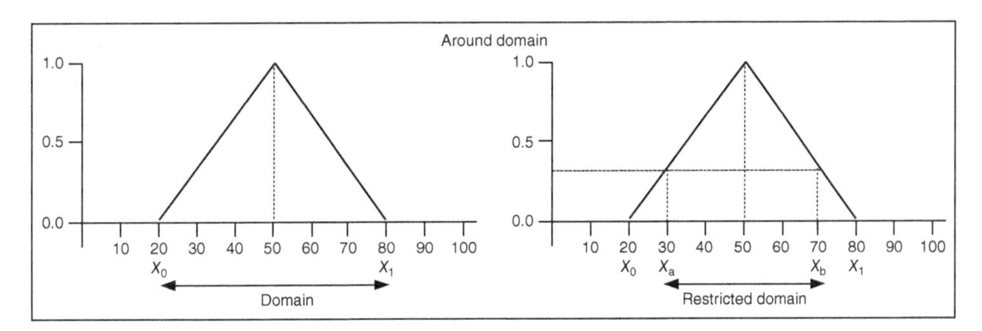

Figure 4.20 Basic linear membership functions around and restricted domains.

Example 4.11 Membership Function Illustrating a Universe of Discourse
When we pursue a new design, often one of the early steps is to design and build a working prototype model of the system or a portion of the system. As the design evolves and grows, so does the number of variables that are part of that design. When we work in a fuzzy world, those variables are described in terms of spaces in that world. As a design evolves, so does the number of spaces, which may often also be overlapping. Such overlaps reflect the fact that the associated variables may be interacting with one another.

The diagram in Figure 4.21 reflects such a situation. Observe that the fuzzy spaces for the variables A, B, and C in Figure 4.21 are partially overlapping, that is, shared. The sources for both fuzzy subsets A and B have control of signals in the range 20–30, and B and C have control of signals in the range of 70–80.

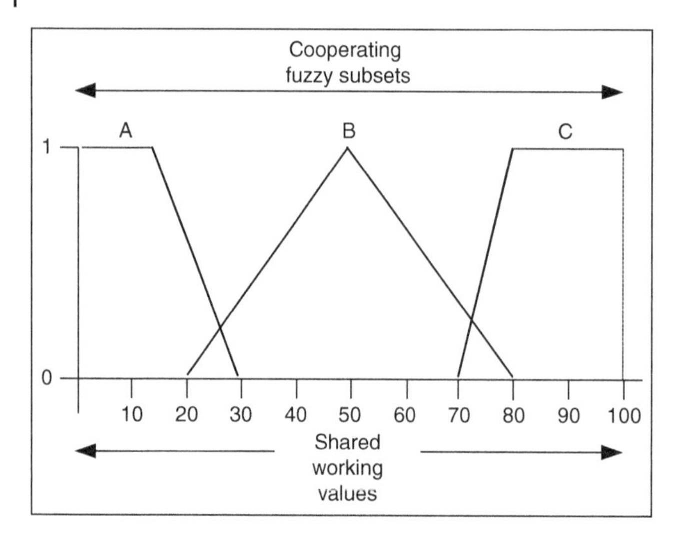

Figure 4.21 Universe of discourse.

The working range from the smallest to the largest value is the *Universe of Discourse*. We encountered this term earlier when we first explored fuzzy subsets of a larger set.

4.9.2 Curved Membership Functions

The previous examples illustrated the straight-line graph, which, as we indicated, is generally among the simplest and easiest fuzzy set or subset to approximate unknown concepts. We now continue our study of graphic membership functions with the basic curved model.

Example 4.12 Membership Function Illustrating Set of Support
Observe that the domain for the fuzzy set in Figure 4.22 spans from 20 to 110 as circumscribed by the large dashed rectangle. However, the nonzero portion of the fuzzy set, circumscribed by the smaller dashed rectangle, starts at 40 and terminates at 90. As a result,

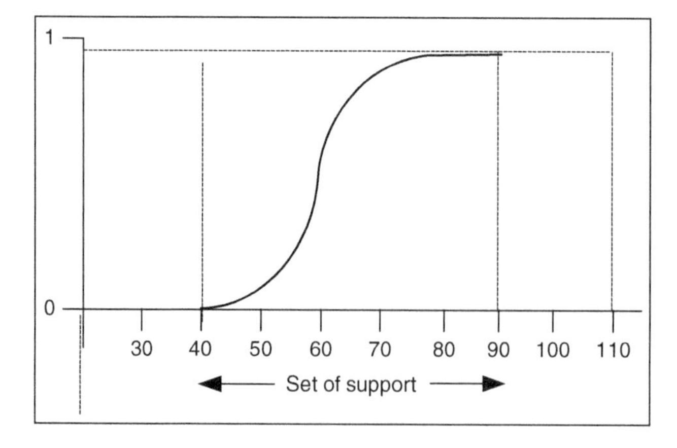

Figure 4.22 Set of support.

it does not cover the entire, potentially active, domain. The portion of the domain that is covered by the smaller rectangle is called the *set of support* or *support set*. The portions that remain uncovered potentially can have a serious impact on the range of operation of the system under design.

Example 4.13 Membership Function Illustrating Sigmoid Curves
In Examples 4.7–4.9, we learned about linear increasing or decreasing membership functions. Such functions that are curved rather than linear are referred to as *growth* and *decline* curves. They are also given the name *sigmoid* or may be, because of their shape, also referred to as *S-shaped curves*.

As we look at the curves in Figure 4.23, we can identify three significant points.

The first two are similar to what we saw in the linear domain. The range of a *growth* (left-hand) curve moves from a beginning membership value of [0] on the left-hand side to an end or full membership value of [1] on the right. The *decline* (right-hand) curve reverses the process, starting with full membership of [1] on the left and [0] membership on the right as also illustrated in Figure 4.23.

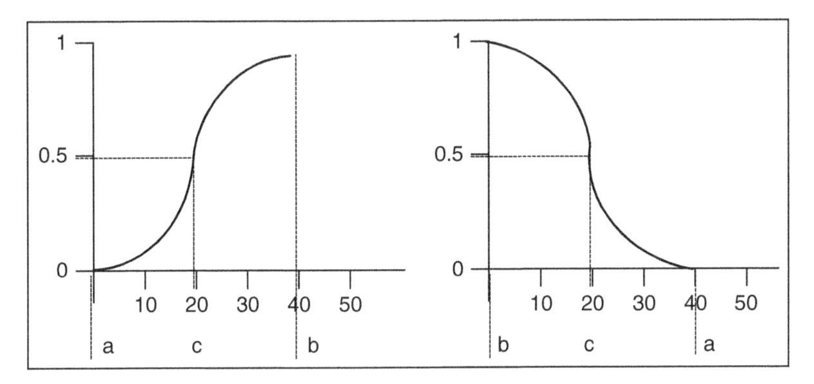

Figure 4.23 Sigmoid membership curves growth and decline.

These first two points on the diagrams are indicated by the symbols (a and b), respectively, for the start and end of the growth curve and by the symbols (b and a), respectively, for the start and end of the decline curve. The third point is called the *inflection point,* which is identified in both diagrams by the symbol (c), which identifies the location (center) where the domain value has a value of 50%.

Example 4.14 Membership Function Illustrating the PI Membership Curve
In Example 4.12, we took a first step toward representing a fuzzy number. The PI curve, illustrated in Figure 4.24, presents a preferred and typically default alternative.

Looking at the characteristics of the curve, first, we note that the membership curve is symmetric. The center of the curve is marked by the symbol (c) and the [0] points on either end are discrete, not asymptotic. As illustrated, the flex point occurs at the 50% point of the degree of membership, and the half width of the base is marked by the symbol (w).

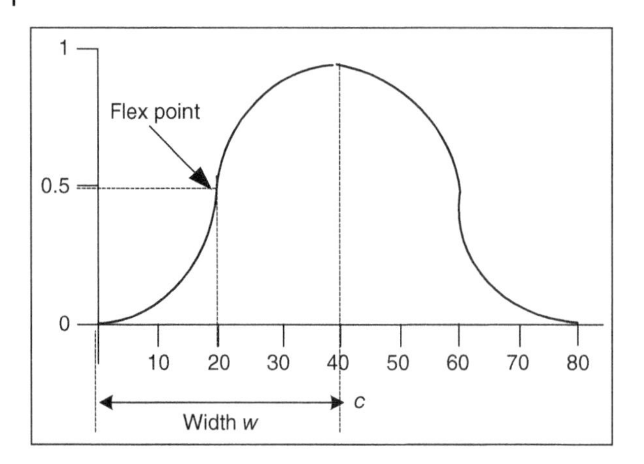

Figure 4.24 PI membership curve.

The name of the curve, PI, derives from the term *photosynthesis-irradiance,* which is a graphical representation of the empirical relationship between the terms *solar irradiance* and *photosynthesis*. Like the sigmoid curve, which may be referred to as an S curve, the PI curve is similarly nicknamed because of its shape as the math symbol π.

Example 4.15 Membership Function Illustrating the Beta Membership Curve
The Beta curve, illustrated in Figure 4.25, is quite similar to the PI version. As the graph indicates, the curve is also built around a center point (c). Looking at the characteristics of the curve, first, we note that the membership curve is similarly symmetric.

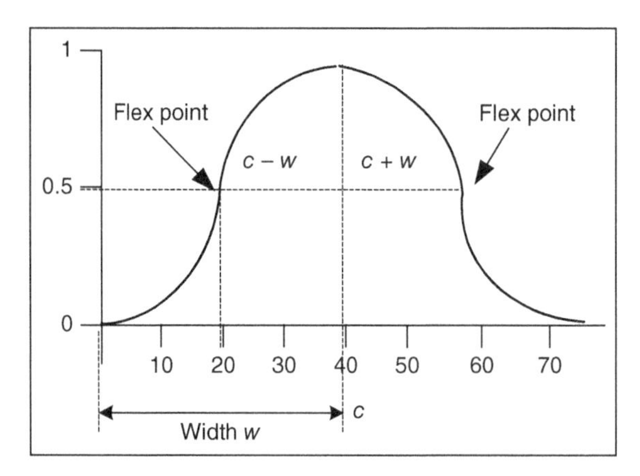

Figure 4.25 Beta membership curve.

Like the PI curve, the flex or inflection point occurs at the 50% point of the degree of membership axis and the half-width of the base is marked by the symbol (w). So, where does this curve come from? Interesting, isn't it? The diagram is taken from the domains of probability and statistics. The distribution is actually a family of distributions defined on the interval [0,1]. The value of the curve B for a particular domain point x is given as:

$$B = 1/\left[1 + \left(\left(x - c\right)/w\right)^2\right] \tag{4.3}$$

The center of the curve is marked by the symbol (c), and the [0] points on either end are discrete, not asymptotic. As illustrated, the flex point occurs at the 50% point of the degree of membership.

Example 4.16 Membership Function Illustrating Gaussian Membership Curves
The Gaussian curves, illustrated in Figure 4.26, follow a model similar to the PI and Beta versions. As the diagram indicates, like the PI and Beta curves, the graphs are built around a center point (c).

Looking at the characteristics of the curve, once again we note that the membership curves are similarly symmetric. Like the PI and Beta curves, the flex or inflection point occurs at the 50% point of the degree of membership axis and the half-width of the base is marked by the symbol (w). The half-width is not constrained to any particular value nor is the shape of the curve constrained to a bell shape as we see in the left-hand curve.

The value of the curve, B, for a particular domain point x is given as:

$$B = e^{-\left(w(c-x)2\right)} \tag{4.4}$$

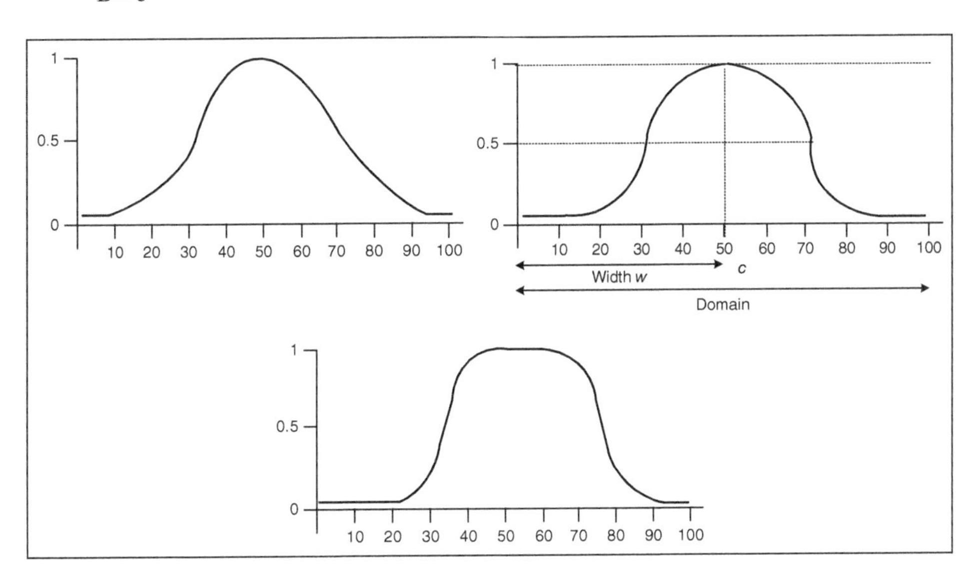

Figure 4.26 Gaussian membership curves.

Example 4.17 Using a Graphic Membership Function – 1
Let

P be the set of people from 0 to 100 years old
Y be the fuzzy subset of people close to 20 years old
M be the fuzzy subset of people close to 50 years old

Using the simple membership functions in Figure 4.27,

Let p be a person who is 40 years old.
The extent to which a person of 40 is:

<u>close to 20</u>	<u>close to 50</u>
$\mu_Y(40) = 0.7$	$\mu_M(40) = 0.8$

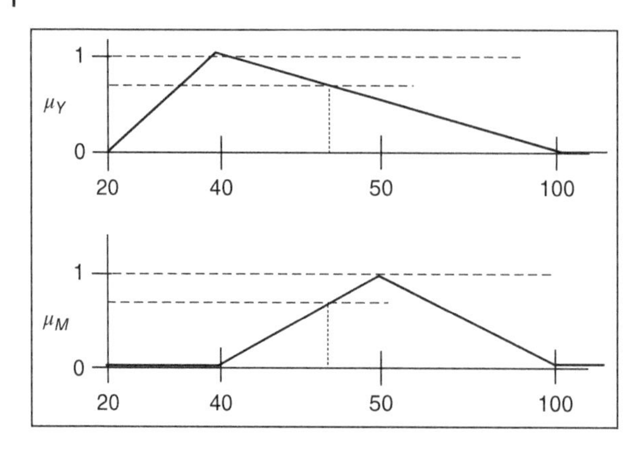

Figure 4.27 Simple membership functions.

not 20

$1 - \mu_Y(40)$

$1 - 0.7 = 0.3$

not 50

$1 - \mu_M(40)$

$1 - 0.8 = 0.2$

close to 20 and 50

$$\min\left(\mu_Y(40), \mu_M(40)\right)$$

$$\min(0.7, 0.8) = 0.7$$

Which intuitively makes sense since 40 is closer to 50 than 20,

close to 20 or 50

$$\max\left(\mu_Y(40), \mu_M(40)\right)$$

$$\max(0.7, 0.8) = 0.8$$

Which again makes sense since 40 is closer to 50 than 20.

Example 4.18 Using a Graphic Membership Function – 2

Let

P be the set of people from 0 to 100 years old

Y be the fuzzy subset of people who are young

Using a simple membership function in Figure 4.28,

The extent to which a person p of 20 is young:

$$\mu_Y(20) = 1$$

The extent to which a person p of 30 is young:

$$\mu_Y(30) = 0.6$$

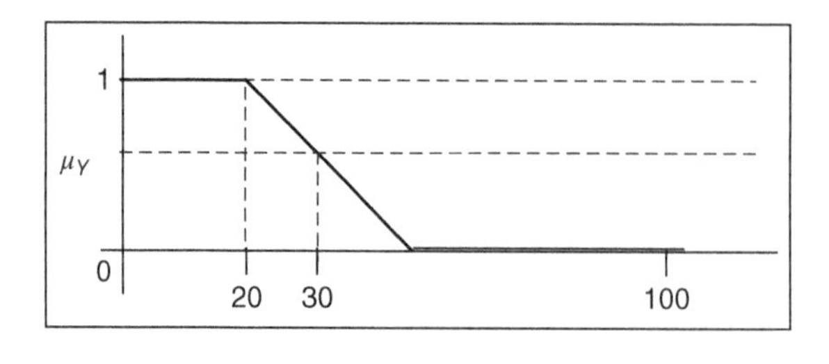

Figure 4.28 Using a membership function.

Example 4.19 Using a Graphic Membership Function – 3

Let

P be the set of people from 0 to 100 years old

O be the fuzzy subset of people who are old

Using a simple membership function in Figure 4.29,

The extent to which a person p of 80 is old:

$$\mu_O(80) = 0.8$$

The extent to which a person p of 60 is old:

$$\mu_O(60) = 0.3$$

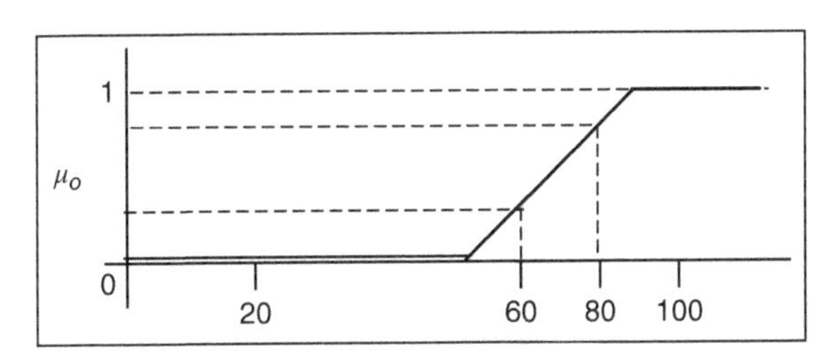

Figure 4.29 Using a membership function.

Example 4.20 Using a Graphic Membership Function – 4

Let

P be the set of people from 0 to 100 years old

M be the fuzzy subset of people who are middle aged

Using a simple membership function in Figure 4.30,

The extent to which a person p of 30 is middle aged:

$$\mu_M(30) = 0.25$$

Figure 4.30 Using a membership function.

The extent to which a person p of 70 is middle aged:

$$\mu_M(70) = 0.25$$

Example 4.21 Using a Graphic Membership Function – 5

Combining the three plots in Figure 4.31,

Let

P be the set of people from 0 to 100 years old

Using three simple membership functions,

young (μ_Y), middle aged (μ_M), old(μ_O):

<u>For a person of 30:</u>

$\mu_Y(30) = 0.6$ $\quad$ $\mu_M(30) = 0.35$

<u>For a person of 60:</u>

$\mu_M(60) = 0.6$ $\quad$ $\mu_O(60) = 0.35$

Thus, a person can be in multiple subsets.

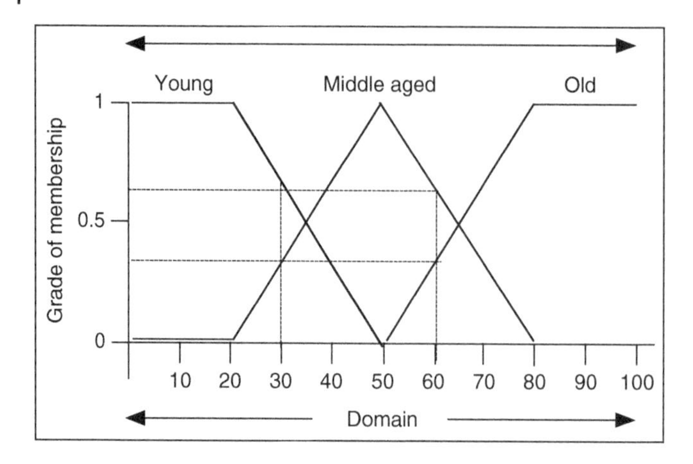

Figure 4.31 Using membership functions.

4.10 Summary

In summary, we examined the concept of set membership in both classical and fuzzy subsets. We reviewed some of the principle definitions and concepts of the theory of ordinary or classical sets and saw how these are identical to fuzzy subsets when the degree of membership in the subset includes all real numbers in the interval [0.0, 1.0].

We worked with the concept of a universe of discourse that was introduced in Chapter 3 as a domain that comprises the total range of ideas, entities, attributes, events, relations, etc. that are expressed, assumed, implied, or hypothesized during the course of a discussion. Such a concept can be a working guide in the process of developing and populating fuzzy sets and subsets.

We reviewed the concepts of crisp sets and subsets and then segued into the corresponding domain of fuzzy sets, subsets, and logic. In the process, we introduced the fundamental/logical operations that can be applied to crisp and fuzzy sets.

We saw that vagueness and imprecision are common in everyday life. Very often, the kind of information we encounter may be placed into two major categories: statistical and nonstatistical. The former we model using probabilistic methods and the later we model using fuzzy methods.

We examined the concepts of fuzzy subsets and fuzzy subset membership and saw that when we restrict the membership function $\mu_A(x)$ so as to admit only two values, 0 and 1, fuzzy subsets reduce to classical sets.

We explored a variety of graphs and graphic shapes that can be used to plot the degree of membership in a set against specific set values. We stipulated that the graphs and graphic shapes are designated by the designer and the design under development or use.

Review Questions

4.1 Explain and describe the term *domain* or *universe of discourse*. Give an example other than that which is in the chapter.

4.2 Do all entities in a fuzzy set or subset have to be identical? Explain the term grade of membership.

4.3 What is the difference between a crisp and a fuzzy subset?

4.4 What does the term support or sup mean? Is this term limited to crisp sets?

4.5 What is the *Law of Excluded Middle*?

4.6 Traditional digital logic design has been governed by the *Law of Excluded Middle.* How have contemporary designs been changing?

4.7 Are the requirements for degrees of set and subset membership the same for classic and fuzzy logic? If not, how are they different?

4.8 Looking at the three membership function graphs in Figure 4.4 for the expression "close to 7," what numeric differences do you see between them?

4.9 Working with the Gaussian curves illustrated in Figure 4.26 as models, create and explain 3 membership functions that could be called Pi, Z, and S curves.

4.10 Explain the difference between fuzzy set membership and probability.

4.11 Looking at the 15 basic fuzzy set properties listed in the chapter, choose 8 of them and illustrate and explain examples that are different from those in the text.

4.12 Repeat Figure 4.12 in the chapter first with 3 friends and then with 4 friends.

4.13 Repeat Figure 4.13 in the chapter first with 3 friends and then with 4 friends.

5

What Do You Mean By That?

THINGS TO LOOK FOR...

- Evolution of language.
- Meaning and purpose of crisp variables.
- Crisp sets revisited.
- Meaning and purpose of linguistic variables.
- Definition of hedges.
- Purpose of hedges.
- Creation and use of hedges.
- Manipulation of hedges.
- Fuzzy sets and membership functions revisited.

5.1 Language, Linguistic Variables, Sets, and Hedges

Have you ever thought about language? Thought about what language is? Thought about where it comes from or why we need it? Look around the world. In general, we all share many of the same concepts, yet at the same time, we all also have different sounds that we call words to refer to those concepts.

Let's reflect back to many thousands of years ago. Our early ancestors learned to convey information to those close to them and to others far apart. Archeologists have discovered that such information was often expressed as collections of sounds, marks, symbols, or graphic images of various sorts. Those marks or symbols eventually evolved and collected into what today might be called an *alphabet*. In isolation, each mark, symbol, or sound probably had only a simple meaning. As the years passed, deeper meaning became associated with the symbols that were eventually interpreted as numbers, letters, or other such primitives. Eventually, our ancestors probably began grouping the symbols and sounds, potentially associating a context with the groupings. They now had data and had taken the first steps toward something very useful.

The symbol 1, paired with a scribbled drawing on the ground or on a cave wall, may have meant a single buffalo or a biological need. The symbols for a leaf may have indicated a tree appendage or a new set of clothes. Creatively applying additional groupings and context to the data gave them *information*. As the process continued over the millennia, associating

Introduction to Fuzzy Logic, First Edition. James K. Peckol.

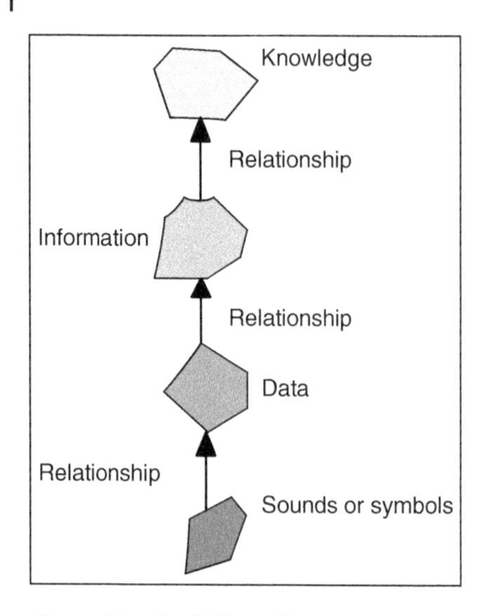

Figure 5.1 Evolution of language.

increased groupings and meanings to the once primitive marks morphed them into what became knowledge and potentially a shared language. The process illustrated in Figure 5.1 shows how we might describe the transition from abstract symbols or sounds into knowledge. At the lowest level, the symbols or sounds exist in isolation. When we attach a meaning to them, they become data. Data collectively becomes information, and information segues into knowledge as relationships are applied. Observe that each level provides a richer expressive power, greater flexibility, and deeper meaning.

One can never be sure where or when language first developed; however, one need not look far to see a wealth of early evidence. Paintings on caves or on bits of pottery clearly show language and meaning in its infancy. Markings on pottery shards suggest primitive account ledgers – one mark may represent one cow, and a different mark may mean five. To early man, communication meant survival. Even in the animal kingdom, various species have developed rather sophisticated "languages" for warning of danger or telling of new sources of food.

As we noted earlier, fundamentally, if we listen to any language or look at a written word, we learn that languages are merely a collection of markings or sounds that by themselves have no significance. To some, "a ka," "rot," "red," or "rouge" are simply sounds, yet to the Japanese, German, English, or French, they describe light at a certain wavelength. Each sound or collection of sounds has an associated concept. By pairing symbols or sounds with concepts, we take the first steps toward evolving and developing a language.

Consider the following example. A woman researcher has been working with a bonobos, also known as a pygmy chimpanzee. She named the chimp Kanzi and worked extensively with him. She contended that Kanzi's aptitude for understanding spoken English and communication with humans using lexigrams made a very strong statement. She further contended that 80% of Kanzi's multiword statements were spontaneous.

Even when paired with a concept, a simple symbol or sound has little more meaning than a single entry in a table of random numbers. One must understand and be able to describe the concept. Is it big or small? Is it heavy or light? How does it compare to other objects? Is it bigger, smaller, heavier, lighter? What do those words really mean? Is the word lighter referring to weight, color, simple, free of worry? What color is it? Are two red objects having the same color or is one darker or lighter than the other? Again, what do those words mean?

People have learned to communicate to exchange information, ideas, or needs. For such an exchange to be meaningful, there must be an agreed-upon relationship between the sound(s) or symbol(s) and the information to be conveyed. Let us consider three cases.

In the simplest case, there is a single *speaker*, a *listener* or *listeners*, a *concept*, and a mutually agreed-upon *sound* or *symbol* associated with that concept to be shared as we see in

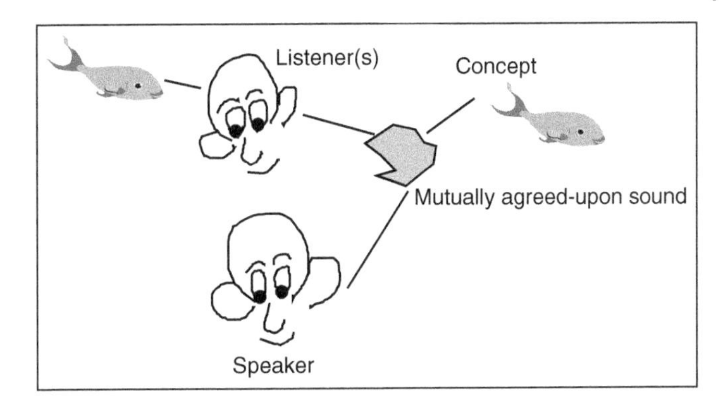

Figure 5.2 Simple communication.

Figure 5.2, both parties have a common understanding of the symbol or sound and the associated concept. As a consequence, the information can be exchanged easily.

Figure 5.3 shows the more complex case. There is still a common understanding of the concept (at least in its abstract sense), but the speaker knows the concept by one sound (or symbol), and the listener(s) know it by a second. Communication is still possible because the listener(s) know that when the speaker utters Sound1, it has the same meaning as or a similar meaning to Sound2.

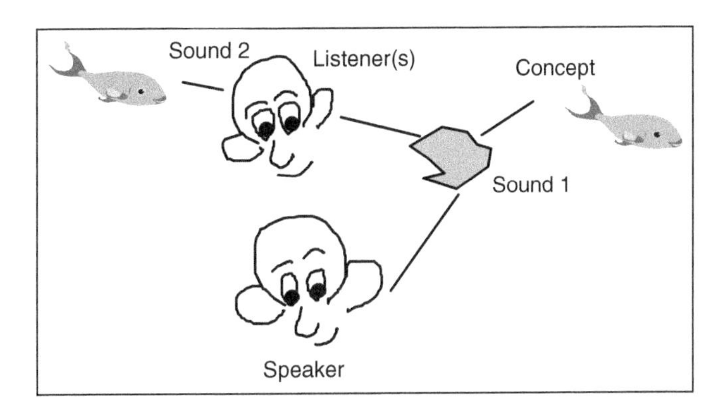

Figure 5.3 Complex communication.

In the most complex case, there is no understanding of the sound or concept by the listener(s), and, as a result, it is not possible to associate a familiar sound or symbol. Such is the case with the Eskimos or folks in other far northern lands. Because snow plays such an important role in their lives, they have several dozen words to describe it and its various, potentially dangerous, conditions. Equivalents for these unfamiliar concepts may not exist or even be necessary in other languages or to those living near the equator, and thus, as Figure 5.4 illustrates, there is no communication.

From a simplistic view, language development is thus a process of associating sounds or symbols (which, in English, we refer to with a set of symbols "*words*" and an associated *sound*) with physical world objects and with the relationships between and among those objects. Such relationships may be real or imagined, concrete or abstract, or more challenging, a different meaning, or double entendre.

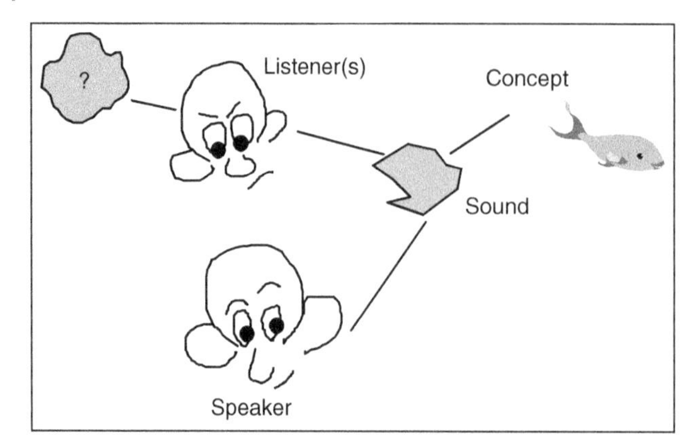

Figure 5.4 No communication.

5.2 Symbols and Sounds to Real-World Objects

The simplest associations between collections of symbols or sounds and physical world objects in the various world languages are names of things or numeric quantities. We generally understand the expression *numeric* to mean a variable whose values are numbers, mumeros, nombres, nummers or many other letter combinations. In some cultures, quantity is limited to one, two, three, many. When one moves beyond such basic notions, formulating associations becomes difficult rather quickly.

Our prehistoric hunter could easily draw or describe one or two buffalo. Having to draw 200 on a cave wall could be a rather time-consuming task at best. By the time he finished drawing all 200 to be able to tell his fellow hunters of his discovery, they would probably all be asleep or gone. Fortunately, for all of us, our ancestors learned that the ability to summarize is vitally important.

One must be able to summarize because it is impossible to have a unique name for each sound, object, characteristic, or combination of characteristics. Through such summaries, our ancestors created variables whose values became words or expressions. Such variables are called *linguistic variables.*

Taking a formal, high-level view, similar to a *numeric variable* whose values are numbers, a *linguistic variable* can be interpreted as a variable whose values are *linguistic*, that is, words or collections of words in a natural or potentially an artificial language.

5.2.1 Crisp Sets – a Second Look

As we continue to learn and develop the concept of *set*, we move increasingly into the worlds of *Boolean algebra* and formal *set theory*. Building on what we learned in the last chapters, once again we pose this question: What is a set?

Formally, a set is defined as a collection of objects called *elements* or *members* that are collectively viewed, from the outside, as a single object. Starting with a set A, we use the symbol € to indicate that a particular element x is a member of set A or the symbol $\notin$ to indicate that it is not a member of set A as in Figure 5.5 when we write as follows:

x € A // x is an element of set A in the left – hand graphic. (5.1)

$x \notin A$ // x is not an element of set A in the right – hand gra

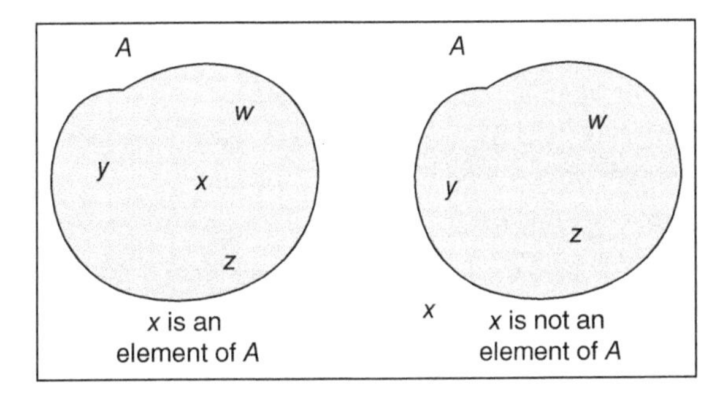

Figure 5.5 Set membership.

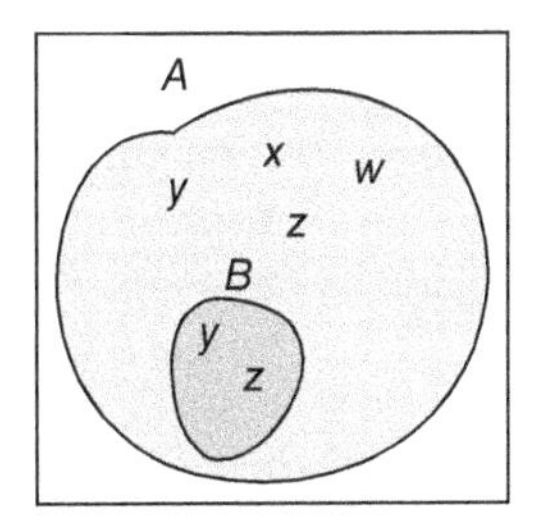

Figure 5.6 Subset membership.

As we have learned earlier, membership in a set or subset can also be indicated using an expression called a *characteristic* or *membership function*. To indicate that a variable x is a member or is not a member of the set A, we write as follows:

$$\mu_A(x) = 1 \text{ // } x \text{ is an element or member of } A. \tag{5.2}$$
$$\mu_A(x) = 0 \text{ // } x \text{ is not an element or member of } A.$$

Such membership is also reflected graphically in Figure 5.5. Depicted in Figure 5.6, a collection of elements B is defined as a *subset* of A if all of the members of B are also members of A as shown for the elements y and z.

Let's examine these and work with Boolean algebra. As noted earlier, Boolean algebra was put forth by George Boole in 1854 in *An Investigation into the Laws of Thought*. From Boolean algebra, when the degree of membership of all elements in a set or subset is binary, that is, either 0 or 1, or TRUE or FALSE, the set or subset is said to be *crisp*. As we have also learned, when the degree of membership of some of the elements in a set is expressed as a fractional value between 0 and 1 such as 0.1, 0.5, or 0.9, the set or subset is said to be *fuzzy*.

As we've learned, two classic relations or operations in Boolean algebra are the *union* and the *intersection* of individual elements or sets. Refer back to Chapters 3 and 4. Such operations on individual elements are given here in Figures (5.6) and (5.7) and in Eqs. (5.1) and (5.2).

The crisp *union* operation expresses the Boolean *OR* of individual elements. The union is computed as illustrated in the matrix or truth table in Figure 5.7 and written as in Eq. (5.1). The + symbol indicates a logical OR, not an addition operation.

+ B A	0	1
0	0	1
1	1	1

Figure 5.7 Boolean union/OR.

Boolean OR +

$$\mathrm{C} = (\mathrm{A} + \mathrm{B}) \text{ or } (\mathrm{A} \cup \mathrm{B}) \text{ // union} \tag{5.3}$$

From Figure 5.7, we have

$$\mathrm{A} = 0, \mathrm{B} = 0 \ \rightarrow 0 + 0 = 0$$
$$\mathrm{A} = 0, \mathrm{B} = 1 \ \rightarrow 0 + 1 = 1$$
$$\mathrm{A} = 1, \mathrm{B} = 0 \ \rightarrow 1 + 0 = 1$$
$$\mathrm{A} = 1, \mathrm{B} = 1 \ \rightarrow 1 + 1 = 1$$

The corresponding membership functions are written as follows:

$\mu_{A \cup B}(x) = 1$ // Element x is an element or member of the set C formed
// by the union of A and B, C = (A ∪ B).
$\mu_{A \cup B}(x) = 0$ // Element x is not an element or member of the set C formed
// by the union of A and B, C = (A ∪ B).

The *intersection* operation expresses the Boolean AND of sets and is also called a *Boolean product*. The intersection is computed as illustrated in the matrix or truth table in Figure 5.8 and written as in Eq. (5.2).

Boolean AND

$$C = (A \bullet B) \text{or} (A \cap B) \, // \text{intersection} \tag{5.4}$$

From Figure 5.8,

$$A = 0, B = 0 \rightarrow 0 \bullet 0 = 0$$

$$A = 0, B = 1 \rightarrow 0 \bullet 1 = 0$$

$$A = 1, B = 0 \rightarrow 1 \bullet 0 = 0$$

$$A = 1, B = 1 \rightarrow 1 \bullet 1 = 1$$

● B A	0	1
0	0	0
1	0	1

Figure 5.8 Boolean intersection/AND.

The corresponding characteristic functions are written as follows:

$\mu_{A \cap B}(x) = 1$ // Element x is an element or member of the set
// formed by the intersection of (A ∩ B), A and B.
$\mu_{A \cap B}(x) = 0$ // Element x is not an element or member of the
// set formed by the intersection of (A ∩ B), A and B.

The term *universe of discourse* generally refers to a collection of entities currently being discussed or analyzed. In the following, our universe of discourse will be the simple concept espoused by the word *number*. Simple in its own right yet the concept of number can quickly be decomposed into two subsets: *integer numbers* and *floating-point* numbers as we see in Figure 5.9.

Within our universe of discourse, we will focus on the subset: integers. In Figure 5.9, the set of numbers is classified into two subsets: subset A all integers and subset B all floating-point numbers. Such sets of entities, from which members of subsets may be drawn, are called M(), the *membership set*. From such a set, we are able to aggregate or group numbers into a variety of *subsets*, each with shared or common properties as illustrated in Figure 5.10.

We can also write the *membership function* of a subset. As noted earlier, such a function indicates whether a member of the *membership set* is also a member of a specific subset.

For subset A, the *characteristic or membership function* is given as $\mu_A(x)$ and takes a value of 0 or false if the variable x is not an integer and 1 or true if it is.

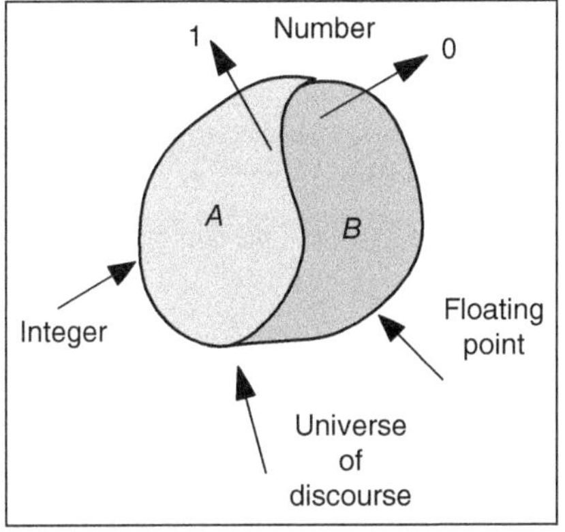
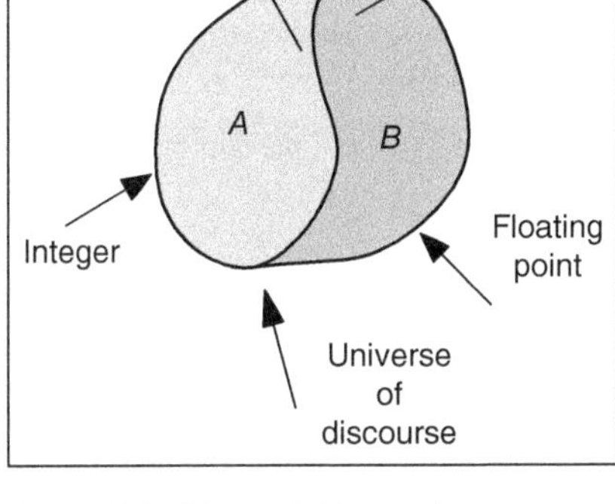

Figure 5.9 The variable number.

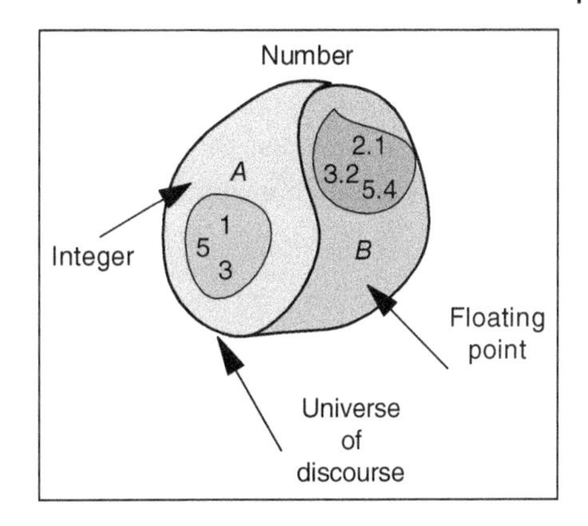

Figure 5.10 Subsets of the set number.

In the foregoing discussion, the *universe of discourse* was limited to integer numbers. Let Y identify the following finite set of integers to represent the *membership set* M().

$$Y = \{1, 2, 3, 4, 5, 6, 7\} \tag{5.5}$$

M(Y) now expresses a *membership set* of numbers within the *universe of discourse*. From such a set, one can aggregate or group members into a variety of *subsets*, each with shared or common properties. Consider the subset A in Figure 5.10 that is comprised of the numbers {1, 3, 5}. In this case, the common properties would be the following:

1) They are all integers.
2) They are all less than seven.
3) They are all odd.

The *membership function* of A is given as:

$$\mu_A(1) = 1,\ \mu_A(2) = 0,\ \mu_A(3) = 1,\ \mu_A(4) = 0,\ \mu_A(5) = 1,\ \mu_A(6) = 0,\ \mu_A(7) = 0 \tag{5.6}$$

The complement of subset A expresses the question "*Which elements do not belong to or are not members of this set?*" and is written as ~*A* and read as *not A*. The complement may also be written as A^c or ~*A*. With respect to the membership set in (Eq. 5.4), the complements can be expressed as:

$$\mu_{\sim A}(1) = 0,\ \mu_{\sim A}(2) = 1,\ \mu_{\sim A}(3) = 0,\ \mu_{\sim A}(4) = 1,\ \mu_{\sim A}(5) = 0,\ \mu_{\sim A}(6) = 1,\ \mu_{\sim A}(7) = 1 \tag{5.7}$$

The complement can also be expressed as:

$$\mu_{\sim A}(x) = 1 - \mu_A(x) \tag{5.8}$$

As we have learned, when the degree of membership in a set or subset is binary, that is, either 0 or 1, or true or false, the set or subset is said to be *classic* or *crisp*. The sets and subsets discussed thus far have all been crisp.

5.2.2 Fuzzy Sets – a Second Look

In the previous section, the degree of membership in crisp sets was characterized as binary. Elements in the membership set did or did not belong to a specified subset as illustrated in Figure (5.5) by the characteristic function of A. Such simple cases are easy to recognize and to quantify. However, the world and its words in its many languages present more of a challenge.

5.2.2.1 Linguistic Variables

The basic concept of the *linguistic variable* was introduced by Lotfi Zadeh in a 1973 paper. Variables, characterized as *linguistic*, provide a means of approximate characterization of phenomena that may be too complex or ill-defined to be easily expressed in standard quantitative terms. Such variables take on a range of possible values representing or expressing the variable's universe of discourse.

Another critical aspect of linguistic variables is related to precision. The concept of precision has two distinct interpretations. The first interpretation, which is traditional, is that of value and is deemed to be precise. The second, in contrast, is deemed imprecise. If someone states that the restaurant is rather close, as far as the distance to the restaurant is concerned, the distance is imprecise in value yet precise in meaning. Such an interpretation is valid if close is interpreted as a fuzzy set with a specified membership function.

If one takes a more formal point of view, each word X in a natural language can be interpreted as a summarized description of a collection of objects. One can then define such a collection to be a fuzzy subset $M(x)$, where $M()$ is a function representing the meaning of X within the universe of discourse.

As an example, the noun *automobile* is the fuzzy subset of objects described by M(automobile) and the adjective *red* is the fuzzy subset of objects described by M(red). The noun phrase *red automobile* can now be described by the fuzzy subset that results from the *intersection* of the subsets M(red) and M(automobile).

If one views color as a variable, one can see at first pass that it can take on different values such as red, blue, or green. Each of these values represents the label of a fuzzy subset within the universe of discourse. That is, color is a variable with values that are labels of fuzzy subsets. To illustrate, let's begin with the color *red* to illustrate such a challenge: *What color is red*?

For linguistic variables such as *color*, the values can be simple terms such as *red*, *green*, or *blue*. Most of the time, such a summary will be understood and is sufficient to convey a concept. At other times, when a concept becomes more obtuse and we need more precision, we add quantifiers and qualifiers such as the phrases *light* red, *dark* red, *slightly* red, *very dark* red, or *not very* red in which the variables are modified by adjectives or adverbs. Such modifiers are called *hedges* and move the terms out of the crisp world into the familiar fuzzy world and to what is called *fuzzy reasoning* or *fuzzy logic*. Such a logic can form a basis for *approximate reasoning*, which is a form of reasoning that is not exact nor very inexact. Such a logic can offer a more realistic framework for human reasoning than the traditional two-valued or Boolean logic.

We may also use the term *red* to speak of the collection of light waves with a wavelength of around 700 nm at a frequency of approximately 4.3×10^{14} Hz near one end of the visible spectrum. For variables such as color, the values are simple terms. Most of the time, such a summary of the light wave characteristics is sufficient for basic understanding.

At those times, when we need greater clarification and precision, we add adjective or adverb quantifiers and qualifiers such as *infra*, *dark*, *light*, or *very*. Other variables, such as *height*, can take on equally complex values with modifiers such as *more* or *less tall, not tall, somewhat tall, very short,* or *very, very short.*

Once again, whether simple or complex, such variables are *linguistic* rather than numeric in character. As noted earlier, their values are words or phrases in a natural (or synthetic) language rather than the numbers one typically associates with a variable. Such variables are defined as *linguistic variables*. We appear to have a problem. How can we express such inexactness or imprecision in the value of a linguistic variable, and how can we use it in an equation?

Back to colors, let's start with two 10-oz-sized cans of paint with colors in two crisp sets, one in a *red* set and one in a *white* set. If we mix 6 oz of the *red* and 4 oz of the *white* together in a third 10 oz can, the resulting color, which we can call *reddish*, has a degree of membership in either the *red* or the *white* fuzzy set.

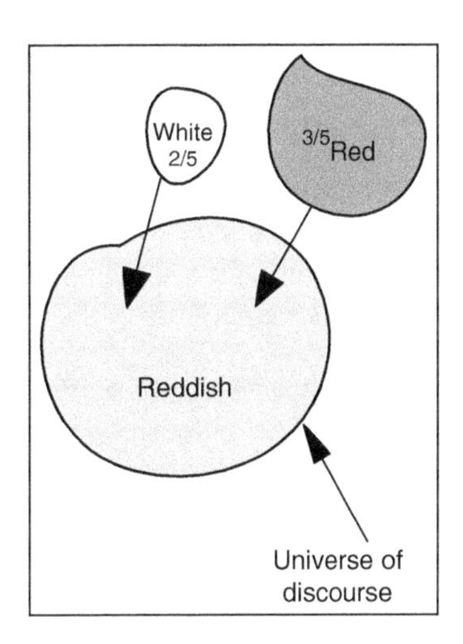

Figure 5.11 Fuzzy set.

The color of the mixture will be 6/10 (0.6) *red* and 4/10 (0.4) *white*. Thus, we can now have a state in which the created color has a degree of membership of 3/5 (0.6) in the *red set* and a degree of membership of 2/5 (0.4) in the *white set*. The resulting set is not crisp; rather, it is *fuzzy* as we see in Figure 5.11.

Each of these values thus represents the label of a fuzzy variable or subset within the universe of discourse. That is, *color* is a term or *linguistic variable* with values that are labels of fuzzy subsets.

Based upon the relative quantities of the two colors mixed together, the membership function or grade of membership for each color in the final mixture or subset can take on any value in the numeric range (0.0, 1.0) including the two endpoints as illustrated using a simple linear graph in Figure 5.12. The domain indicates ounces of red paint in the mixture left-to-right and ounces of white right-to-left both with a resolution is 2.0 oz.

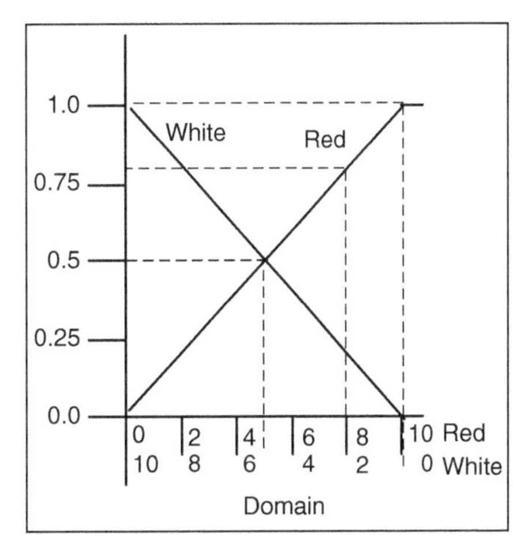

Figure 5.12 Graph of red color linguistic variable.

Whether simple or complex, such variables, as we see, are *linguistic* rather than *numeric* in character. Their values are words or phrases in a natural (or synthetic) language rather than the numbers one typically associates with a numeric variable. In our example, *red* will be such a *linguistic variable* to which we can assign values that are words such as *light red, dark red, reddish, pinkish* or other similar terms rather than integer numbers.

5.2.2.2 Membership Functions

In contrast to crisp variables, for which the truth-values are Boolean or two-valued (0, 1) and represented by the integer values 0 and 1, the truth-values of fuzzy variables are multivalued with truth-values ranging between 0.0 meaning completely false and 1.0 meaning completely true. Also accepted are truth-values that can be partly true and partly false at the same time. At the end of the day, fuzzy logic reflects how people actually think or reason.

In the fuzzy context, M as a subset is a semantic rule that associates with each linguistic variable *X* its meaning. The expression *M(X)* denotes a *fuzzy membership set.*

The *characteristic* or *membership function* of a fuzzy subset takes on a meaning similar to that which it did for the crisp subsets. However, now, the function indicates the *degree* or extent to which a *linguistic variable* is a member of a *membership set* for a named fuzzy subset.

For subset *A*, the *membership function* is written as $\mu_A(x)$.

x : Identifies the fuzzy variable to which the function is applied.
A : Identifies the subset to which the degree of membership applies.
$\mu_A(x)$: Indicates the membership function. The return value will be the degree of membership of x in subset A.

The return value from the function application will be a value between 0.0 and 1.0, including the two endpoints.

We begin with a universe of discourse A and set x_i of seven linguistic variables. The degree of membership of the variables in A might be any of the following:

$$\mu_A(x_1)=0.1,\ \mu_A(x_2)=0.8,\ \mu_A(x_3)=0.2,\ \mu_A(x_4)=0.3 \\ \mu_A(x_5)=0.8,\ \mu_A(x_6)=0.7,\ \mu_A(x_7)=0.4 \tag{5.9}$$

The *complement* of a fuzzy subset mirrors that of the crisp subset as was illustrated in Chapters 3 and 4. However, rather than asking if an element does not belong to a set, it poses this question: "*How much* or *to what degree do elements belong* or *not belong to the set?*"

Working with the set in (5.7), we have the fuzzy *complement*:

$$\mu_{\sim A}(x_1)=1-0.1=0.9,\ \mu_{\sim A}(x_2)=1-0.8=0.2,\ \mu_{\sim A}(x_3)=1-0.2=0.8 \\ \mu_{\sim A}(x_4)=1-0.3=0.7,\ \mu_{\sim A}(x_5)=1-0.8=0.2,\ \mu_{\sim A}(x_6)=1-0.7=0.3 \\ \mu_{\sim A}(x_7)=1-0.4=0.7 \tag{5.10}$$

5.3 Hedges

In our studies thus far, we have learned that *crisp variables* are quantitative, whereas *linguistic variables* are variables with more qualitative values comprising words or phrases in a natural or potentially an artificial language rather than numerical. Such variables are generated from a set of primary terms such as *young* or its antonym *old* or *tall* with its antonym *short.*

The meaning or value of a linguistic variable can be further enhanced or altered via collections of adjective or adverb modifiers such as *not*, *very*, or *more or less,* often combined with the connectives *and* and *or*. For example, when asked about a person's height, the response may be that they are *not very short* and *not very tall*.

Because the value of a linguistic variable such as *reddish* is the label of a fuzzy subset, it may be augmented by one or more modifiers and accompanied by its *membership function*. Earlier we saw how this is done with primary terms such as *young* or *old* as values for the linguistic variable *Age*.

By using the intersection and union of fuzzy subsets, we have seen how to express more complex phrases such as *tall and blond*. Now we shall see how to include modifiers as well. In the vocabulary of fuzzy logic, as stated earlier, such modifiers are called *hedges* and are typically adjectives or adverbs (not bushes in front of a building).

During the discussion of set operations, we saw that the modifier or hedge *not* is represented mathematically by $1-\mu_A(x)$, where $\mu_A(x)$ expresses the membership of the variable X in the fuzzy subset A. Thus, the value *not tall* is given by the expression $1-\mu_{\text{Tall}}(x)$. Let us now begin with the natural language sentence *Jean is a very tall man* to see how we might represent other hedges as well. In this sentence, *tall* is the primary term, and *very* is the hedge.

To understand how one might want to represent a hedge such as *very*, let's consider the intent of the phrases *tall* and *very tall* from a fuzzy logic perspective. To begin to develop a feeling for such sets and their relation to each other, we begin with two sets of people. We'll call one set *tall* people and the other set *very tall* people. We'll list people's heights, in meters, in both sets: 1.7 will mean 1.7 meters.

Tall People {1.7, 1.8, 1.9, 2.0, 2.1}

Very Tall People {1.7, 1.8, 1.9, 2.0, 2.1}

Should an individual's grade of membership in the fuzzy subset *Tall* be larger, smaller, or the same as their membership in the fuzzy subset *very Tall*? Intuitively, one might think that it should be smaller.

To be able to accurately reflect the difference between the subsets *Tall* and *very Tall*, one would want a person with a degree of membership of 1.8 in the fuzzy subset *Tall* to have a somewhat lower degree of membership in the subset *very Tall*.

Also, if one looks at the *Tall* membership function in Figure 5.13, one can see that the taller members of the subset *Tall* are concentrated in the right most portion of the plot. It is clear, then, that a representation of *very Tall* should emphasize the taller members of the subset and to de-emphasize those who are less tall. That is, we wish to compress the plot toward the right.

Ideally, we would like a means to mathematically modify the *Tall* membership function in such a way that there will be a large effect on those members with a high grade of membership and little or no effect on those with a lower grade of membership.

Figure 5.13 Tall membership function.

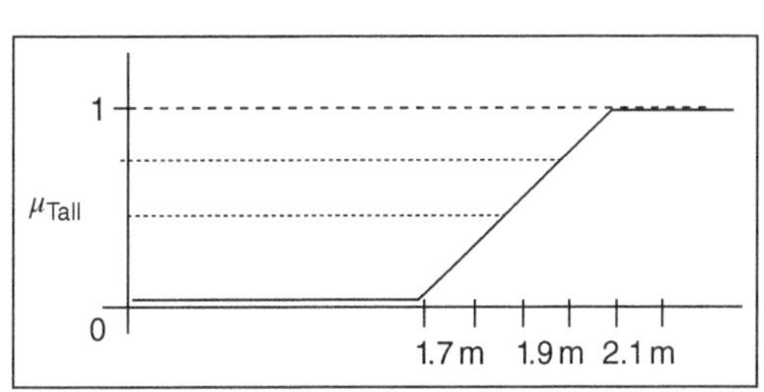

The *square law* is one such function. With the square law, small values are increased by a small amount and large values are increased by a large amount as can be seen from the squares of the numbers 3 and 5, which give $3^2 = 9$ and $5^2 = 25$, respectively.

From Figure 5.13, the fuzzy subset *Tall* is written, with the associated grades of membership, μ_{tall}, as:

$$\text{Tall} = \{(1.7/0.0), (1.8/0.25), (1.9/0.5), (2.0/0.75), (2.1/1)\} \quad (5.11)$$

from which Tall^2 becomes

$$\text{Tall}^2 = \{(1.7/0.0), (1.8/0.625), (1.9/0.25), (2.0/0.56), (2.1/1)\} \quad (5.12)$$

which is plotted as Figure 5.14.

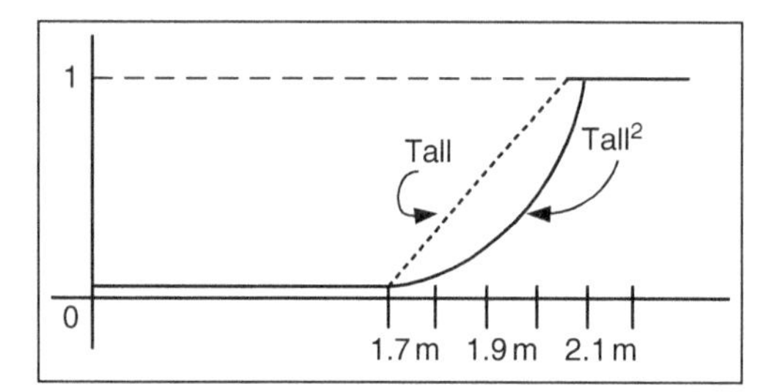

Figure 5.14 Tall and very tall membership functions.

As Figure 5.14 shows, we have accomplished our objective. Our new distribution emphasizes members with a high degree of membership in the subset Tall while de-emphasizing those with a lower degree of membership.

Based upon these results, we can make the following definition.

Definition 5.0

very: The fuzzy subset R (Result) that results from applying the hedge *very* to the fuzzy subset Tall (T) is the subset of ordered pairs (z_{1k}, z_{2k}) such that for all ordered pairs (x_{1i}, x_{2i}) in Tall, $z_{1k} = x_{1i}$ and $z_{2k} = x_{2i}$ in the subset R or equivalently $Z_{2k} = \mu_T(x_{1i})^2$.

One may follow a similar line of reasoning to develop the hedge *more or less.* In the sentence *Jean is more or less tall,* the apparent intent of the hedge is to relax or dilute membership in the subset *Tall,* i.e., to emphasize those members with a lower degree of membership. The desired effect can be achieved with the square root function that increases small values by a large amount and large values by a small amount.

From Figure 5.15, the fuzzy subset *Tall* is written as:

$$\text{Tall} = \{(1.8/0.25), (1.9/0.5), (2.0/0.65), (2.1/1)\}$$

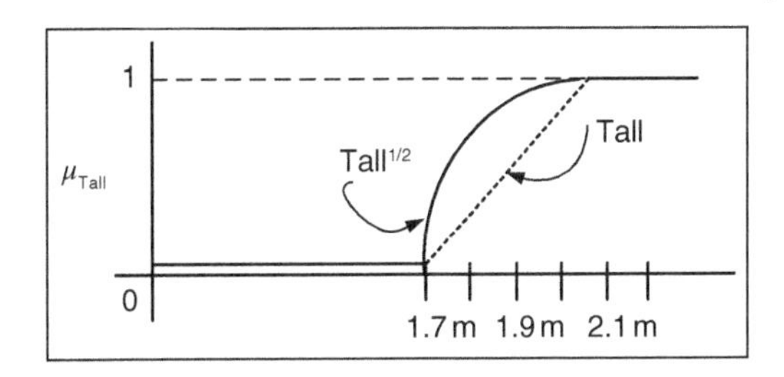

Figure 5.15 Tall and more or less tall membership functions.

from which Tall$^{1/2}$ becomes:

$$\text{Tall}^{1/2} = \{(1.8/0.5), (1.9/0.7), (2.0/0.8), (2.1/1)\}$$

which is plotted as Figure 5.15.

As Figure 5.15 shows, the new distribution has indeed emphasized those members with a lower degree of membership in *Tall*.

From these results, we can make the following definition:

Definition 5.1

> **more or less:** The fuzzy subset R (Result) that results from applying the hedge *more or less* to the fuzzy subset Tall (T) is the subset of ordered pairs (z_{1k}, z_{2k}) such that for all ordered pairs (x_{1i}, x_{2i}) in T, $z_{1k} = x_{1i}$ and $z_{2k} = {x_{2i}}^{1/2}$ in the subset R or equivalently $Z_{2k} = \mu_T(x_{1i})^{1/2}$.

One can also develop hedges by combining existing ones.

Example 5.0 Combining Hedges
Begin with the phrase *X is A.*

$\mu A(x)$

Then *X is very A.*

$\mu A^2(x)$

Then *X is very very A.*

$\mu A^4(x)$

Then somewhere in between *X is highly A.*

$\mu A^3(x)$

Example 5.1 Combining Hedges
Begin with the phrase *X is exact.*

$\mu Exact(x)$

Then *X is not exact.*

$1\text{-}\mu Exact(x)$

Then *X is very not exact* – poor grammar

X is very inexact – better grammar

$(1\text{-}\mu Exact(x))^2$

Then *X is not very exact.*

very exact

$\mu Exact(x)^2$

not very exact

$1 - \mu Exact(x)^2$

Notice that *very inexact* is different from *not very exact.*

The notion of exponentiation can be generalized to give the following definitions. See Figures 5.14 and 5.15.

In the following definitions, P is a linguistic hedge.

Definition 5.2 In the representation of linguistic hedges, the operation of *concentration* is given by P^n.

Definition 5.3 In the representation of linguistic hedges, the operation of *dilution* is given by $P^{1/2}$.

We can now formalize the definition of hedges.

Definition 5.4 Let *X* be a word in a natural language and let the function *M* be defined such that *M(X)* is the meaning of *X* within the universe of discourse. Further, let *hX* be the composition of h, a hedge, with *X* such that *M(hX)* is the meaning of *hX* within the universe of discourse.

With such an interpretation, *h* can be regarded as an operator that transforms the fuzzy subset *M(X)* representing the meaning of *X* into the fuzzy subset *M(hX)* representing the meaning of *hX*.

5.4 Summary

We began this chapter with a look at early symbols and sounds and their evolution to language and knowledge. Building on such origins, we introduced the concept of *sets* and moved into the worlds of formal *set theory*, *Boolean algebra*, and introduced *crisp variables*. From these basics, we defined and examined sets, subsets, set membership, and the concept of a characteristic function to assess if a variable is or is not a member of a set.

From the crisp world, we migrated into the fuzzy world and introduced the linguistic variable as one whose values are words or phrases in a natural (or synthetic) language. Such words are interpreted as representing labels on fuzzy subsets within a universe of discourse.

We learned that the values for a linguistic variable are generated from a set of primary terms, a collection of modifiers called *hedges*, and a collection of *connectives*. Hedges affect the value of a linguistic variable by either concentrating or diluting the membership distribution of the primary terms.

We concluded with a discussion of the purpose, creation and use, and manipulation of hedges.

Review Questions

5.1 Briefly describe what we call an alphabet and describe its origins.

5.2 Briefly describe the relationship between symbols, sounds, and concepts.

5.3 How do symbols, sounds, concepts, and an alphabet relate to communication?

5.4 What is a *linguistic variable*?

5.5 Identify the requirements for one set to be a subset of another.

5.6 What is the meaning of the expression: *universe of discourse*?

5.7 Let X identify the following set of letters for the membership set J()

$X = \{$a, B, c, D, E, f, H, i, O, P u$\}$. Identify and label the possible subsets of J(). Identify the membership functions for each of the subsets.

5.8 Express the complement of each of the fuzzy subsets in 5.7.

5.9 In the chapter, we created a fuzzy set and membership graph as a combination of amounts of two colors of paint. Create a similar fuzzy set and a membership graph as the combination of three different musical notes played at different amplitudes. What might that sound like?

5.10 Looking at the 15 basic fuzzy set properties listed in the chapter, choose 8 of them and illustrate and explain examples that are different from those in the text.

5.11 Explain the term hedge and illustrate with an example.

5.12 Consider that we have two groups with six people each hiking and climbing a tall mountain. Those in one of the groups are getting various distances close to the edge of a cliff and those in the other group getting similarly close to the edge are considered to be getting very close. Create and explain a *close membership function* and a *very close membership function*.

5.13 You are designing a subsystem that will be integrated into a larger system that you are also designing. The subsystem has a four-channel bus input that connects to an internal logic block. It is important that there is minimal delay across the four signals when they arrive at the internal logic block. Assume that the channel is 15 mm long and there is a propagation delay of 0.1 ps (picosecond) per mm per megahertz. The frequency of signals on Channel 0 averages 40 MHz, signals on Channel 1 average 30 MHz, signals on Channel 2 average 20 MHz, and those on Channel 4 average 10 MHz. Give the design for the subsystem.

6

If There Are Four Philosophers...

THINGS TO LOOK FOR...

- Classical and fuzzy reasoning.
- Fuzzy inference and approximate reasoning.
- Relations between fuzzy subsets.
- Notion of equality.
- Concepts of containment and entailment.
- Concepts union and intersection in the fuzzy world.
- Concepts of conjunction and disjunction in the fuzzy world.
- Conditional relations.
- Min-max composition.
- Modus ponens and modus tollens.

6.1 Fuzzy Inference and Approximate Reasoning

As we noted earlier, fundamentally, a language is a collection of markings or sounds that by themselves have little or no significance. Nonetheless, it does not seem to discourage us. We routinely use sounds and visual images to express a concept. By pairing sounds and images with concepts and ideas, we take the first step toward evolving and developing a language. With a language for expressing concepts and ideas, we take our next step and slowly evolve a process that we call reasoning. That process is a very important and critical step in the design process.

We use the term *reasoning* to refer to a set of processes by which useful conclusions are deduced from a collection of concepts, premises, or data. In classical logic, such premises are precise and certain, whereas in fuzzy logic, they can be ambiguous, possibly vague, imprecise, or perhaps even conflicting. Trying to describe fuzzy reasoning is a fuzzy process in its own right. Fuzzy or approximate reasoning is often more qualitative than quantitative. Nonetheless, fuzzy logic underlies our ability to understand languages, play games, solve ill-defined problems, or make rational and sophisticated decisions in complex, uncertain, or unfamiliar environments.

We noted in an earlier chapter that as engineers, our main goals are to recognize, to understand, and to solve problems. To aid in our quest, we continually seek tools that will

Introduction to Fuzzy Logic, First Edition. James K. Peckol.

help us to understand and to facilitate that process. We also learned that Boolean algebra brings us some of those tools in the form of variables and equations. Variables give us a tool to identify and to quantify real-world entities. Equations provide us with a means to specify and to work with relations among those variables as we pursue our goals and the challenges of understanding and accurately solving problems.

Also, essential in large part to solving problems is our ability to reason. Such a skill derives significantly from our ability to describe and to understand relationships between or among objects that we have identified and potentially named. We have seen that those names represent a summarized description that can be interpreted as linguistic variables whose values correspond to the labels of fuzzy subsets. The operations defined earlier between such (fuzzy) subsets now form the basis for defining relationships between (fuzzy) propositions, i.e. relationships between (fuzzy) propositions are reflected in the corresponding operations on their underlying subsets.

As we move forward, we will see that fuzzy logic also adds, through variables and equations, to our growing collection of tools. The tools we will introduce in the following pages are often similar in name but richer in function compared with those found in classic logic. In our tool chest, sometimes a power saw is preferred over a handsaw although both can be very effective in their respective domains.

In the discussions that follow, we will also be working with graphs of membership functions. The graphs are created primarily for illustration purposes. Such graphs in real-world applications are a very handy tool; however, to be effective, they will need to reflect the data and values specific to those applications.

6.2 Equality

Let's start out with the basics and begin with *equality*. In its simplest form, the term *equality* is expressed by a statement such as X *is* A. This phrase usually implies that some object X is the same as some other object A. For example, *the component is an operational amplifier* – or that some object X has property B. For example, *the operational amplifier has a high slew rate.*

When equality is expressed in a classical way, its meaning is precise. The phrase

"The size of the resistor is 36 ohms"

equates the attribute *magnitude* of a specific property, that is, "*the number of ohms*" of a particular resistor to the number 36.

The formula:

$$\text{size}(\text{resistor}) = 36$$

expresses the relation as a proposition equating the value of the linguistic variable *size* to the single integer 36. In this example, there is no uncertainty about the value of the *size* (number of ohms) of the resistor. By using such a relationship, we can now convey certain specific information about the resistor to someone else.

However, when expressing such a concept, the speaker or author of the expression is assuming that the listener or reader knows and understands the property of the resistor to which the expression *size* is referring. That is, "*the number of ohms*" is not the length, diameter, or some other property of the device.

If the meaning of equality is extended to include more ambiguous referents, the relation can become blurred. When we modify the phrase given above to read

The size of the resistor is low,

The corresponding proposition becomes

size (resistor) = *low*.

In the sentence above, the term *low* is a *linguistic variable*. Unless there is a specific agreed-upon value for the variable *low*, its grade of membership in a classic or crisp subset cannot be asserted as *false* or *true* (0, 1), i.e. it has a specific value or it does not.

We can express the resistor size along the horizontal axis of a linear fuzzy *membership function* $\mu_{(R)}$ graph and the grade of membership along the vertical axis as in Figure 6.1. From the graph, we can see that a 36 ohm resistor would have a grade of membership of approximately 0.75 in the fuzzy subset *low*.

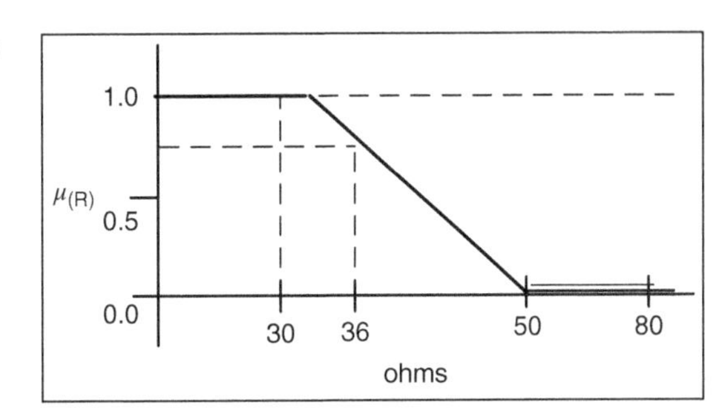

Figure 6.1 Membership function graph for resistor size.

The value of the linguistic variable *size* is now described by a fuzzy subset of numeric values rather than a single value (which may also be interpreted as a crisp subset, as described above, with a single member).

The equality process can be expressed formally as:

Equal

Let

S – Universe of discourse

A, B – (Fuzzy) subsets of S

X – Linguistic variable with values in S

We equate X to A:

X is an A // X is a member of the subset A.

If we state (know) that A and B are equal

A = B

We can further conclude, therefore, that

X is also a B.

Several examples will illustrate how such reasoning proceeds.

Example 6.1

We start with a classical or crisp view. In each of the following propositions, we articulate a *quantitative* description (20 years, 100 kph, 20 ns) to describe the entity being classified (wine, car, risetime).

Classical

a) X is Age(wine).
   Age(wine) = 20 years
   X is 20.
   We know the wine is 20 years old.
b) X is Speed(car).
   Speed(car) = 100 KPH
   X is 100.
   We know the car is moving at 100 KPH.
c) X is Rise time(signal).
   Rise time(signal) = 20 ns
   X is 20 ns.
   We know the rise time of the signal is 20 ns.

In a classic logic application, the value of X (Age, Speed, or Rise time) would be tested or assessed for concurrence with the specific value stated or required for the application and designated as (FALSE, TRUE) or (0.0, 1.0), accordingly.

We now move to a fuzzy view.

Fuzzy

a) X is Age(wine).
   Age(wine) = *very old*
   X is very old.
   We know the wine is very old.
b) X is Speed(car).
   Speed(car) = *fast*
   X is fast.
   We know the car is moving fast.
c) X is Rise time(signal).
   Rise time(signal) = *slow*
   X is slow.
   We know the rise time of the signal is slow.

Observe that in each expression, we have articulated *qualitative* rather than *quantitative* values or descriptions to convey aspects of the entities under discussion. Recall that we examined the *very* modifier and the square operation when we introduced the concept of *hedges* in Chapter 5. Notice now that in stating the age of the wine we have also used the hedge *very*.

Do we know specific values for any of the variables? No. However, in each case, we have a sufficient intuitive feel for or understanding of the intended application and meaning of the expression.

For each of the fuzzy logic applications, we can also develop a grade of membership graph reflecting the ranges of specific values corresponding to appropriate meanings for each of the applications. Figure 6.2 represents a Pi-shaped graph illustrating the degree of membership in the subset *Quality* on the vertical axis corresponding, along the horizontal axis, to the years of aging wine.

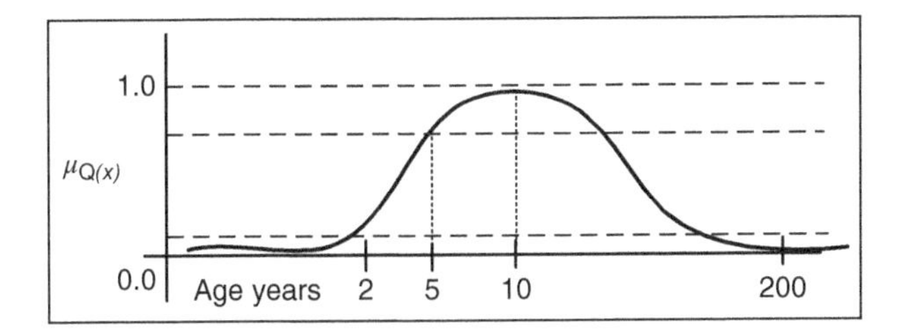

Figure 6.2 Membership function for wine Quality vs. Aging.

Are these articulations incorrect or do they provide sufficient knowledge to permit an understanding of a concept?

6.3 Containment and Entailment

As we learned earlier, *containment* and *entailment* are weaker forms of equality. Once again, articulation is qualitative rather than quantitative. The *containment* relation expresses the idea that one subset is contained within a second.

The term *entail* or *entailment* can be a bit more casual. The term *entailment* expresses *containment* from the opposite perspective. That is, the second subset contains or *entails* (we could also use the word *implies*) the first.

Thus, for example, the graphic in Figure 6.3 contains a set of integers {1..20} that we can call set Y. The set Y then becomes our universe of discourse.

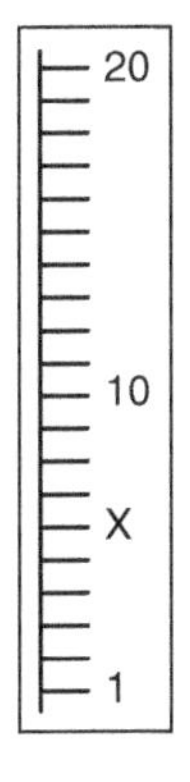

Figure 6.3 Entailment and containment.

Looking closer, we see that Y necessarily *contains* or *entails* the set of integers {1..10}, which we can call S(Y) to indicate a subset of Y. Consequently, we may also then reason that if X is an integer that is less than 10 (greater than 1 being implied), then X is also an integer that is less than 20.

As with equality, extending *entailment* or *containment* to fuzzy subsets introduces a degree of "qualitativeness" or vagueness. The reasoning process expressed formally in the following examples follows traditional lines.

Example 6.2

Let

S – Universe of discourse
A, B – (Fuzzy) subsets of S
X – Linguistic variable with values in S, then

X is A // X is a variable of type A.
A ⊆ B // Subset A is contained in subset B.
X is B // Therefore, X is also a variable of type B.

Example 6.3

Consider that we are designing a set of algorithms we'll call *SortFile* that can be used to place a random collection of entities in a file in a specific order according to some criteria. One of the important characteristics of algorithms in *SortFile* is the speed at which they can perform a required task.

We can have the statement:

quicksort is a fast algorithm.

Proposition:

Speed (quicksort) = Fast

Fast is a fuzzy subset of *SortFile,* and *quicksort*'s membership in that subset is expressed as:

$$\mu_{\text{Fast}}\left(\text{quicksort}\right)$$

Slow is also a fuzzy subset of *SortFile,* and *quicksort*'s membership in that subset is given as:

$$\mu_{\text{Slow}}\left(\text{quicksort}\right)$$

NotSlow is another fuzzy subset of *SortFile*. Quicksort's membership in the subset *NotSlow* will be the complement or negation of *Slow*:

$$1-\mu_{\text{Slow}}\left(\text{quicksort}\right)$$

We can now see that the set hierarchy (*SortFile*) in Figure 6.4 *contains* or *entails* the subsets (Fast), (Slow), and (*NotSlow*) with (quicksort) being a member of each of the subsets.

Now, if *Fast* is also a subset of the fuzzy subset *NotSlow*, i.e. if *NotSlow* entails *Fast,* then

$$1-\mu_{\text{Slow}}\left(\text{quicksort}\right)\supseteq\mu_{\text{Fast}}\left(\text{quicksort}\right)$$

from which, one can infer: Quicksort is a *not slow* algorithm or that Quicksort is *not slow*.

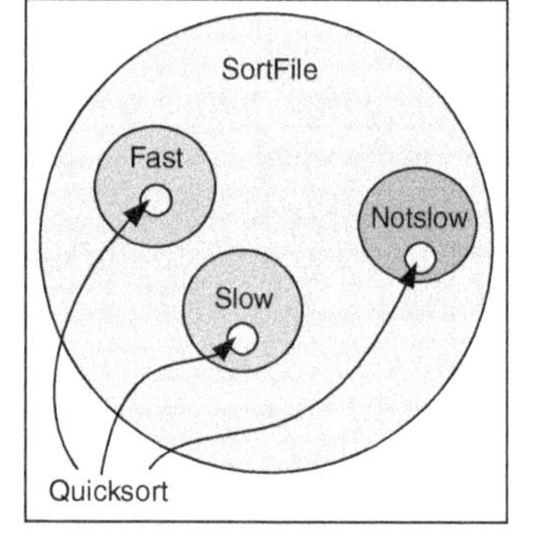

Figure 6.4 Set, subset, member hierarchy.

In essence, *entailment* permits one to infer a proposition that is less specific from one that is more specific. One may also regard *entailment* as a loose generalization of the notion of inheritance applied to fuzzy subsets.

If one interprets statements of the form X is A to also mean X has property A, then given that B is a superset of A, one may also reach the conclusion that X is B as X has property B. Thus, A inherits property B.

The important thing to remember here is that we are working with fuzzy logic not crisp. In the fuzzy world, an entity is not constrained to membership in a single set. In contrast, an entity can have a grade of membership in multiple subsets as illustrated in Figure 6.5, where we have the subsets, *Slow* and *Fast*, of the set *SortFile* and the subset *NotSlow* of the set *Fast*.

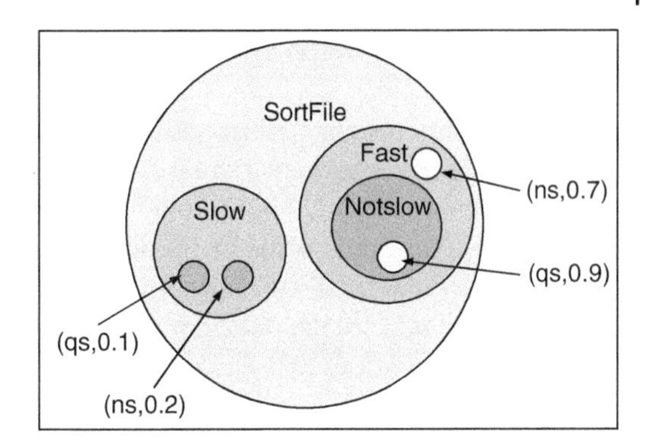

Figure 6.5 Entailment and containment.

For some tasks, *quicksort* (qs) and *NotSlow* (ns) will be slow and will have grades of membership (0.1 and 0.2) in the *Slow* subset. For other tasks, *NotSlow* will be a subset of *Fast* and *quicksort* will be a subset of *NotSlow*.

Example 6.4

We are designing a module within a larger system, and one of our requirements is the need for several capacitors of a specific value that are subject to a constraint that their tolerance is less than 1.0%.

Statement:

The capacitor's tolerance is very tight.

Proposition:

Tolerance (capacitor) = very tight

Once again, we are using the hedge *very*.

Here, we start with membership in the subset *tight*:

$$\mu_{\text{Tight}}(\text{capacitor})$$

Next, *very tight* is interpreted as a fuzzy subset, and the capacitor's membership in the subset *very Tight* is given by:

$$\left(\mu_{\text{Tight}}(\text{capacitor})\right)^2$$

In this example, we are using the square law operation that we discussed when we introduced the concept of *hedges* in Chapter 5.

Since the fuzzy subset *Tight* entails the subset *very Tight*, one may write:

$$\mu_{\text{Tight}}(\text{capacitor}) \supseteq \left(\mu_{\text{Tight}}(\text{capacitor})\right)^2$$

from which one can infer:

The tolerance on the capacitor (which is very tight) is also *tight* as well.

Thus, when building a circuit, one could certainly substitute a higher quality (*tight tolerance*) part for one of lower quality if needed.

6.4 Relations Between Fuzzy Subsets

In isolation, the concept of fuzzy subsets is curious, interesting, and sometimes potentially confusing. However, to do real work, as we did with Boolean algebra, we need to be able to identify, understand, and apply relations among such entities. To that end, we will now introduce and examine some of the vocabulary and logical relations that we can have between fuzzy subsets and that also will work with and among fuzzy subsets. In retrospect, once again, we will see that much of the vocabulary and many such relations mirror those between similar entities that we are familiar with from our work in the crisp world.

6.4.1 Union and Intersection

Following our adventures with George Boole in the crisp world, we learned about the logical OR and the logical AND operators and relationships. We'll begin our study of relationships between fuzzy sets and subsets with the familiar things that we know. In Chapter 4, we introduced the implementation of the concepts *union* and *intersection* as Properties 5 and 6 in the world of fuzzy set theory. We will now introduce their application.

6.4.1.1 Union

As a first step, it is important to understand that the fuzzy union of two subsets, as in Figure 6.6 is not the same as combining two glasses of water, as in Figure 6.7. Several subsets combined do not overflow their containing set. More appropriately, it's like sets Y and X in Figure 6.3. If I have set Y (1..20) and set X (1..10) and bring them together, the resulting subset members will be the total from Y(1..20).

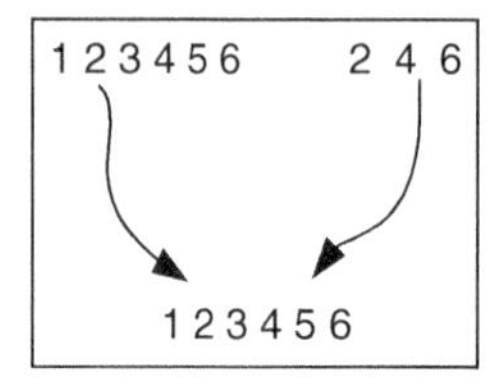

Figure 6.6 A union.

Figure 6.7 Not a union.

Let's now define two subsets $R_1(x, y)$ and $R_2(x, y)$ as expressed in Figure 6.8. The *union* of two such subsets is expressed as:

$R_1 \cup R_2$ or $R_1 + R_2$ // the + indicates the OR operation, not *add*.

R_1	y_1	y_2	y_3	y_4
x_1	0.2	0.4	0	1
x_2	0.6	0.3	0.5	0
x_3	0.7	0.4	1	0.5
x_4	0	0	0.5	0.4

R_2	y_1	y_2	y_3	y_4
x_1	0.2	1	0.3	0.7
x_2	0.5	0.4	0.5	0.8
x_3	0.1	0.5	0.9	0.5
x_4	1	0.8	0.6	0.5

$R_1 \cup R_2$	y_1	y_2	y_3	y_4
x_1	0.2	1	0.3	1
x_2	0.6	0.4	0.5	0.8
x_3	0.7	0.5	1	0.5
x_4	1	0.8	0.6	0.5

Figure 6.8 The union of two fuzzy subsets.

The membership function for the resulting *union* subset is given as Eq. (6.1).

$$\mu_{R_1 \cup R_2}(x, y) = \text{MAX}\{\mu_{R_1}(x, y), \mu_{R_2}(x, y)\} \tag{6.1}$$

The two example subsets for R_1 and R_2 are given in the tables in Figure 6.8.

Each row in the subset resulting from the *union* of the original two subsets is computed as follows:

For the row x_1 from R_1 we have,

(0.2, 0.4. 0.0. 1.0)

and row x_1 from R_2,

(0.2, 1.0. 0.3. 0.7)

We compute row x_1 for the *union* $R_1 \cup R_2$ as:

MAX {(0.2, 0.2), (0.4, 1.0), (0, 0.3), (1.0, 0.7)}

which gives us row x_1 for the *union* $R_1 \cup R_2$ as {0.2, 1.0, 0.3, 1.0}.

We repeat the same process for the remaining three rows to complete the table for the *union* $R_1 \cup R_2$.

6.4.1.2 Intersection

Define two subsets $R_1(x, y)$ and $R_2(x, y)$.

The *intersection* of two such subsets is expressed as:

$R_1 \cap R_2$ or $R_1 \bullet R_2$ // the • indicates the intersection operation
// not *multiply*

The membership function for the resulting *intersection* subset is given as Eq (6.2).

$$\mu_{(R_1 \cap R_2)}(x, y) = \text{MIN}\{\mu_{R_1}(x, y), \mu_{R_2}(x, y)\} \tag{6.2}$$

The two example subsets R_1 and R_2 are given in the tables in Figure 6.9.

R_1	y_1	y_2	y_3	y_4
x_1	0.2	0.4	0	1
x_2	0.6	0.3	0.5	0
x_3	0.7	0.4	1	0.5
x_4	0	0	0.5	0.4

R_2	y_1	y_2	y_3	y_4
x_1	0.2	1	0.3	0.7
x_2	0.5	0.4	0.5	0.8
x_3	0.1	0.5	0.9	0.5
x_4	1	0.8	0.6	0.5

$R_1 \cap R_2$	y_1	y_2	y_3	y_4
x_1	0.2	0.4	0	0.7
x_2	0.5	0.3	0.5	0
x_3	0.1	0.4	0.9	0.5
x_4	0	0	0.5	0.4

Figure 6.9 Intersection of two fuzzy subsets.

Each row in the subset resulting from the *intersection* of the original two subsets is computed as follows:

For the row x_1 from R_1 and row x_1 from R_2, we compute row x_1 for $R_1 \cap R_2$ as:

MIN {(0.2, 0.2), (0.4, 1.0), (0,0.3), (1.0, 0.7)}

which gives us row x_1 for $R_1 \cap R_2$ as {0.2, 0.4, 0, 0.7}.

We repeat the same process for the remaining three rows to complete the table for the intersection $R_1 \cap R_2$.

6.4.2 Conjunction and Disjunction

We will now examine the operations of *conjunction* and *disjunction*. Reasoning using either of these propositions is reflected in statements of the following form:

X is A *and* X is B => *X is A and B.*

or

X is A *or* X is B. => *X is A or B.*

Such combinations are interpreted according to the *intersection* or *union* of their underlying (fuzzy) subsets. By such means, one may thus conclude that the first phrase is equivalent to *X is A and B* and the second is equivalent to *X is A or B*. Reasoning using *conjunctive* or *disjunctive* relationships may be formally expressed in the fuzzy world as follows:

Conjunction / Disjunction

Let

S – Universe of discourse

A, B – (Fuzzy) subsets of S

X – Linguistic variable with values in S

then

We can write:

X is A *and* X is B => X is A ∩ B.

X is A *or* X is B => X is A ∪ B.

Let's now put these two concepts to work.

Example 6.5

In our study of fuzzy set properties in Chapter 4, we worked with a universe of discourse comprising five people and two fuzzy subsets, tall people and blond people, of that universe. In this example, our universe of discourse will be a finite set of six ordinary transistors: T_0. . .T_5.

Following our example under Property 6 in Section 4.6.2, we will consider two properties of a transistor: its beta and its leakage. Based on these properties, we will specify two fuzzy subsets: B (beta) and L (leakage) within our universe of discourse, S. Consider the diagram in Figure 6.10.

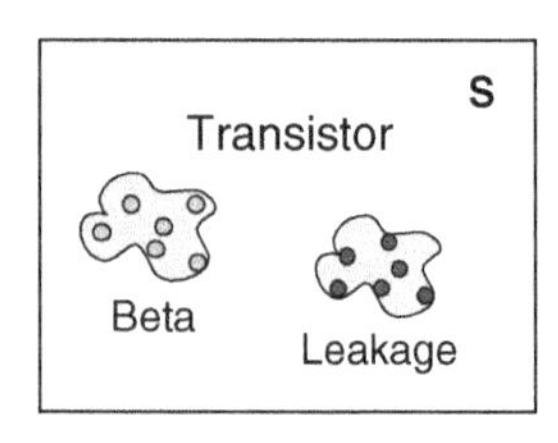

Figure 6.10 Universe of discourse S.

The grades of individual device membership in subsets beta and leakage will fall within the range of 0.1–0.9. Our technical goal is high beta and low leakage. Therefore, potential grades of membership for beta will be 0.1–0.9 for low to high beta. We will also specify a grade of membership of 0.1 for high leakage (we don't want it) and 0.9 for low leakage (very good).

Now consider that we are designing a circuit and that we need a transistor with high beta *and* low leakage so we evaluate a number of different devices.

Statements:

The beta of the transistor is *High.*

The leakage of the transistor is *Low.*

Propositions:

Beta (transistor) = High

Leakage (transistor) = Low

High and Low are fuzzy subsets with the transistor's membership given by:

μ_{high} (transistor) and μ_{low} (transistor), respectively

The implied *and* connecting the two statements indicates the intersection of the underlying subsets. Thus,

Beta (transistor) and Leakage (transistor) ⇒ High beta and Low leakage, and we conclude that the transistor has high beta and low leakage. We express that requirement as an *intersection* operation with membership in the combined subset given by:

X is B *and* X is L→ X is B ∩ L.

Example 6.6

Let S (the universe of discourse) be a finite set of six different transistors. Now, let B and L be fuzzy subsets of S containing the transistors with high beta and the devices with low leakage, respectively, of set S given in Figure 6.11.

Figure 6.11 Basic universe of discourse S.

$S = \{T_0, T_1, T_2, T_3, T_4, T_5\}$

$B = \{(T_0,0.8), (T_1,0.9), (T_3,0.5), (T_3,0.7), (T_4,0.4), (T_4,0.6)\}$

$L = \{(T_0,0.5), (T_1,0.9), (T_3,0.7), (T_3,0.9), (T_4,0.8), (T_4,0.7)\}$

The implied (*and*) connecting the two statements indicates the *intersection* of the underlying subsets. Thus,

Beta (transistor) *and* Leakage (transistor) => High beta *and* Low leakage and we conclude that:

The preferred transistor has high beta and low leakage.

With membership in the combined subset given by:

min (μ_{high} (transistor), μ_{low} (transistor))

From our earlier studies in Chapter 4, Fuzzy Logic Property 6, we learned that the result of the intersection of two fuzzy subsets T (tall) and B (blond), expressed as T ∩ B, is the fuzzy subset of ordered pairs:

(z_{1k}, z_{2k}) formed from all ordered pairs

(x_{1i}, x_{2i}) in T and

(y_{1j}, y_{2j}) in B

such that

$(x_{1i} = y_{1j}) = z_{1k}$ and

$z_{2k} = \min(x_{2i}, y_{2j})$

or equivalently,

$\min(\mu_T(x_{1i}), \mu_B(y_{1j}))$

Once again, the ordered pairs within each subset comprise the name and the grade of membership of the entity in a fuzzy subset.

Like the fuzzy subsets *Tall* and *Blond*, in the universe of discourse S discussed earlier, *Beta* (B) and *Leakage* (L) are the fuzzy subsets within the universe of discourse, *Transistor*.

Membership values within those subsets are given by:

$$B = \Sigma_{\sup x} \mu_B(x)/x$$

$$L = \Sigma_{\sup y} \mu_L(y)/y$$

Transistor beta (β), which expresses the gain or amplification factor of the transistor, typically ranges from 20 to 500. For the case of beta, we are looking for high values. Thus, the degree of membership in the fuzzy subset B should be high for high beta values and small for small beta values. Therefore, the range of values of our degree of membership in the fuzzy subset B will correspond to the typical beta values.

Leakage or leakage current is the current due to the minority charge carriers flowing in a transistor. Such a current can shunt signals or affect bias voltages and thereby potentially have a destructive effect on circuit operation.

Depending upon the type of transistor, leakage (L) values can range from 0 μ_A to over 2500 μ_A. Here, we read μ_A as microamps. For the case of leakage, we are looking for small values ideally approaching 0 μ_A. Thus, the degree of membership in the fuzzy subset L should be high for small leakage values and small for high leakage values. A possible leakage grade of membership function is given in Figure 6.12.

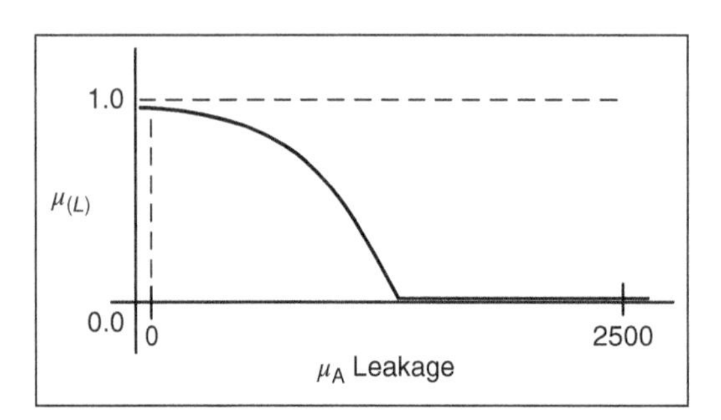

Figure 6.12 Membership function for transistor leakage.

Let us now examine a collection of the six different transistor types T_0–T_5 and determine which type best suits our needs. We test first for beta and then leakage.

$$B = 0.9 / T_0 + 0.6 / T_1 + 0.8 / T_2 + 0.5 / T_3 + 0.7 / T_4 + 0.4 / T_5$$

$$L = 0.8 / T_0 + 0.7 / T_1 + 0.3 / T_2 + 0.6 / T_3 + 0.6 / T_4 + 0.7 / T_5$$

Repeating, our goal is high beta *and* low leakage.

Thus, we form the conjunction of the two sets:

X is B *and* Y is L → Z is B ∩ L.

For each device, we compute: $\min((\mu_B(x), \mu_L(y))$.

$$B \cap L = \{ (T_0, \min(0.9, 0.8)), (T_1, \min(0.6, 0.7)), (T_2, \min(0.8, 0.3)), (T_3, \min(0.5, 0.6)), (T_4, \min(0.7, 0.6)), (T_5, \min(0.4, 0.7)) \}$$

$$= 0.8/T_0 + 0.6/T_1 + 0.3/T_2 + 0.5/T_3 + 0.6/T_4 + 0.4/T_5$$

Based on these results, it appears that transistor T_0 is our best choice. Why do we choose the min rather than max operation?

6.4.3 Conditional Relations

One may express a *conditional* or *conditioned* relationship between two variables with statements of the form *if X is A, then Y is B,* in which the values of the linguistic variables are grades of membership in underlying (fuzzy) subsets.

Typical examples are as follows:

1) If the temperature of the device exceeds x degrees, then cooling must be enabled.
2) If Kyleen is over 6 feet tall, then she will play volleyball.
3) If the weather is stormy, then sailing is dangerous.
4) If I increase the emitter resistor, then the transistor's gain will decrease.

Let's examine the second statement of this list. Kyleen is potentially a member of two subsets: tall people and people playing volleyball. Kyleen's membership in the second subset is *conditioned* on her being a member in the first. Similar *conditioned* relations are routine in complex and high-performance digital and embedded systems. They are also common and routinely potentially essential in automatic control systems.

Statements such as those examples describe a *conditioned* relationship between two variables; therefore, any subset representing that relationship must contain all possible combinations of the values of those variables. One such subset is given by the *Cartesian product* of the variables.

The Cartesian product is a mathematical operation that yields a single set from multiple sets. Given the sets A and B, the Cartesian product, written as A x B (read as A cross B), yields the set of all ordered pairs (a_i, b_j) for which a_i € A and b_j € B.

A Cartesian product can also be expressed in a tabular form in which the cells of the table contain the ordered pairs formed by computing the product of the elements in the rows and columns. Let A be defined as the set {1, 2, 3} in the left-hand vertical column and B be defined as the set {4, 5, 6} in the top horizontal row. The table, A × B, will now be computed and illustrated as shown in Figure 6.13.

A \ B	4	5	6
1	(1,4)	(1,5)	(1,6)
2	(2,4)	(2,5)	(2,6)
3	(3,4)	(3,5)	(3,6)

Figure 6.13 Cartesian cross product A × B.

Specifying the grade of membership of each resulting pair in the consequent subset to be the fuzzy *min* of the individual grades of membership intuitively satisfies the intent of the relation. As we will learn, numerous other interpretations have also been proposed.

Conditioned Fuzzy Subsets

A *conditioned* relationship between subsets may be expressed formally as follows:

Let

S_1 and S_2 be two fuzzy sets

A(x) be a fuzzy subset contained in set S_1

then

$A(x) \subset S_1$

If the membership function for the subset A(x) depends upon x € S_2 as a parameter, then the subset A(x) will said to be conditioned on the subset S_2.

The conditional membership function for A(x) will be written as follows:

$\mu_A(x||y)$ where x € S_1 and y € S_2 // the || is pronounced "given."

We now have a mapping of S_2 into the collection of fuzzy subsets defined on S_1.

A fuzzy subset B € S_2 will induce a fuzzy subset $A \subset S_1$ whose membership function will be:

$$\mu_A(y) = \max\left(\min\left[\mu_A(x \,||\, y), \mu_B(y)\right]\right)$$

Example 6.7

Let

$S = \{1, 2, 3\}$

$A = 0.9/1 + 0.7/2$

$B = 0.6/1 + 0.8/2 + 1.0/3$

from which

$$
\begin{aligned}
(A \times B) &= \min(0.9, 0.6)/(1,1) + \min(0.9, 0.8)/(1,2) + \min(0.9, 1.0)/(1,3) + \\
&\quad \min(0.7, 0.6)/(2,1) + \min(0.7, 0.8)/(2,2) + \min(0.7, 1.0)/(2,3) \\
&= \underbrace{0.6/(1,1) + 0.8/(1,2) + 0.9/(1,3)}_{A1} + \underbrace{0.6/(2,1) + 0.7/(2,2) + 0.7/(2,3)}_{A2}
\end{aligned}
$$

which can also be written as in Figure 6.14.

	B_1	B_2	B_3
A_1	0.6	0.8	0.9
A_2	0.6	0.7	0.7

Figure 6.14 Conditioned fuzzy subsets.

The (fuzzy) conditioned statement of the form if A *then* B can be seen to be a special case of the more general statement *if* A *then* B *else* C, which can also be rewritten as *if* A *then* B *or if not* A *then* C. This latter statement describes the union of the two conditional relations: *if* A *then* B is expressed by $A \times B$ and *if not* A *then* C is expressed by $A^c \times C$. The complete relation is then given by:

$$A \times B + A^c \times C$$

Here the "+" is the logical OR operator, and the exponent c of *A* indicates the complement or negation. It should be clear that if the subset *C* is (repeatedly) replaced by a similar *if-then-else* statement, then the conditions may be nested indefinitely.

6.4.4 Composition Revisited

Real-world or commonsense reasoning is very often a process of taking the solution to one problem, possibly modifying it slightly, then applying it to a second. A variation on such reasoning is often found in hardware or software design or in troubleshooting where a variety of test stimuli are used as inputs to a failing or improperly operating system, and the consequences (resulting outputs) of each are observed and assessed.

For example, given conditional statements such as the following:

If I *compile* the program late at night, it works.

If I *drive* fast, there is a noise in the engine.

If the *temperature* is low, the circuit oscillates.

Can anything be said about compiling *somewhat* or *very, very* late at night, or about the engine noise if driving is *more or less* fast, or about the circuit if the temperature is *very* low? In each case, the value of the premise linguistic variable X (*compile, drive, temperature*) has been modified by a *hedge h* (*late, fast, low*) as we discussed in Chapter 5.

Since the original expressions stated above imply a relation R between two fuzzy subsets, a *premise* subset X and a *consequent* subset Y, applying R to members of the *premise* subset X will affect members of the *consequent* subset Y. Taking the application of R a step further.

Let the subset X entail *h* X. Thence, applying R to X will also apply R to the subset *h* X. As we see in Figure 6.15, applying R to the subset *h* X will yield a subset of Y that we'll call Y′ (Y prime) that contains the image of *h* X under the mapping.

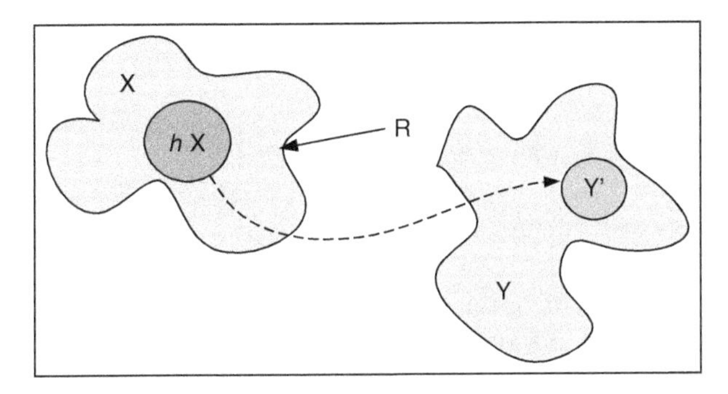

Figure 6.15 Applying Relation R to Set X.

We may now write if X ⊇ (*h* X), that is, if (X) is a superset of (h X), then RoX ⊇ Ro(*h*X). Said another way, if R is composed with X, then (Y) will contain the image of (*h* X) under the mapping. See Figure 6.15. The symbol "o" indicates *composition*.

Two of the more common fuzzy relations are denoted *max-min* and *max-product composition*. Composition in such relations is indicated by the symbol "o". The min operator may be replaced by a different operator, provided that the replacement is associative and monotonically nondecreasing in each argument. Later we will have occasion to define the max-product composition and see some of its particular uses.

6.4.4.1 Max-Min Composition

Formally, fuzzy reasoning using the composition of relations may be expressed as follows.

Let

S – Universe of discourse
A – Fuzzy subset of S
R – Fuzzy binary relation, X × Y // X cross Y
X, Y – Linguistic variables with values in S
X is A.
(X, Y) is R.
Y is A o R.

The membership function with respect to the resulting subset is given as Eq. (6.2) and illustrated in Figure 6.16:

$$\mu_{\mathrm{AoR}}\left(\mathrm{Y}\right) = \max_{\mathrm{X}}(\min\left(\mu_{\mathrm{A}}\left(\mathrm{X}\right), \mu_{\mathrm{R}}\left(\mathrm{X},\mathrm{Y}\right)\right) \tag{6.3}$$

where $\max_{\mathrm{x}}$ is taken over all x in X.

A		
0.1	0.5	0.9

R	5	10	15
20	0.1	0.3	0.6
30	0.2	0.5	0.8
40	0.3	0.7	0.9

A o R		
0.3	0.7	0.9

Figure 6.16 MAX-MIN composition of a fuzzy subset and a fuzzy relation.

Working with Figure 6.16, we take the row A cross the column R:

AR_1: max{min(0.1, 0.1), min(0.5, 0.2), min(0.3, 0.9) }
= max{0.1, 0.2, 0.3} = 0.3

AR_2: max{min(0.1, 0.3), min(0.5, 0.5), min(0.9, 0.7)}
= max{0.1, 0.5, 0.7} = 0.7

AR_3: max{min(0.1, 0.6), min(0.5, 0.8), min(0.9, 0.9) }
= max{0.1, 0.5, 0.9} = 0.9

We now generalize the max-min composition operation to support its application to two fuzzy relations. Define two subsets $R_1(x,y)$ and $R_2(y,z)$.

Let

S – Universe of discourse
X, Y, Z – Linguistic variables with values in S
R_1 – Fuzzy binary relation of X,Y → $R_1(X,Y)$
R_2 – Fuzzy binary relation of Y, Z→ $R_2(Y, Z)$
R_3 – Subset of X × Z, $R_3(X,Z)$ // read *x* as cross
(X,Y) is R_1.
(Y,Z) is R_2.
Z is $R_1 * R_2$.
* is an operator that is associative and monotonically nondecreasing in each operand.

The *max-min composition* of the two subsets is expressed as:

R(X,Z) = R_1(X,Y) o R_2(Y,Z) // the o indicates composition

The membership function with respect to the resulting subset is given as Eq. 6.4 and illustrated in Figure 6.17:

$$\mu_{R1} o R_2(x,z) = \max_{y \in Y} \min\left[\mu R_1(x,y), \mu_{R2}(y,z)\right] \tag{6.4}$$

From Figure 6.17, each row in a subset resulting from the max-min composition of the original two subsets is computed as follows:

X_1Z_1: max{min(0.2, 0.2), min(0.4, 0.5), min(0.7, 0.1), min(1,1)}
= max{0.2, 0.4, 0.1, 1} = 1

X_2Z_1: max{min(0.6, 0.2), min(0.3, 0.5), min(0.5, 0.1), min(0,1)}
= max{0.2, 0.3, 0.1, 1} = 1

X_1Z_2: max{min(0.2, 1), min(0.4, 0.4), min(0.7, 0.5), min(1,0.8)}
= max{0.2, 0.4, 0.5, 0.8} = 0.8

X_2Z_2: max{min(0.6, 1), min(0.3, 0.4), min(0.5, 0.5), min(0,0.8)}
= max{0.6, 0.3, 0.5, 0} = 0.6

R_1	y_1	y_2	y_3	y_4
x_1	0.2	0.4	0.7	1
x_2	0.6	0.3	0.5	0

R_2	Z_1	Z_2
Y_1	0.2	1
Y_2	0.5	0.4
Y_3	0.1	0.5
Y_4	1	0.8

R_1 o R_2	Z_1	Z_2
x_1	1	0.8
x_2	1	0.6

Figure 6.17 MAX-MIN composition of two fuzzy subsets.

6.4.4.2 Max-Product Composition

Define the two subsets $R_1(x,y)$ and $R_2(y,z)$.

The *max-product composition* of the two subsets is expressed as:

$$R(X,Z) = R_1(X,Y) \circ R_2(Y,Z)$$

R_1	y_1	y_2	y_3	y_4
x_1	0.1	0.7	0.4	0.2
x_2	0.3	0.6	0.5	0.9

R_2	Z_1	Z_2
Y_1	0.2	1
Y_2	0.5	0.4
Y_3	0.1	0.5
Y_4	1	0.8

$R_1 \circ R_2$	Z_1	Z_2
x_1	0.35	0.28
x_2	0.9	0.72

Figure 6.18 Max-product composition of two fuzzy subsets.

The membership function with respect to the resulting subset is given as Eq. 6.4 and illustrated in Figure 6.18. The symbol × is multiply.

$$\mu_{R_1 \times R_2}(x,z) = \max_{y \in Y}\left[\mu_{A(x,y)} \times \mu_{B(y,z)}\right] \tag{6.5}$$

Each row in the subset resulting from the composition of the original two subsets is computed as follows:

X_1Z_1: max{ (0.1×0.2), (0.7×0.5), (0.4×0.1), (0.2× 1)}
= max{0.02, 0.35, 0.04, 0.2} = 0.35

X_2Z_1: max{ (0.3×0.2), (0.6×0.5), (0.5×0.1), (0.9×1)}
= max{0.06, 0.3, 0.05, 0.9} = 0.9

X_1Z_2: max{ (0.1×1), (0.7×0.4), (0.4×0.5), (0.2×0.8)}
= max{0.01, 0.28, 0.2, 0.16} = 0.28

X_2Z_2: max{ (0.3×1), (0.6×0.4), (0.5×0.5), (0.9×0.8)}
= max{0.3, 0.24, 0.25, 0.72} = 0.72

Example 6.8

Let's now return to our transistor and see how we can apply these concepts. We know that transistor X and transistor Y are approximately the same and that X has very small leakage. What can we say about the leakage of transistor Y?

We now compare four transistors.

Let

S = {1, 2, 3, 4}, a set of possible *nanoamp* leakages for a particular transistor type. The possible leakage grades of membership will be (1.0, 0.5, 0.0, 0.0).

R = Fuzzy relation *Approximately Equal* over the elements of S, which can be written as:

R = the set of elements in Figure 6.19

	1	2	3	4
1	1.0	0.5	0	0
2	0.5	1.0	0.5	0
3	0	0.5	1.0	0.5
4	0	0	0.5	1.0

Figure 6.19 Approximately Equal over a Set.

A = Fuzzy subset *Small*
= 1.0/1 + 0.6/2 + 0.2/3 +0.0/4

1	2	3	4
1.0	0.6	0.2	0.0

From basic mathematics, we know that multiplying a fraction by itself (squaring) will produce a smaller fraction. Thus, since *Small* is given by A, *very Small* can be given by A^2.

$A^2 = (1.0)^2/1 + (0.6)^2/2 + (0.2)^2/3 + (0.0)^2/4$

1	2	3	4
1.0	0.36	0.04	0.0

Thus, $A^2 \times R =$ // A^2 cross R

1.0	0.36	0.04	0

1.0	0.5	0	0
0.5	1.0	0.5	0
0	0.5	1.0	0.5
0	0	0.5	1.0

Using *max-min composition*,
A2 x R =
{
max (min(1.0, 1.0), min(0.36, 0.5), min(0.04, 0.0), min(0.0, 0.0)),
max (min(1.0, 0.5), min(0.36, 1.0), min(0.04, 0.5), min(0.0, 0.0)),
max (min(1.0, 0.0), min(0.36, 0.5), min(0.04, 1.0), min(0.0, 0.5)),
max (min(1.0, 0.0), min(0.36, 0.0), min(0.04, 0.5), min(0.0, 1.0))
}

= {
max (1.0 0.36 0 0), max(0.5 0.36 0.04 0), max(0 0.36 0.04 0), max(0 0 0.04 0)
}
= {1.0, 0.5, 0.36, 0.04}
Y is given as:
Y = 1/1 + 0.5/2 + 0.36/3 + 0.04/4
and A^2 – very small is given as:
= 1/1 + 0.36/2 + 0.04/3 + 0/4
which suggests that the leakage for the sample of transistor type Y is somewhat larger than *very small.*

Example 6.9
We know that sailing in stormy weather can be dangerous. We now want to find out how dangerous it can be if we wish to sail when it is *very* stormy.

Let

S = *Universe of discourse*

A = Fuzzy subset *Stormy*, which is based on the *wind velocity* of 20, 30, or 40 mph or more of the storm *Paulette*. The grade of membership is based on the seriousness of the wind velocity.

$= 0.1/20 + 0.5/30 + 0.9/40$

D = Fuzzy subset *Dangerous* based on *wave size* of 5, 10, or 15 feet.

R = Fuzzy relation *Stormy-Danger* over elements of S relating stormy weather (*wind velocity*) and D dangerous sailing (*wave size*).

Specify R from Figure 6.16 as:

R =

R	5	10	15
20	0.1	0.3	0.6
30	0.2	0.5	0.8
40	0.3	0.7	0.9

Let us pose the following question:

If the weather is very stormy, what is the danger of sailing?

Since *stormy* is given by A, *very stormy* will be given by A^2.

$A = 0.1/20 + 0.5/30 + 0.9/40$

$A^2 = 0.01/20 + 0.25/30 + 0.81/40$

If we first compose A, then A^2 with R, then using max-min composition, we get the following:

Stormy

$A \times R = A$

0.1	0.5	0.9

R	5	10	15
20	0.1	0.3	0.6
30	0.2	0.5	0.8
40	0.3	0.7	0.9

$$
\begin{aligned}
A \times R = \{ & \\
& \max(\min(0.1, 0.1), \min(0.5, 0.2), \min(0.9, 0.3)), \\
& \max(\min(0.1, 0.3), \min(0.5, 0.5), \min(0.9, 0.7)), \\
& \max(\min(0.1, 0.6), \min(0.5, 0.8), \min(0.9, 0.4)) \\
\} & \\
= \{ & \\
& \max(0.1, 0.2, 0.3), \max(0.1, 0.5, 0.7), \max(0.1, 0.5, 0.4) \\
\} & \\
= [& 0.3, 0.7, 0.5]
\end{aligned}
$$

$$D_{Stormy} = 0.3/5 + 0.7/10 + 0.5/15$$

Very Stormy

$A2 \text{ x } R = A^2$

0.01	0.25	0.81

R	5	10	15
20	0.1	0.3	0.6
30	0.2	0.5	0.8
40	0.3	0.7	0.9

$$
\begin{aligned}
A2 \text{ x } R = & \\
\{ & \\
& \max(\min(0.01, 0.1), \min(0.25, 0.2), \min(0.81, 0.3)), \\
& \max(\min(0.01, 0.3), \min(0.25, 0.5), \min(0.81, 0.7)), \\
& \max(\min(0.01, 0.6), \min(0.25, 0.8), \min(0.81, 0.9)) \\
\} & \\
= \{ & \\
& \max(0.01, 0.2, 0.3), \max(0.01, 0.25, 0.7), \max(0.01, 0.25, 0.81) \\
\} & \\
= \{ & 0.3, 0.7\ 0.81\}
\end{aligned}
$$

$$D_{\text{very Stormy}} = 0.3/5 + 0.7/10 + 0.81/15$$

The expression for danger inferred for *very stormy* shows some compression with respect to that for *stormy*, which is consistent with the hedge *very*.

Example 6.10

As noted earlier, the *Cartesian product* is a relation between two sets that yields a set of all possible ordered pairs for which the first element is from the first set and the second element is from the second set. The procedure can generalize to an infinite number of sets.

In this example, we will work with two sets and build the relation R based on the *Cartesian product* of the two fuzzy subsets A and D and the *min* of each of the pairs.

S – Universe of discourse
A – Fuzzy subset *Stormy,* which is based on *wind speed*
= 0.1/20 + 0.5/30 + 0.9/40
D – Fuzzy subset *Dangerous* based on *wave size* of 5, 10, or 15 feet
= 0.1/5 + 0.5/10 +0.9/15
R – Fuzzy relation *Stormy-Danger* over elements of S relating stormy weather (*wind speed*) and dangerous sailing (*wave size*)

A x D

	5	10	15
20	min (0.1, 0.1)	min (0.1, 0.5)	min (0.1, 0.9)
30	min (0.5, 0.1)	min (0.5, 0.5)	min (0.5, 0.9)
40	min (0.9, 0.1)	min (0.9, 0.5)	min (0.9, 0.9)

Let R = A x D:
R =

	5	10	15
20	0.1	0.1	0.1
30	0.1	0.5	0.5
40	0.1	0.5	0.9

Let us pose the following question:
If the weather is *very stormy*, what is the danger in sailing?
Since *stormy* is given by A, *very stormy* will be given by A^2.
A = 0.1/20 + 0.5/30 + 0.9/40
A^2 = 0.01/20 + 0.25/30 + 0.81/40

If we first compose A, then A^2 with R, we get the following:
D_{Stormy}

A×R

	5	10	15
20	min (0.1, 0.1)	min (0.1, 0.5)	min (0.1, 0.9)
30	min (0.1, 0.1)	min (0.5, 0.5)	min (0.5, 0.9)
40	min (0.1, 0.1)	min (0.5, 0.5)	min (0.9, 0.9)

$D_{veryStormy}$
$A^2 \times R$

	5	10	15
20	min (0.1, 0.1)	min (0.1, 0.25)	min (0.1, 0.81)
30	min (0.1, 0.1)	min (0.5, 0.25)	min (0.5, 0.81)
40	min (0.1, 0.1)	min (0.5, 0.25)	min (0.9, 0.81)

which gives us:

$$\begin{aligned} D_{Stormy} = \max\,(&(\min(0.1, 0.1), \min(0.1, 0.5), \min(0.1, 0.9) + \\ &\min(0.1, 0.1), \min(0.5, 0.5), \min(0.5, 0.9) + \\ &\min(0.1, 0.1), \min(0.5, 0.5), \min(0.9, 0.9)) \\ = \{&\, 0.1/5 + 0.5/10 + 0.9/15 \,\} \quad \text{// Fuzzy subset D} \end{aligned}$$

$$\begin{aligned} D_{veryStormy} = \max\,(&(\min(0.1, 0.01), \min(0.1, 0.25), \min(0.1, 0.81) + \\ &\min(0.1, 0.01), \min(0.5, 0.25), \min(0.5, 0.81) + \\ &\min(0.1, 0.01), \min(0.5, 0.25), \min(0.9, 0.81)) \\ = \{&\, 0.1/5 + 0.5/10 + 0.81/15 \,\} \end{aligned}$$

This last example reveals two interesting points. First, the relation R was developed as the Cartesian product of the two fuzzy subsets A and D. Although mathematically accurate, the Cartesian product does not inherently introduce causality into the resulting relation. Consequently, R does not embody any notion of the actual physical relationship (in this case, between stormy weather and dangerous sailing) that may exist between the corresponding subsets. Therefore, one should be aware that the fuzzy subset resulting from such a composition may not always reflect what is intended.

Second, observe that when the fuzzy subset *stormy* (A) is composed with R, the result is the fuzzy subset *dangerous* (D). This consequence holds in general, that is, when a relation is formed by taking the Cartesian product of two operands, composing one of the operands with that relation will yield the other operand.

This result may be expressed more formally as follows:

Let

S – Universe of discourse
R – Fuzzy binary relation, X x Y
A, B – Fuzzy subsets of S
X, Y – Linguistic variables with values in S

X is A.
(X, Y) is R.
Y is $A \times R$.

If R is $A \times B$, then $Y = B$

permits one to view such a composition as an approximate extension of the modus ponens, which we'll discuss next.

6.5 Inference in Fuzzy Logic

In any logic or process of reasoning, inference is the formal process of deducing a proposition C (consequent or conclusion) from a set of premises P on the basis of evidence and reasoning. Stated rules of inference govern the process by which such deductions are legally made within that logic or reasoning process.

Classical propositional calculus most frequently uses two sets of rules for governing inference. The rules for *deduction* forward involve reasoning from a set of premises (antecedent) to a conclusion (consequent). Those for deduction backward involve reasoning from a consequent to a premise.

The inference rules governing reasoning in the forward direction (*deduction*) are called the *modus ponens.*

> If P, the premises, imply C, and if the premises P are declared to be valid or true, therefore, then C must be true.

The inference rules governing backward implication (*abduction*) are called the *modus tollens.*

> If P, the premises, imply C and if the consequent C is invalid or false, P are declared to be invalid or false.

These sets of implication rules upon which these two forms of reasoning are based are formally stated as follows:

Let

S – Universe of discourse
A, B – A and B are fuzzy subsets of S
X, Y – Variables with values in S

Modus Ponens

Premises

P_1 X is A.
P_2 If X is A, then Y is B.

Conclusion

C Y is B.

Modus Tollens

Premises

P_1 Y is B.
P_2 If X is A, then Y is B.

Conclusion

C X is A.
P_1 Y IS B.
P_2 IF X IS A THEN Y IS B.
C X IS A.

One must be cautious when interpreting the conclusions drawn using abductive reasoning (based on *modus tollens*) although such reasoning is often used in troubleshooting hardware or software, in medical diagnostic procedures, or as justification for many myths or folk tales. Such reasoning contains a potential fatal flaw. The flaw lies in the unjustified (yet often made) presumption of causality between premise variables A and B.

Above, premise P_2 states that if A is true, then B will follow. However, it makes no assertion about other possible causes of B. Thus, the truth of B does not imply the truth of A, only the potential of truth.

The deductive and abductive rules given above can be extended from classical logic to fuzzy logic with only minor reinterpretation of the premises and special consideration given to implication.

Let

S – Universe of discourse

A, B – A and B are fuzzy subsets of S

X, Y – Variables with values in S

Fuzzy Modus Ponens

P_1 X is A.

P_2 <u>If X is A, then Y is B.</u>

C Y is B.

Fuzzy Modus Tollens

P_1 Y is B.

P_2 <u>If X is A, then Y is B.</u>

C X is A.

Because fuzzy reasoning uses fuzzy subsets and fuzzy implication, the propagation of such fuzzy subsets through the reasoning chain is not crisp. Consequently, some alteration in the expected conclusion is permitted as given in Table 6.1 for modus ponens and Table 6.2 for modus tollens. The conclusions shown are based on P_1 as shown and P_2 as stated above.

Recall that we encountered such an unexpected result in the *Dangerous Sailing Dilemma* problem.

Modus Ponens

Table 6.1 Modus ponens fuzzy conclusions.

Premise P_1: X is A	Consequent Y is B
X is A	Y is B
X is very A	Y is very B
X is very A	Y is B
X is more or less A	Y is more or less B
X is more or less A	Y is B
X is not A	Y is unknown
X is not A	Y is not B

Modus Tollens

Table 6.2 Modus tollens fuzzy conclusions.

Premise P_1: Y is B	Consequent X is A
Y is not B	X is not A
Y is not very B	X is not very A
Y is not more or less A	X is more or less A
Y is B	X is unknown
Y is B	X is A

In practice, when a reasoning system is designed based on either the modus ponens or modus tollens, the truth of premise P_1 is implicit.

As we have seen, fuzzy logic may be viewed as an extension of classical logic. We will now see that inference in fuzzy logic also differs from that in classical logic in several important ways.

1) Traditional logic systems are grounded in formal classic or crisp set theory; thus, a proposition must be either true or false. Because fuzzy logic is based on fuzzy set theory, the truth values of a proposition may range over all values within the truth interval that is usually specified as the closed interval [0.0..1.0].
2) The predicates in traditional logic systems are constrained to non-fuzzy subsets of the universe of discourse, whereas predicates in fuzzy logic may be crisp or fuzzy.
3) Traditional logic systems permit only the qualifiers *all* and *some*. In contrast, fuzzy qualifiers include a variety of others such as *many*, *most*, *few*, *very*, *several*, or *about*.
4) Fuzzy logic provides a means of modifying the meaning of a predicate by the use of hedges.

6.6 Summary

Working with the classic and fuzzy logic tools that we've studied, we now begin to bring them together in processes called *fuzzy inference* and *approximate reasoning*.

We use the word *reasoning* to refer to a set of processes by which we are able to deduce useful conclusions from a collection of concepts, premises, or data. In classical logic, such premises are precise and certain, whereas in fuzzy logic, they can be ambiguous, possibly vague, imprecise, or perhaps even conflicting.

As part of the reasoning process, we introduced and demonstrated the fundamental concepts and various relationships of *equality*, *containment* and *entailment*, *conjunction* and *disjunction*, and *union* and *intersection* among sets and subsets.

We noted earlier that as engineers, our main goals are to recognize, to understand, and to solve problems. We now note that a large part of solving problems is our ability to reason. To facilitate that process, we worked with graphs of membership functions. We note here also that an important part of solving problems is conducting failure modes and criticality analysis: What can potentially go wrong with our design during operation and how serious is it if something does fail.

We introduced and demonstrated a variety of relationships between and among fuzzy sets and subsets as an important aspect of the reasoning process. We learned and applied an essential component of such a process called *inference* in which we deduce a conclusion from a set of premises. We also learned and demonstrated that inference rules governing reasoning in the forward direction (*deduction*) are called the *modus ponens* and those governing backward implication (*abduction*) are the *modus tollens.*

Review Questions

6.1 Briefly describe what we mean by the term equality in fuzzy reasoning? Give several examples.

6.2 Briefly describe and develop a membership function and graph for the movement of a surgical knife in performing a delicate operation.

6.3 What do the terms containment and entailment mean, and how are they related? Give an example of each different from that in the text.

6.4 Give examples of the union and intersection of three fuzzy subsets that are different from those in the text.

6.5 Give examples of the conjunction and disjunction of three fuzzy subsets that are different from those in the text.

6.6 Explain and give two examples of a conditional or conditioned relationship between two variables that are different from those in the text.

6.7 Explain the term max-min composition and give an example that is different from those in the text.

6.8 Explain the term modus ponens and give an example different from those in the text.

6.9 Explain the term modus tollens and give an example different from those in the text.

7

So How Do I Use This Stuff?

THINGS TO LOOK FOR...
• A formal approach to fuzzy logic design. • Formulating a design methodology. • Practicing a formal design methodology. • Recognizing and dealing with design hazards. • Fuzzification and defuzzification strategies. • Fuzzy implication functions. • Fuzzy inference. • Fuzzy control and expert systems. • Membership functions. • First steps toward system design.

7.1 Introduction

Fuzzy logic has been increasingly applied in the control of real-world processes and in the manipulation of inexact knowledge. We have learned how to map a problem from the crisp world into the fuzzy world and then to use fuzzy logic in the control and management of real-world processes and to work with the supporting tools in the manipulation of inexact knowledge.

Two of the major attractions of fuzzy logic are: it permits one to express problems in (familiar) linguistic terms, and it can be applied where the numerical mathematical model of a system may be too complex or impossible to build using conventional techniques.

The next step, called *defuzzification*, is to move our working fuzzy solution back into the world of real numbers. Such a process maps a fuzzy set into or onto a crisp set.

A fuzzy system usually consists of three major components:

1) A mechanism for mapping from a linguistic expression of a problem to a degree of membership in a fuzzy subset. This process is called *fuzzification*.
2) A mechanism for processing. Such a mechanism is usually composed of a collection of fuzzy implications (referred to as *IF-THEN* rules) and a means for manipulating such rules based on system inputs and internal knowledge state. The former is called a *knowledge base* and the latter is an *inference engine*. The result of processing is a *fuzzy solution set*.

Introduction to Fuzzy Logic, First Edition. James K. Peckol.

3) A mechanism for mapping from a fuzzy solution set to a single output value. This is called *defuzzification*.

A simplified model for a fuzzy system is shown in Figure 7.1.

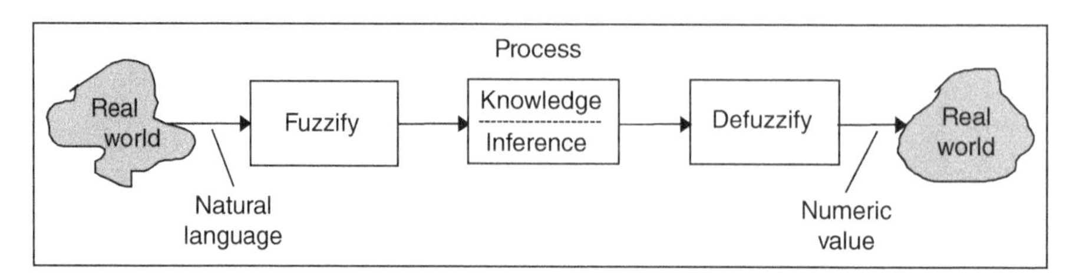

Figure 7.1 A simplified fuzzy system.

7.2 Fuzzification and Defuzzification

We have had a brief review of Boolean algebra, which is classified as a crisp logic. We learned that it is so designated because member statements and variables have either one of two values. As we learned, such variables or statements are either true (1) or false (0). Figure 7.2 illustrates possible variable membership truth values for a Boolean algebra/crisp logic control system.

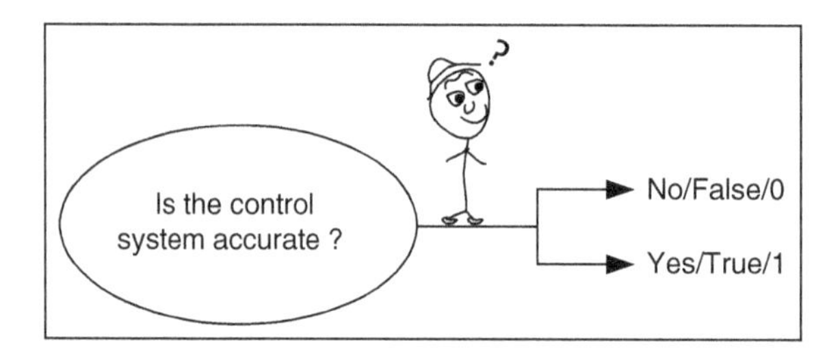

Figure 7.2 Possible membership values in a crisp logic system.

7.2.1 Fuzzification

As we moved into what became known as the fuzzy world, the here-to-fore crisp constraints were relaxed. Such a change allowed statements or variables to express the degree to which something was true or false through *membership functions*. We are changing a variable from a real scalar value into what is known as a *fuzzy value*.

Such a modification enables the conversion of crisp variables into *linguistic variables* and the capture and utilization of the vagueness and imprecision often found in the entities in the surrounding world into the means to support fuzzy reasoning. Such variables can then be used to express quantities or values that cover a specific range. Such a transformation from crisp to fuzzy is denoted as *fuzzification*.

Figure 7.3 illustrates possible variable membership (linguistic variable) truth values for a fuzzy logic control system.

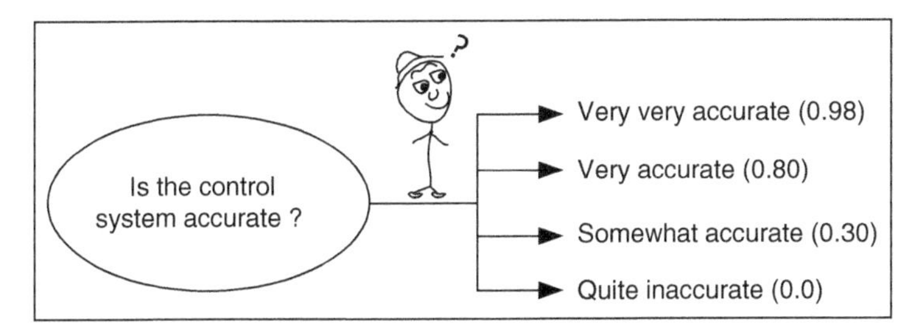

Figure 7.3 Possible membership values in a fuzzy logic system.

Now consider the following example: Someone said that he was often riding his bicycle though crowded neighborhoods. Of concern would be the rate at which the vehicle was being operated, which could be expressed as the following subset:

$$\text{Velocity}: \left(\text{slow, moderate, fast, very fast, too fast}\right) \tag{7.1}$$

Someone might then ask: What was your speed?

$$\text{Response}: \text{It was pretty fast.}$$

Someone could then ask: How fast is pretty fast?

A different response might be: "I was going 45 mph." That crisp input may be quickly converted into a linguistic variable "very fast" or "too fast" expressing the context or potential danger.

Earlier we looked at the graphical representation and potential associated grades of membership of the expression "around 7." Let's now look at a first cut of graphically expressing such degrees of membership or membership functions in a fuzzy subset. Such data can be graphically expressed in numerous ways, of which the simplest one is straight lines. We'll start with the linguistic variable subset from Eq. (7.1) and illustrated in Figure 7.4.

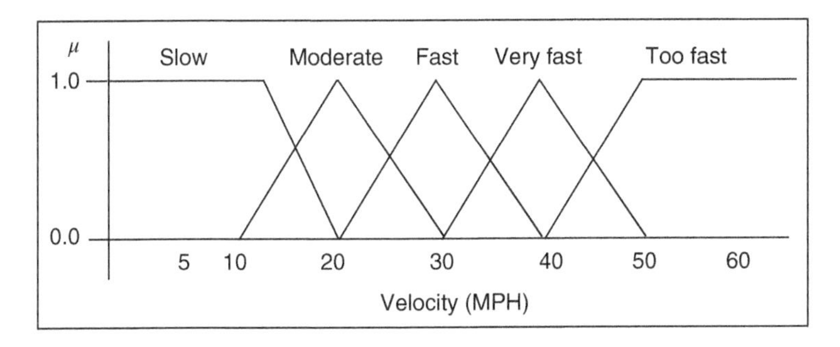

Figure 7.4 Membership functions.

In the diagram, the degree of membership: (0.0 . . . 1.0) is presented on the vertical axis and velocity: (0 . . . 60) on the horizontal axis.

Moving further into the fuzzy world, we pose the following question: "How fast is 12 mph in that world?" From Figure 7.5, a 12 MPH velocity is termed 50% *moderate* with a degree of membership of 0.5 in the *moderate* velocity set and 80% *slow* with a membership of 0.8 in the *slow* velocity set. Similarly, the earlier response of 45 MPH is indicated as 50% *very fast* and also 50% *too fast*. In both cases, we have a 0.5 membership in the two sets.

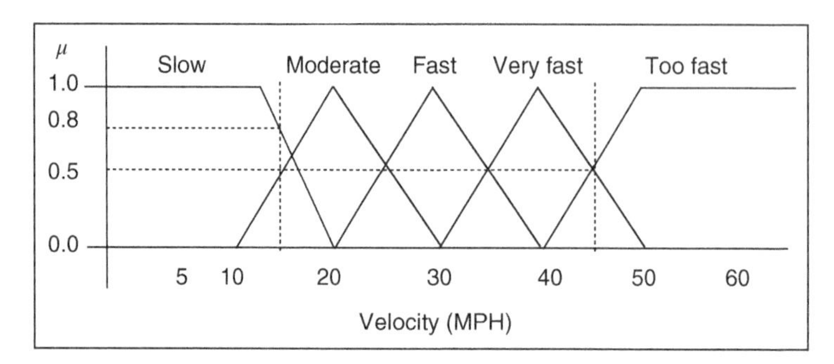

Figure 7.5 Velocity degree of membership.

As we noted, the simplest membership functions are constructed using straight lines as illustrated in Figure 7.5 and using triangles and trapezoids in Figure 7.6. Such diagrams are not constrained to isosceles triangles or symmetric trapezoids. The membership function shapes are driven by the system being designed or the problem being solved. Two possible alternatives are illustrated by the slopes of the lines in Figure 7.6.

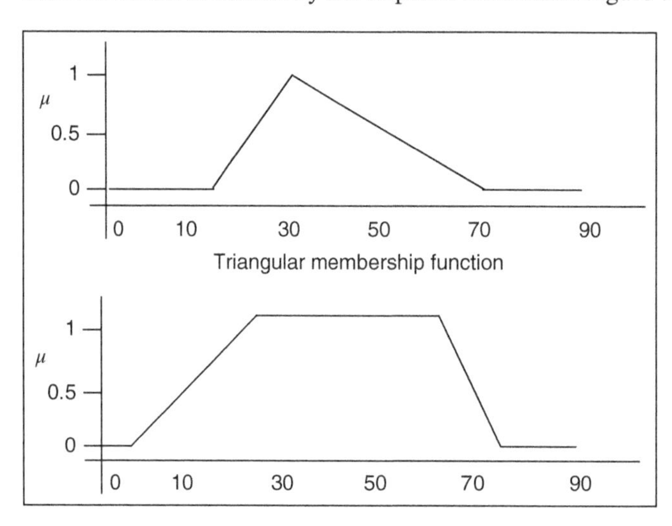

Figure 7.6 Trapezoidal membership function.

Let's now look closer at the basic features of graphical membership functions *core*, *support*, and *boundary*, which are identified in Figure 7.7. We begin with a universe of discourse, a fuzzy set F comprising members x from that universe, and with a degree of membership y in that set. We learned earlier that the degree of membership y of the variable x in the fuzzy set F is given by Eq. (7.2).

$$\mu_F(x) = y \tag{7.2}$$

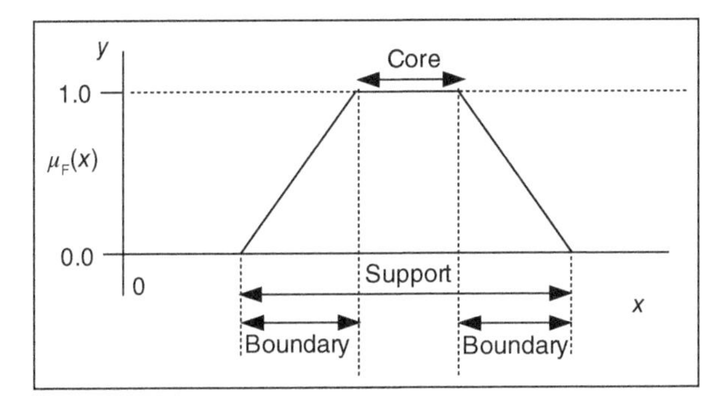

Figure 7.7 Membership function features.

7.2.1.1 Graphical Membership Function Features

Core The region of the membership function labeled *core* in Figure 7.7 comprises of all elements “x” in the universe of discourse with a degree of membership of “1.0” in the fuzzy set F.

$$\mu_F(x) = 1.0$$

Boundary The region of the membership function labeled *boundary* in Figure 7.7 comprises those elements in the universe of discourse with a degree of membership in the range between “0.0” and “1.0” in the fuzzy set F.

$$0.0 <= \mu_F(x) <= 1.0$$

Support The region of the membership function labeled *support* in Figure 7.7 comprises those elements in the universe of discourse with a degree of membership in the range greater than “0.0” in the fuzzy set F.

$$\mu_F(x) > 0.0$$

As shown in Figure 7.1, the fuzzy results generally and eventually must be converted into a crisp, real-world output signal (or signals) using a process called *defuzzification*, which we will introduce next. The defuzzified signals specify the action to be taken to manage the real-world output devices.

7.2.2 Defuzzification

The diagram in Figure 7.8 presents a high-level view of a typical fuzzy system. Up to this point, we have learned and worked with the portion of the system enclosed within the dashed line on the left. Ultimately, we must get crisp data back to the real world. The tool to do that is illustrated in the defuzzifier on the right, which we will examine now.

As we have learned, fuzzification supports a mapping to convert the crisp values into fuzzy (linguistic) values. Such values become the inputs to the algorithms implemented

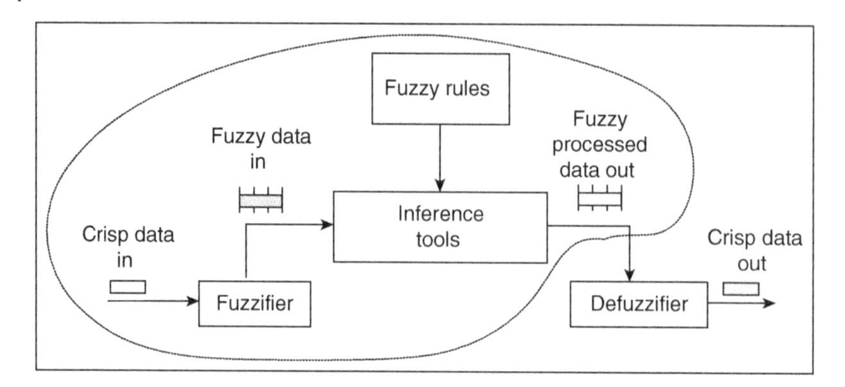

Figure 7.8 Fuzzy system – high-level view

and supported by the underlying analog and digital logic to affect the desired behavior and the intended operation of the system(s) under design.

Ultimately, returning to the crisp world is generally required. Such a return is accomplished through what is called the *defuzzification* process. Defuzzification is interpreting the membership degrees of the fuzzy sets into a specific decision or real value. Thus, through the process of defuzzification, a second mapping converts the fuzzy signals into their real-world crisp counterparts.

Defuzzification is the inversion of fuzzification. The defuzzification process is capable of generating non-fuzzy (crisp) control actions that illustrate the possibility distribution of an inferred fuzzy control action or actions.

As we've seen, probably the most commonly used fuzzy set membership function has the graph of a triangle. If, based on design requirements, such a triangle were to be cut in a straight horizontal line somewhere between the top and the bottom, and the top portion were to be removed, we would now have a trapezoid.

The first step of defuzzification typically repeats such a process to the parts of the membership graphs to form trapezoids (or other shapes if the initial shapes were not triangles). For example, if the output has a "Reduce Temperature (20%)" triangle, then this graphic will be cut 20% of the way up from the bottom. In the most routinely used technique, all of these trapezoids are then superimposed to form a single *fuzzy centroid* geometric figure. The centroid of this shape, called the *fuzzy centroid*, is calculated. The x coordinate of the centroid is the defuzzified value.

7.3 Fuzzy Inference Revisited

In the last chapter we introduced modus ponens, modus tollens, and the concept of fuzzy inference. Following Zadeh's introduction of the compositional rule of inference into approximate reasoning, almost 40 different fuzzy implication functions have been proposed.

The four preferred alternatives are given in Table 7.1. The most commonly used are those by Mamdani and Larsen.

Table 7.1 Fuzzy implication functions.

Mamdani	$\int_{X \times Y} (\mu_A(X) \wedge \mu_B(Y))/(x, y)$ $\wedge$ is the min operator
Larsen	$\int_{X \times Y} (\mu_A(X) \bullet \mu_B(Y))/(x, y)$ $\bullet$ is the arithmetic product
Zadeh 1	$\int_{X \times Y} 1 \wedge (1 - \mu_A(X) + \mu_B(Y))/(x, y)$
Zadeh 2	$\int_{X \times Y} (\mu_A(X) \wedge \mu_B(Y)) \vee (1 - \mu_A(X))/(x, y)$ $\vee$ is the max operator

7.3.1 Fuzzy Implication

Note: Σ_{XY} denotes a discrete membership function. The $\int_{XY}$ denotes a continuous membership function, not an integral. The terms (x, y) denote the respective degrees of membership. An implication infers that a conclusion can be drawn from something although the conclusion might not explicitly stated or identified.

Let's now examine Mamdani's (min) implication function from Table 7.1.

Let

S – Universe of discourse
A, B – Fuzzy subsets of S, illustrated below
a,b – Linguistic variables with values in S
R() – Fuzzy rule
^ – min

Implication

$$A \Rightarrow B$$

$$R(a, b) = A \times B$$
$$= \int \mu_A(a) \wedge \mu_B(b)/(a,b)$$

We know A and B and the grades of membership in each.
We can find $R(a,b) = A \times B$.

a	2	3	4
$\mu_A(a)$	0.1	0.5	0.9

Subset A of S

b	2	5	7	9
$\mu_B(b)$	0.2	0.6	0.7	1

Subset B of S

$A \times B =$

min (0.1, 0.2) min (0.1, 0.6) min (0.1, 0.7) min (0.1, 1)
min (0.5, 0.2) min (0.5, 0.6) min (0.5, 0.7) min (0.5, 1)
min (0.9, 0.2) min (0.9, 0.6) min (0.9, 0.7) min (0.9, 1)

$$\begin{matrix} (0.1) & (0.1) & (0.1) & (0.1) \\ (0.2) & (0.5) & (0.5) & (0.5) \\ (0.2) & (0.6) & (0.7) & (0.9) \end{matrix} \Rightarrow$$

	b	2	5	7	9
a	2	0.1	0.1	0.1	0.1
	3	0.2	0.5	0.5	0.5
	4	0.2	0.6	0.7	0.9

Subset A × B

Our fuzzy rule, implication, will be: "If 'a' then 'b'."

One can view Mamdani's implication function as a clipping operation in the sense that the value of the consequent membership function is limited (by the min operator) to the value of $\mu_A(X_1)$, which is shown in Figure 7.9. The right-hand graphic expresses the defuzzified result.

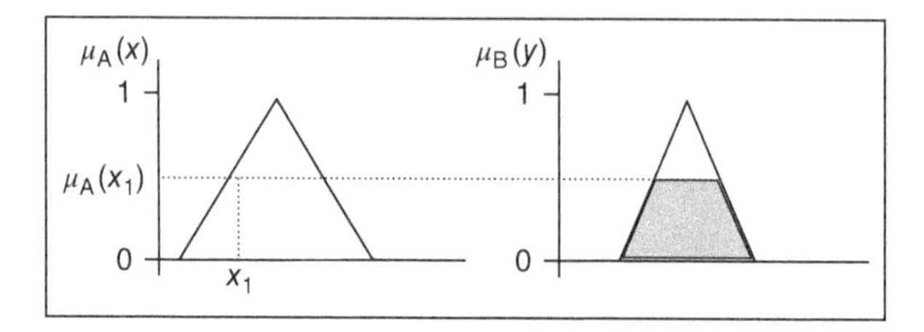

Figure 7.9 Mamdani's implication function.

Similarly, one can view Larsen's implication function as a scaling operation because the value of the consequent membership function is arithmetically limited to the value of $\mu_A(X_1)$, shown in Figure 7.10.

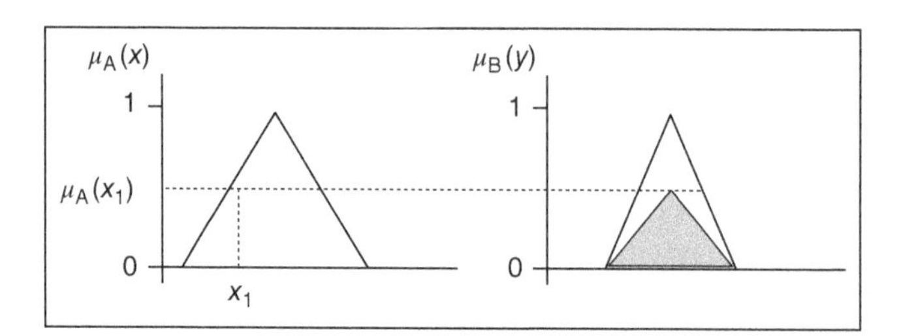

Figure 7.10 Larsen's implication function.

7.4 Fuzzy Inference – Single Premise

In most of the interesting real-world problems, we must contend with numerous and often contradictory requirements. Consider the following problem. Let us devise a simple set of rules for driving in the winter in a snowy climate.

1) If the road is *slippery*, then decrease speed.
2) If you are *impeding* traffic, then increase speed.

This rule set contains two conflicting requirements – one must simultaneously increase and decrease speed. Using traditional means – crisp logic – one cannot arrive at a solution that will simultaneously satisfy both requirements.

Let us now look at the problem from a fuzzy perspective. We can specify two values, *slow* and *medium* for the linguistic variable *speed*. These values will be the labels of two (possibly overlapping) fuzzy subsets.

The fuzzy concepts *slippery* and *impeding* can be modeled by some metric (with appropriate fuzzy values) such as the coefficient of friction in the former case and the length of the tailback in the latter. The specific shapes and values for the membership functions are not important at the moment.

The essence of the problem can be distilled into two propositions:

Surface(slippery) ⇒ Speed(slow)
Impeding(moderately) ⇒ Speed (medium)

for which the membership functions are shown in Figure 7.11.

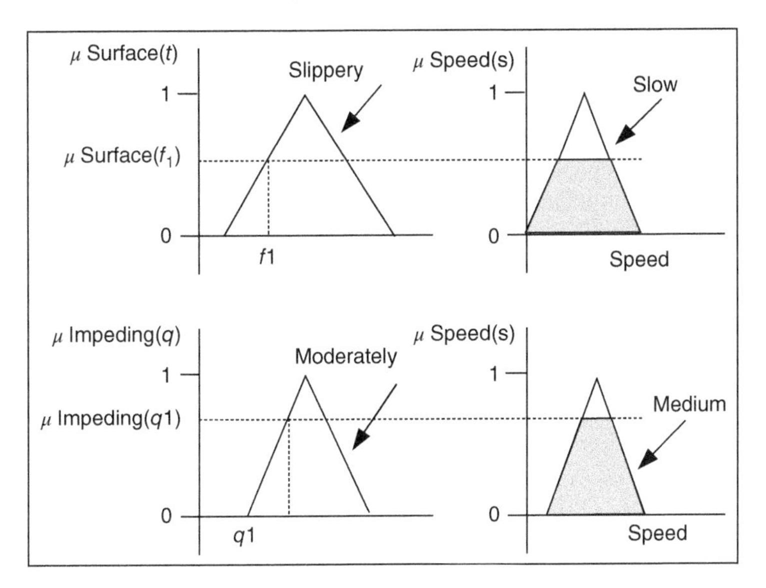

Figure 7.11 Winter driving dilemma.

After proceeding through the formal logic, one arrives at two conflicting consequents:

Speed(slow)
Speed(medium)

The membership functions for the consequents are reproduced and superimposed in Figure 7.12.

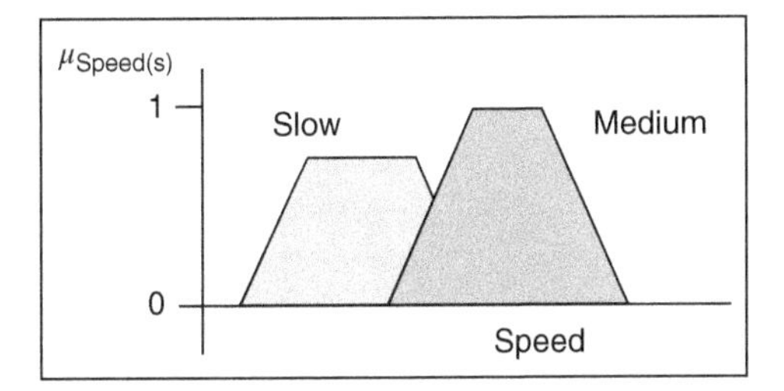

Figure 7.12 Winter driving dilemma continued.

As is often the case, restating a problem in different terms does little to resolve it; however, it may give one a different view of the problem and perhaps some insight into how to approach solving it.

The distribution in Figure 7.12 represents a fuzzy consequent space: *premise to consequence*. That space must be mapped into the (subset of the) space of crisp actions that best represents the intent of the fuzzy consequent.

Such a mapping, as we've learned, is called a *defuzzification strategy*. Regrettably, there is no systematic procedure for choosing the best strategy to achieve such a mapping. Optimistically, however, this lack offers a rich area for future research. At present, three such strategies are commonly used: *max criterion*, *mean of maximum*, and *center of gravity*.

7.4.1 Max Criterion

The *max criterion* chooses, as the crisp output, the point at which the (fuzzy) distribution reaches a maximum value. If there is more than one instance at which $\mu()$ reaches the maximum value, some form of conflict resolution is necessary. One common solution is to opt for the conservative alternative by selecting the value of the output corresponding to the first occurrence of the maximum.

As Figure 7.13 shows, the maximum grade of membership for the composite distribution is that of the medium distribution. We shall opt for the conservative strategy (ice is dangerous as you know, unless it's in a glass) and select the value S_2 as the crisp consequent.

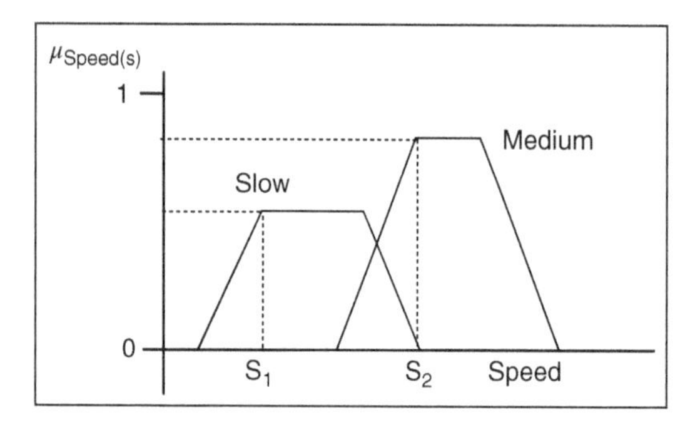

Figure 7.13 Winter driving dilemma continued max criterion defuzzification.

7.4.2 Mean of Maximum

The *mean of maximum* strategy attempts to resolve the conflict inherent in the max criterion by returning a crisp consequent that is the mean of all individual consequents whose membership functions reach the maximum value in the composite distribution. For the discrete case, each membership function is a singleton value. Such a singleton can be interpreted as the fuzzy equivalent of a delta function.

Defuzzification for the discrete case of the mean of the maximum method is given as:

$$C = \sum_{R} \frac{Ymi}{n}$$

C – Crisp consequent

Y_{mi} – Value of Y for which $\mu_R(y_i) = \max$

R – The fuzzy singletons

n – The number of singletons for which $\mu_R(y_i) = \max$

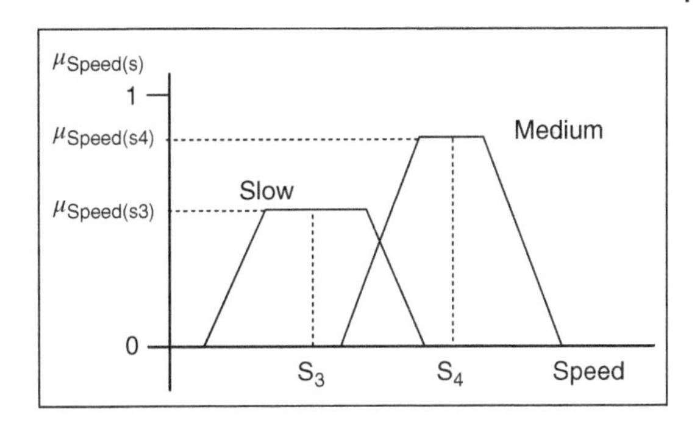

Figure 7.14 Winter driving dilemma continued mean of maximum criterion defuzzification.

For the continuous case, Y_{mi} is the value of Y_i at the midpoint of the maximum of each of the appropriate distributions. With only the medium distribution reaching the maximum value, the mean of maximum strategy will return the value s_4 as seen in Figure 7.14.

7.4.3 Center of Gravity

The center of gravity strategy takes a different approach to eliminate the ambiguity encountered with the max criterion. Rather than dealing with edges, this strategy returns the center of gravity for the complete distribution as its crisp consequent.

For a discrete distribution, the method gives:

$$C = \frac{\sum_R \mu_i(Y_i) \bullet Y_i}{\sum_R \mu_i(Y_i)}$$

C – Crisp consequent
Y_i – Value of Y at the ith singleton
R – The fuzzy singletons
μ_i – The grade of membership for the ith singleton

For the continuous case, the sum is taken over the individual distributions. Y_i is the value of Y at the center of gravity of the ith distribution, and $\mu_i(Y_i)$ is the grade of membership at Y_i.

We now return to the *Winter Driving Dilemma*. If Mamdani's implication function is used (as has been done in Figure 7.12), the three defuzzification strategies give the following results:

With the *center of gravity* strategy, the contributions of both the slow and the medium distributions are considered when producing the final result as shown in Figure 7.15.

$$S_5 = \frac{\mu_{Speed}(s_3) \cdot s_3 + \mu_{Speed}(s_4) \cdot s_4}{\mu_{Speed}(s_3) + \mu_{Speed}(s_4)}$$

It should be clear that, in this example, Larsen's implication function will give the same results as Mamdani's for the *mean of maximum* and *center of gravity* defuzzification strategies and will give a speed of s4 for the max criterion.

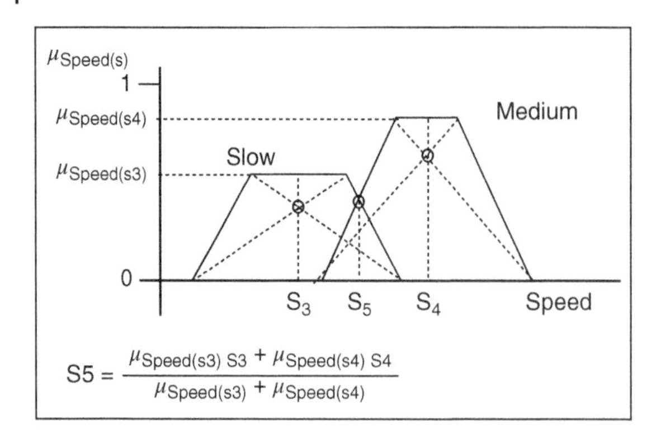

Figure 7.15 Winter driving dilemma continued center of gravity defuzzification.

7.5 Fuzzy Inference – Multiple Premises

The fuzzy implication can be extended rather simply to antecedents with multiple premises. From the basic form:

If A then B
redefine A as
$A = A_1 \cap A_2$ // AND
or
$A = A_1 \cup A_2$ // OR

in which AND is the fuzzy intersection and OR is the fuzzy union. Now one can reason about problems such as the following:

If the road is dry and you think you are Nigel Mansell, then speed up.
If you are poor and you have many tickets, then slow down.

Nigel Mansell is treated as a value for the linguistic variable *raceCarDriver*. The membership functions for the *Grand Prix Driver Dilemma* are now given in Figure 7.16.

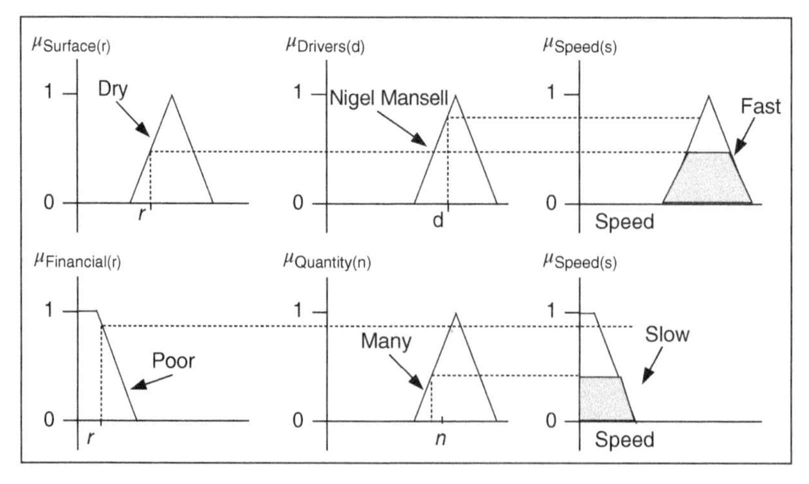

Figure 7.16 Grand prix driver dilemma.

Terms:

r road surface	d driver	s speed
r rich	n number	

Because the antecedents in each of the above implications are expressed as a conjunction of premises, the min operator is used to determine the grade of membership in the combined subset. The crisp result may be obtained by any of the defuzzification strategies.

7.6 Getting to Work – Fuzzy Control and Fuzzy Expert Systems

A fuzzy system is called an *expert system* when it is used to advise, diagnose, or reason and a *control system* when it must control the behavior of physical world processes or systems. An expert system is designed to work with inexact, incomplete, or differing kinds of information and a fuzzy controller functions by replacing or augmenting a traditional PID (Proportional Integral Derivative) controller.

The implementation methods for fuzzy controllers and expert systems are similar as well. One of the few differences is their ultimate use. Eventually, the controller must be connected to a physical process, whereas the expert system usually operates autonomously and may include self-modifying or learning capability.

Let us now examine a simple traditional closed-loop feedback controller such as that shown in Figure 7.17. In this model,

- The inputs specify the desired behavior of the system.
- The outputs drive the physical process, perhaps the heater in a reflow solder oven.
- The sensors measure the state of the physical process.
- The controller combines data from the inputs and sensors to generate proper output(s).

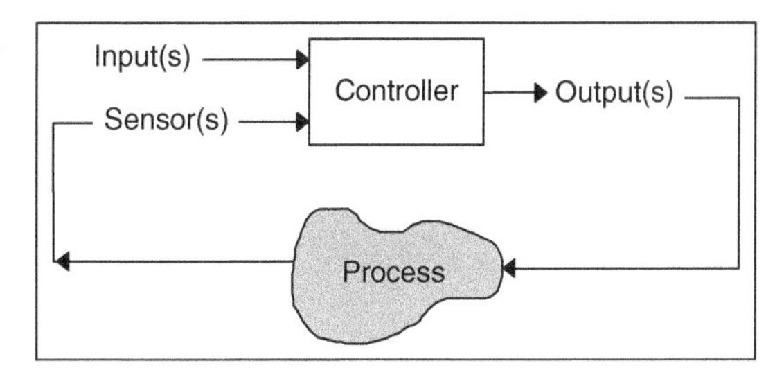

Figure 7.17 A simplified controller system.

One of the simplest controllers is that used to control the temperature in an oven or in a house. As in any such system, the control occurs with respect to what is called a *set point*. In a traditional system, such as that shown in Figure 7.18, if the temperature is too hot, then the heater is turned off, and if the temperature is too cool, then the heater is turned on.

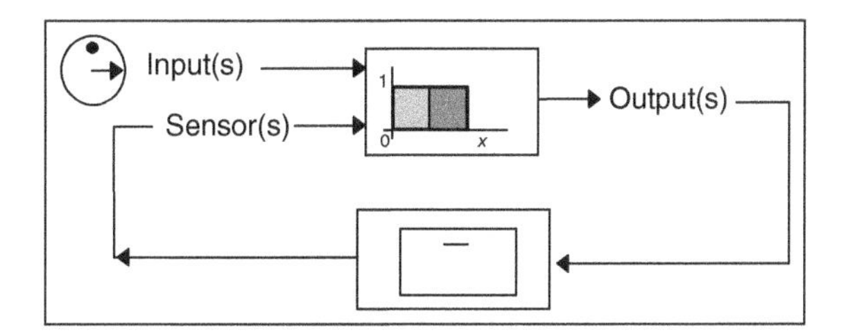

Figure 7.18 A temperature controller block diagram.

All possible temperatures fall into two nonoverlapping (crisp) sets: HOT and COOL as seen in Figure 7.19.

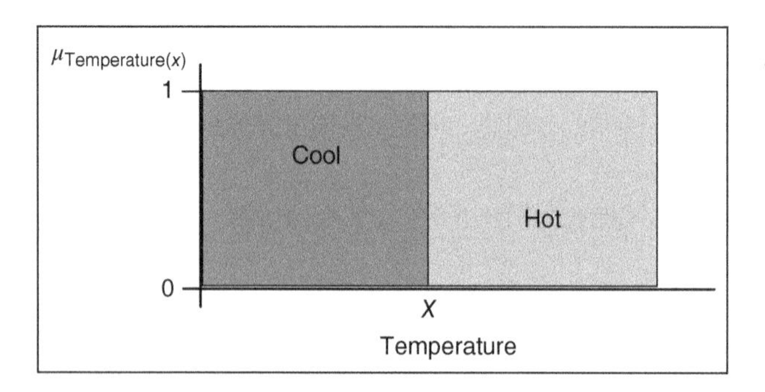

Figure 7.19 Traditional temperature control membership function.

In this context, the control algorithm is given by the two basic rules:

C1: If Temperature is COOL, then turn Heater ON.
C2: If Temperature is HOT, then turn Heater OFF.

A real system may incorporate a dead band or hysteresis around the set point, but the fact remains that the heater is either ON or OFF to one extent or the other.

Let us now consider an alternative fuzzy logic implementation as shown in Figure 7.20

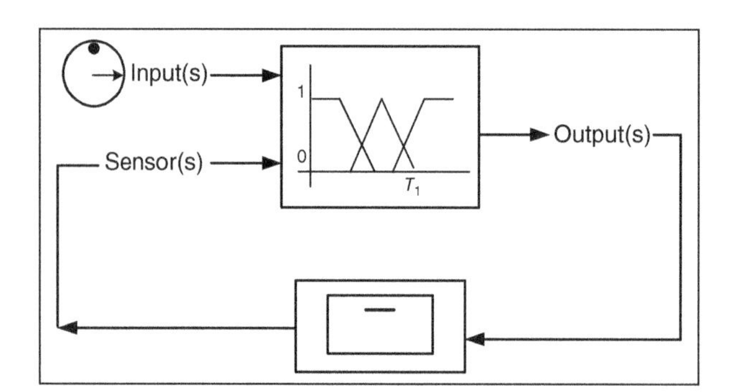

Figure 7.20 A temperature controller block diagram.

By using linguistic variables and hedges, such as *very hot/cool, on_high, on_medium*, or *warm*, we can give a more accurate (in a practical sense) description of the temperature operating range and of how much heat needs to be supplied.

In this example, we will cover the temperature range and the extent to which the heater is on (0–100%) with the linguistic variables *temperature* and *heat* and assign values corresponding to the fuzzy subsets given in Table 7.2.

Table 7.2 Heater ON temperature and extent.

Temperature	HOT	WARM	COOL
Heat	ON	ON_LOW	OFF

To ensure a smooth transition between regions, we will allow WARM to overlap HOT and COOL and ON_LOW to overlap ON and OFF. The membership functions for the two linguistic variables are given in Figure 7.21.

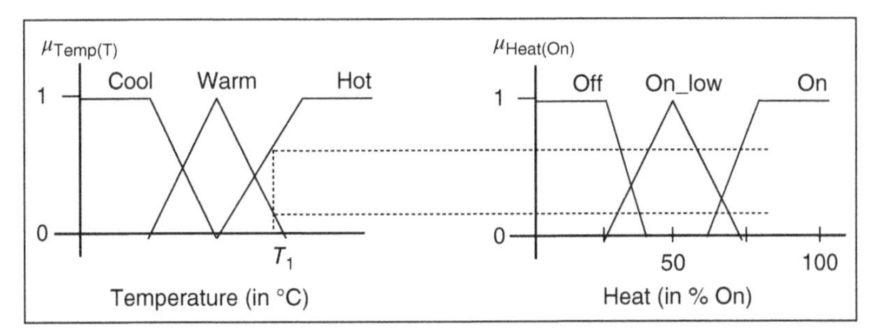

Figure 7.21 Temperature control membership functions.

The controller output, *Heat*, is given as a percentage of full ON, and *control* is implemented according to the following basic rules:

F1: If Temperature is COOL, then Heater is ON.
F2: If Temperature is WARM, then Heater is ON_LOW.
F3: If Temperature is HOT, then Heater is OFF.

If the sensor *temperature* is at a value of T1, then the antecedents of the implications or rules F2 and F3 will be true. If we use Mamdani's implication function and the center of gravity defuzzification strategy, the membership function distributions for the consequents and centers of gravity (cog) will appear as in Figure 7.22.

Figure 7.22 Temperature control membership functions.

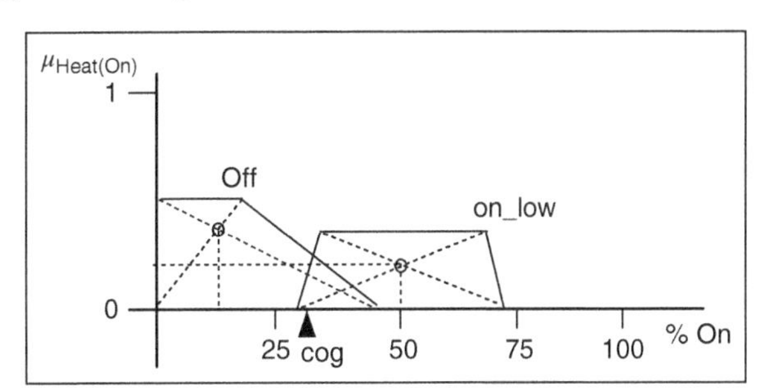

From the distributions $\mu_{ON\_LOW}(\%On) = 0.20$ and $\mu_{OFF}(\%On) = 0.5$, the corresponding heater controls are 50 and 18% ON, respectively.

The crisp output is now computed as:

$$\begin{aligned} \% \text{ On} &= \frac{0.20 \bullet 0.5 + 0.5 \bullet 0.18}{0.2 + 0.5} \\ &= 0.27 \text{ or } 27\% \end{aligned} \tag{7.3}$$

Unlike traditional control, fuzzy control does not demand that the heat be full ON or full OFF. Because fuzzy logic can support partial membership in contradictory subsets, one is able to combine conflicting requirements to produce the appropriate output. Such ability often leads to more efficient and lower cost solutions. Fuzzy controllers also have many other advantages over their traditional counterparts. Typically, they are less expensive to develop and build. Because they are able to work over a wider range of operating conditions, they are generally more robust. Finally, because their control algorithm (encoded as implications or rules) is expressed in a natural language, they are easier to understand and can be easily customized.

Let us now look at the problem of optimizing the flow of people through a queue at a rock concert. Although this is a rather simple example, it is representative of the kinds of problems one encounters and must consider when trying to regulate the flow of items such as either the raw material or finished goods on a production line or traffic on congested motorways.

Example 7.1

We have the problem of regulating the rate at which people enter the arena for a rock concert. If people enter too quickly, someone could be injured in the rush. On the other hand, if they enter too slowly, it will take too long to get them all seated, and the concert will be delayed.

We will control the rate at which people enter by placing turnstiles with a variable delay at all of the entrances to the arena as shown in Figure 7.23.

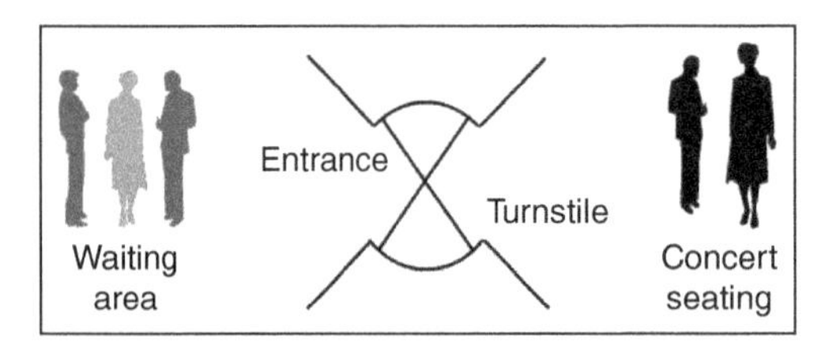

Figure 7.23 Queuing and delay at a turnstile.

The turnstile is designed to admit 5–10 people with each advance. If the crowd density is low, we will allow free access. As the density increases, the time between advances will also increase.

Let us now look at the design of the system.

System Design

The system will have two inputs, *crowd-density* and *rate-of-seating*, and a single output, the *turnstile-delay*. We must choose appropriate linguistic variables and corresponding values for each of these. Next, we must design the membership function for each value. Finally, we need to write the implication functions specifying the behavior of the system and select the defuzzification strategy.

Thus,

Inputs

Crowd-Density
Rate-of-Seating

Outputs

Turnstile-Delay

Definitions

For this example, each linguistic variable is specified to have 3 values. Although the specific number is arbitrary, a greater number of values provides smoother transitions between regions. The common number is from 5 to 9. Also, typically, an odd number of values is chosen so as to provide symmetry about the center of the distribution.

Crowd-Density gives an indication of the number of people within a certain area. To quantify crowd density, we define *Density* as a linguistic variable to specify the number of people in a given area. The values of *Density* will be the fuzzy subsets *light*, *medium_heavy*, and *heavy*. We shall define unit density to be 2 people/m^2.

Rate-of-Seating provides a measure of how quickly people inside the concert hall get to their seats. We will define the linguistic variable *Rate* as a measure of the average number of people being seated within a given time. The values for rate will be *slow*, *medium_fast*, and *fast*. We shall specify that 10 people per second is the maximum seating rate.

Turnstile-Delay is the system output. By controlling the delay, we will be able to regulate the rate at which an average number of people will be admitted to the concert hall. As the delay increases, it will take longer to admit people, thus, reducing crowd density. We will define *Delay* as the linguistic variable with values *short*, *medium*, and *long*, to specify the delay between advances.

Membership Functions

We will use the simple straight-line membership functions as shown for the values of each of the variables. We will allow the regions to overlap, once again, to ensure smoother transitions between the regions as shown in Figure 7.24.

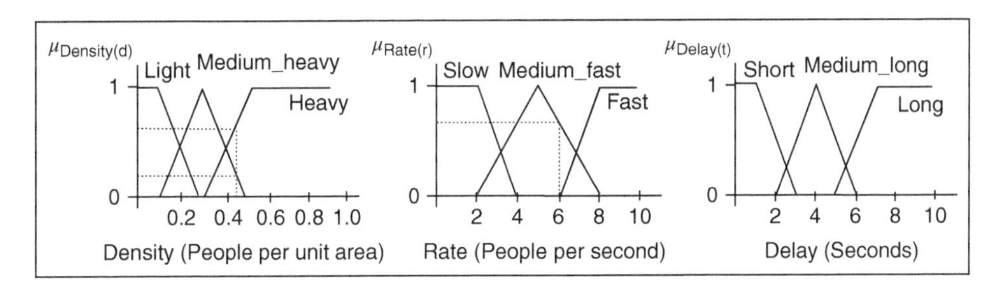

Figure 7.24 Membership functions.

7.6.1 System Behavior

We can specify the desired behavior of the system in Table 7.3 relating the system inputs and the desired system outputs. The headings of the rows and columns will specify the input values, *Rate* (slow, medium_fast, fast) and *Density* (light, med_heavy, heavy), respectively, and the table entries will specify the *Delay* values (short, med_long, long).
Rate Density

Table 7.3 System behavior.

	Slow	Medium_Fast	Fast
Light	SHORT	SHORT	SHORT
Med_Heavy	MED_LONG	SHORT	SHORT
Heavy	LONG	MED_LONG	SHORT

From Table 7.3, we can now write the implication functions or rules.

R1: If Density is LIGHT and Rate is SLOW, then Delay is SHORT.
R2: If Density is MED_HEAVY and Rate is SLOW, then Delay is MED_LONG.
R3: If Density is HEAVY and Rate is SLOW, then Delay is LONG.
R4: If Density is LIGHT and Rate is MED_FAST, then Delay is SHORT.
R5: If Density is MED_HEAVY and Rate is MED_FAST, then Delay is SHORT.
R6: If Density is HEAVY and Rate is MED_FAST, then Delay is MED_LONG.
R7: If Density is LIGHT and Rate is FAST, then Delay is SHORT.
R8: If Density is MED_HEAVY and Rate is FAST, then Delay is SHORT.
R9: If Density is HEAVY and Rate is FAST, then Delay is SHORT.

7.6.2 Defuzzification Strategy

For this example, we will use the center of gravity method.

7.6.2.1 Test Case

If we assume a (Crowd) *Density* of 0.45, which is little less than 1 person/m^2, and a rate of 6 s to seat a person, then rules R5 and R6 will fire. Because multiple rules have fired, their outputs must be combined to give a single crisp output.

The resulting grade of membership for each rule is given as:

$$\begin{aligned}
&\text{density} = (\text{d}), \text{rate} = (\text{r}) \\
&\text{R5} = \min\left(\mu_{\text{Med_Heavy}}(\text{d}), \mu_{\text{Med_Fast}}(\text{r})\right) \\
&\quad = \min(0.25, 0.6) \\
&\quad = 0.25
\end{aligned} \tag{7.4}$$

$$\begin{aligned}
&\text{R6} = \min\left(\mu_{\text{Heavy}}(\text{d}), \mu_{\text{Med_Fast}}(\text{r})\right) \\
&\quad = \min(0.8, 0.6) \\
&\quad = 0.6
\end{aligned} \tag{7.5}$$

Using Mamdani's implication function, the output distribution in Figure 7.25 becomes:

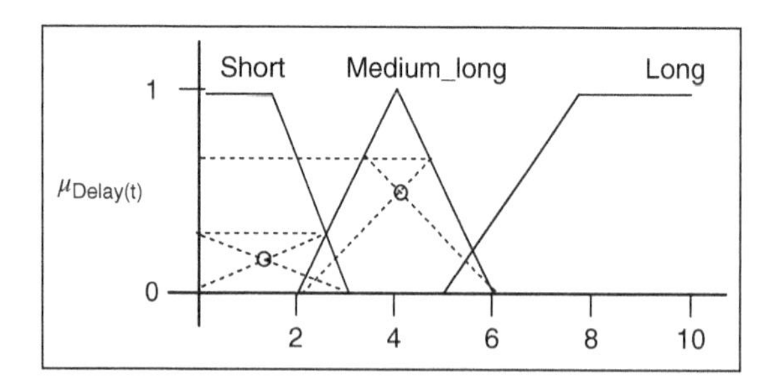

Figure 7.25 Output distribution.

Using the center of gravity defuzzification strategy, the delay is given as:

$$\begin{aligned}
\text{delay} &= \frac{0.25\text{g}1.6 + 0.6\text{g}4.0}{0.25 + 0.6} \\
&= 3.3\,\text{sec}
\end{aligned} \tag{7.6}$$

Example 7.2

Let us now look at one final example in which we will use fuzzy logic in the design of an automatic transmission to help in selecting the best driving gear.

We need to select the best gear in an automatic transmission for various vehicle speeds and load conditions. When the driver is leisurely cruising along the motorway, the transmission should be in a higher gear. On the other hand, when the driver is trying to pull a heavy load, a lower gear is more appropriate.

The block diagram for the system is given in Figure 7.26.

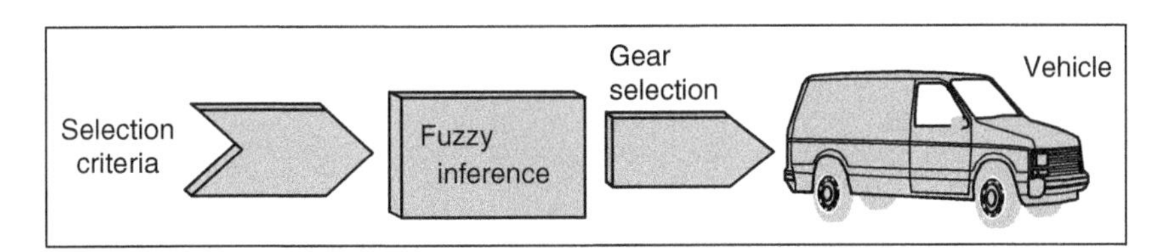

Figure 7.26 Block diagram – automatic transmission design.

When choosing the criteria for selecting the best gear, we will consider the *speed* of the vehicle, the *demands* on the engine, and *requests* from the driver for additional speed or power.

System Design

The system will have four inputs, the *vehicle speed*, the *engine speed* and *load*, and the *throttle position* and a single output, the *gear selected*. We must choose appropriate linguistic variables and corresponding values for each of them. Next, we must design the membership function for each value. Finally, we need to write the implication functions specifying the behavior of the system and select the defuzzification strategy.

Thus,

Inputs

Vehicle
  Speed
Engine
  Speed
  Load
Throttle

Outputs

Gear 1–4

Definitions

Vehicle Speed indicates the forward velocity of the vehicle in kmph. We will define *Speed* as the linguistic variable describing the velocity and will specify a numeric range of 0–250 kmph. For greater resolution, we will use five values of *Speed* distributed as follows:

slow	0–40 kmph
med_slow	30–90 kmph
medium	80–130 kmph
fast	120–170 kmph
very_fast	above 160 kmph

Engine Speed identifies the rate of engine rotations in RPM. We will define *Rotation* as the linguistic variable and specify a numeric range of 0–8000 RPM. Four values should be sufficient for quantifying rpm.

low	0–2500 rpm
medium	1500–4500 rpm
med_fast	3500–6500 rpm
fast	greater than 5500 rpm

Engine Load will provide a measure of the horsepower the engine is expected to deliver. We will define *Load* as the linguistic variable and specify a numeric range of 0–500 hp and will assign the following five values:

light	0–150 hp
med_light	50–250 hp
medium	150–350 hp
med_heavy	250–450 hp
heavy	above 350 hp

Input from the *Throttle* will reflect the driver's desire to increase or decrease the speed or to pull a heavier load. We define *Throttle* as the linguistic variable and specify a numeric range of 0–100%. We will cover the range with only four values since we do not need very high resolution.

low	0–30%
medium	10–50
high	30–70%
very_high	60–100%

The single output, *Gear*, identifies which gear has been selected. We will define *Gear* as the linguistic variable. Because present-day transmissions cannot select fractions of a gear, the values of the linguistic variable will be the four discrete values one to four.

7.6.3 Membership Functions

We will use the simple straight-line membership functions as shown for the values of each of the input variables. We will allow the regions to overlap to ensure smoother transitions between the regions as shown in Figure 7.27.

The output membership function will consist of four discrete values in Figure 7.28.

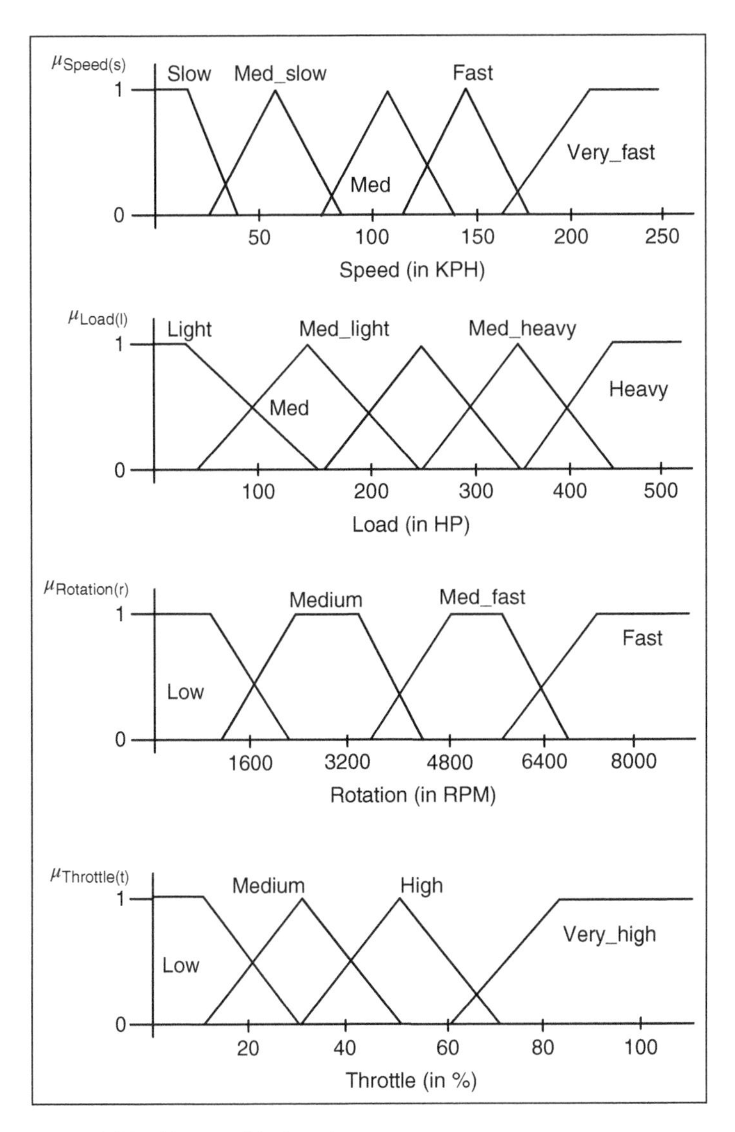

Figure 7.27 Membership functions.

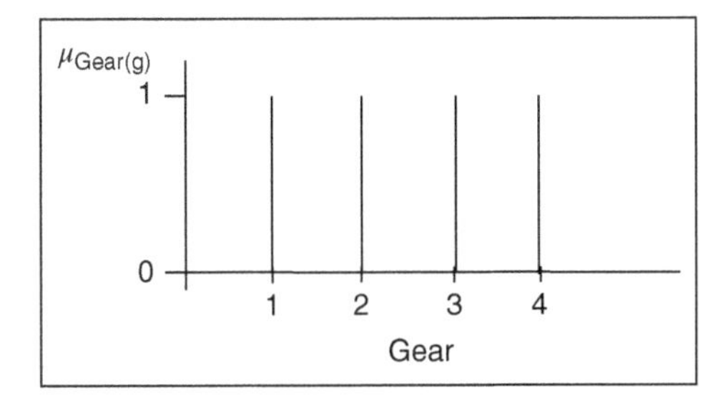

Figure 7.28 Gear selection.

7.6.4 System Behavior

We can now write a variety of rules to cover many different driving situations.

If we are hauling a boat, we would write:

R1: If Throttle is LOW and
Speed is SLOW and
Rotation is LOW and
Load is HEAVY,
then Gear is 1.

or if we are driving on a country road,

R2: If Throttle is HIGH and
Speed is FAST and
Rotation is MED_FAST and
Load is LIGHT,
then Gear is 4.

7.6.4.1 Defuzzification Strategy

We will again use the center of gravity defuzzification strategy. Note that the center of gravity for a singleton is its value.

The system we have designed accommodates most typical or normal driving conditions. We would now like to be able to extend it to a number of different contexts such as climbing hills, driving in ice or snow, or descending winding roads. We will do this by incorporating a simple expert system to infer the vehicle context from several additional inputs. The output of the expert system will then be used to influence the gear selection strategy. The block diagram for the system now appears as in Figure 7.29:

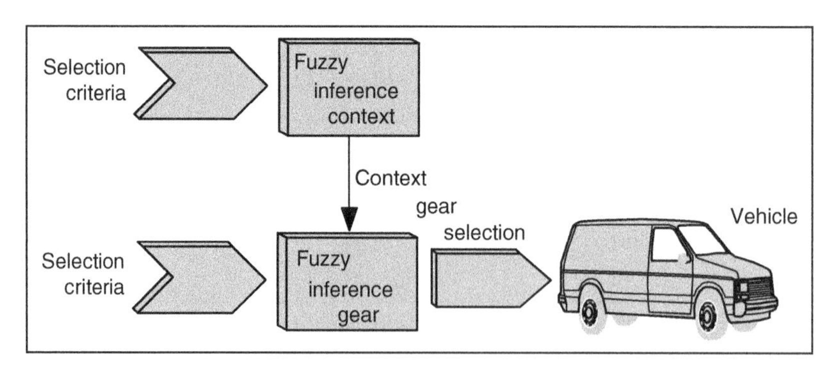

Figure 7.29 Block diagram – automatic transmission design.

Context now becomes an additional criterion for selecting a gear. Information that could be useful for inferring the vehicle context includes vehicle speed and the short-term variation in speed, the angle of inclination of the vehicle with respect to the horizontal or the throttle position.

Context rules would take the form:

C1: If Speed is MED_FAST and
Variation is HIGH and

Inclination is LARGE_NEG and
Throttle is LOW,
then Context is STEEP_DOWNHILL,
from which rules of the form:
R1: If Context is STEEP_DOWNHILL
then Gear is 2

may be written.

7.7 Summary

We have come to the fuzzy world with a background in the design and development of digital and embedded systems in the crisp world coupled with some basic knowledge of the fundamentals in the fuzzy world. In this chapter, we worked to extend that knowledge into the domain of fuzzy logic system design. As we have seen, fuzzy logic may be viewed as an extension of classical logic.

We have learned that inference in fuzzy logic differs from that in classical logic in several important ways.

1) Traditional logic systems are grounded in formal set theory; thus, a proposition must be either true or false. Because fuzzy logic is based on fuzzy set theory, the truth values of a proposition may range over all values within the truth interval that is usually specified as the closed interval (0.0..1.0).
2) The predicates in traditional logic systems are constrained to non-fuzzy subsets of the universe of discourse, whereas predicates in fuzzy logic may be crisp or fuzzy.
3) Traditional logic systems permit only the qualifiers *all* and *some*. In contrast, fuzzy qualifiers include a variety of others such as *many*, *most*, *few*, *several*, or *about*.
4) Fuzzy logic provides a means of modifying the meaning of a predicate by the use of hedges.

Our first step toward formal system design was to learn to move from the crisp to the fuzzy world. We learned to execute the process of *fuzzification* starting with linguistic variables, set membership functions, and associated graphs. We then introduced four of the major fuzzy implication functions proposed by Mamdani, Larson, and Zadeh and worked with those from Mamdani and Larson.

Once in the fuzzy world, we applied the fuzzy reasoning process, then we took the tools and worked through the fuzzy design process. Our targets were three basic systems: a temperature controller, a rock concert turnstile, and an automotive automatic transmission. Recognizing that, ultimately, we needed to return crisp data to the real world, we introduced and learned the process of *defuzzification*.

Review Questions

7.1 Explain the term fuzzification.

7.2 Give an example of a fuzzy logic system and explain potential membership values for such a system.

7.3 Identify and explain the basic features of a fuzzy logic membership function.

7.4 Explain the term defuzzification and explain why such a process is necessary.

7.5 Propose and solve a fuzzy logic single-premise problem and a set of rules like the Grand Prix Driver Dilemma. Include all the appropriate diagrams.

7.6 Propose and solve a fuzzy logic multiple-premise problem and a set of rules like the Winter Driving Dilemma. Include all the appropriate diagrams.

7.7 Apply Larson's implication function to the problem in Section 7.2.1.

8

I Can Do This Stuff!!!

THINGS TO LOOK FOR...

- Potential applications for fuzzy logic.
- Formulating a design methodology.
- Preliminary steps in a formal design methodology.
- Identifying and understanding the system requirements.
- Recognizing and dealing with failure modes and design hazards.
- Executing the steps in formal system design methodology and design.
- Simulation and test of the design.

8.1 Introduction

Fuzzy logic is finding increased application in the control of real-world processes and in the work with and the manipulation of inexact knowledge. The significance of the impact that fuzzy logic has had can be seen from the fact that in recent years in Japan alone over 2000 patents related to fuzzy logic have been granted. Two of the major attractions of fuzzy logic are: it permits one to express problems in (familiar) linguistic terms and it can be applied where the numerical mathematical model of a system may be too complex or impossible to build using conventional techniques.

8.2 Applications

Fuzzy systems have been found to give excellent results in the following general areas.

1) Systems for which complete or adequate models are difficult or impossible to define or develop.
2) Systems or tasks that are usually controlled or executed by a human expert.
3) Systems with inputs and/or outputs that are continuous and complex and that have a nonlinear transfer function.
4) Systems or tasks that use human observations as input, control rules, or decision rules.

Introduction to Fuzzy Logic, First Edition. James K. Peckol.

5) Systems or tasks such as economics or the natural sciences in which vagueness is common.
6) Tasks using devices such as air conditioners, washing machines, and vacuum cleaners.
7) With today's advances in sundry kinds of transportation, fuzzy logic finds its way on the ground into the automotive world and in the air into helicopter autopilots.
8) Applications in the fields of AI (artificial intelligence), expert systems, machine learning, and neural networks.

8.3 Design Methodology

A good methodology is as important in the design of fuzzy systems as it is in the design of the more traditional systems. Fuzzy logic does not offer a magic or quick and easy solution. Like other tools such as the C or C++ programming languages, integrated circuit catalogs, or personal computers, fuzzy logic allows one to create a "design" quickly. However, by neglecting or improperly using such tools, one can create (serious) problems just as quickly. One must spend time during the early stages of any design to understand the requirements, to analyze the problem and possible solutions, and then to evaluate each alternative solution to determine if it meets the initial requirements.

The fundamental and essential key to any successful design process is knowing and understanding the problem and possible potential hazards that might be identified during the design process and potentially appear in the postproduction delivered product. Without such knowledge, one cannot and should not proceed with a design.

To *identify the hazards*, one must first identify the context or environment in which the system is intended to operate postproduction. The hazards encountered by a spacecraft or undersea vehicle are different from those seen by a pacemaker and are different again from those that might occur on a cell phone. Once the hazards and consequences that must be dealt with are known and thoroughly understood, the design process can proceed. However, knowing the potential hazards and consequences is only part of the process. A second part of the operating context is the *potential risk(s)* to and from the environment(s) in which the system is to operate. Risk assessment must also include effects on people, equipment, and those to and from the natural environment.

Focus then shifts from the environment to the system(s) being designed. If the initial analysis has determined that the potential risks are minimal or nonexistent, then it is probably not necessary to incorporate additional or extraordinary *safety measures* into the design. If such is not the case, then methods to mitigate or eliminate the risks must be examined and added.

On a space station, for example, a failure to the system that provides a breathable atmosphere for the astronauts represents significant risk. Compensation alternatives outside of the system such as providing a space suit may be a reasonable safety measure. Others might include an emergency oxygen generator that is activated immediately on failure of the controller for the primary system. On the ground, a rocket capsule without an inside door handle can potentially lead to fatalities in the event of a fire. Nonetheless, risks deemed minimal should still be resolved.

With the preliminary analysis completed, the results of that analysis must be incorporated into *System Requirements* and the *System Design* specifications. Every specification that we write should also include a list of regulatory agencies, safety standards, and proprietary guidelines with which the system complies or must comply. Examples of these two documents can be found in Appendix A of this book.

The next step in the process of developing safe and robust designs is to focus on the design itself. Start with the considerations at the project and detail levels, then move to the high-level design aspects of the system. After such considerations are understood, examine potential failure modes and consequences of such failures in the major functional components of the system and propose methods by which those failures can be detected and addressed.

Requirements definition is the process of identifying and understanding the needs of all interested parties and then documenting those needs as written definitions and descriptions. At this stage, the focus should be on what problem the system has to solve, not how. The initial emphasis is on the world in which the system will operate, not on the system itself. The emphasis is on the role that such specifications play in the safety and reliability of the system. The subsequent design process then shifts the focus from what problem the system has to solve to how the problem will be solved and how and for what the system will be tested.

A typical application is made up of a series of translations that begin with the customer's requirements and lead to the required system hardware and machine code that runs on that hardware. Errors arise whenever any one of these translations fails to accurately reflect the initial intention or goal. The single largest cause of such errors and ultimately of unreliable or unsafe systems is simply one of translation, misunderstanding, or neglect. A specification given in feet misread or misinterpreted as meters is an example.

At each step through the development process, it is important that the current deliverables be critically reviewed for conformance with requirements and specifications as well as for proper or potentially improper and unsafe operation.

8.4 Executing a Design Methodology

The following are a general set of specification and design guidelines. They do not guarantee a perfect solution to every problem, but if used carefully and consistently, they can help to produce a better and potentially safer solution.

1) All parties involved in the development of the system should spend time during the early stages of the development cycle to thoroughly understand the problem, the environment in which one is working, and the associated requirements then put together a formal *Requirements Specification*. Such a document presents a qualitative outside view of the system. An example of such a document can be found in Appendix A.

   Then, increasing the formality and based upon the stipulated requirements, as the next step, one should develop a formal *Design Specification* that presents a detailed quantitative outside view of the system. An example of such a document can also be found in Appendix A.

2) With the two high-level specifications in hand, the next step is to move inside the system, increase the detail, and define and specify the functional and operational characteristics of the system. Then, based upon the system characteristics, identify and analyze possible solutions, and evaluate each alternative solution to determine if it fully meets the initial requirements. Such characteristics establish the architecture of the system.
3) An important task that should occur at several times during the development cycle is periodic design reviews. Someone who is knowledgeable but is not directly involved in the product's design process should go over the design and look for potential problems with the design as it progresses. Any problems found should be corrected, not simply ignored or dismissed.
4) One can model any system according to the very simple high-level external diagram in Figure 8.1.

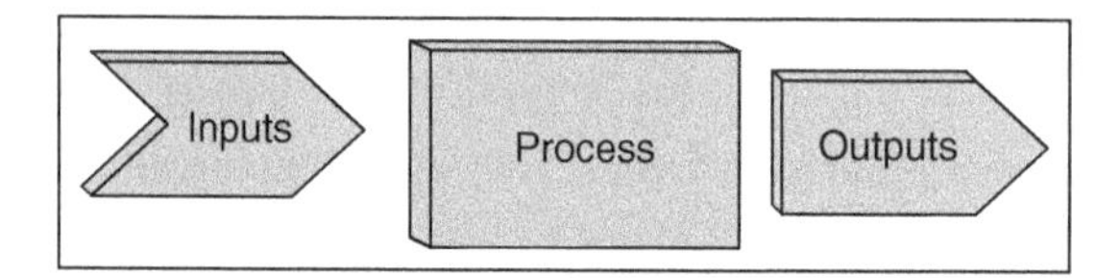

Figure 8.1 System model – high level external view.

A fuzzy system usually consists of three major components:

a) A mechanism for mapping from a linguistic expression of a problem to a degree of membership in a fuzzy subset. As we've learned, this process is called *fuzzification.*

   At a high level, the same process occurs when mapping from a problem statement to a problem solution implemented using such tools as Verilog or VHDL Boolean algebra, or the C, C++, or Python languages. Among the alternatives specifically designed for fuzzy programing are HALO and FCL (Fuzzy Control Language), which is a domain-specific programming language with some limitations for targeting fuzzy control. Numerous other possibilities also exist.

b) A mechanism for processing.

   Such a mechanism is usually composed of a collection of fuzzy implications (referred to as IF-THEN rules) and a means for manipulating such rules based upon system inputs and the internal knowledge state. The former is called a *knowledge base* and the latter is an *inference engine.* The result of the processing is a *fuzzy solution set.*

c) A mechanism for mapping from a fuzzy solution set to a single output value. As we've learned, this is called *defuzzification.*

   Again, the same process occurs when mapping from a software design to a hardware implementation.

5) A simplified, high-level model for a fuzzy system design is given in Figure 8.2. The system may also include feedback of some form from an output signal or signals to the input of the system.

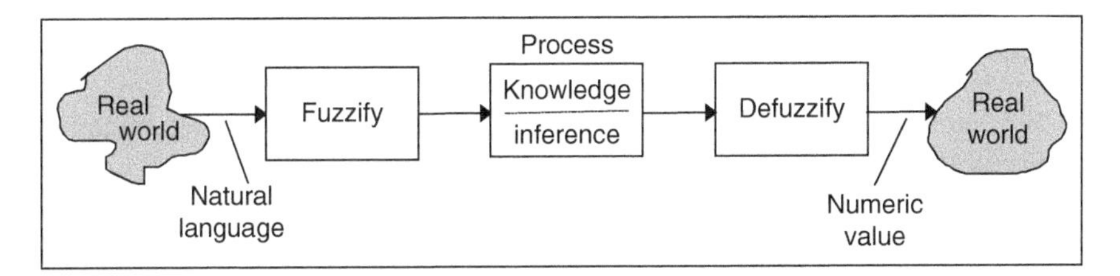

Figure 8.2 A simplified fuzzy system.

Using such a model, one must specify, in detail, what flows into the system, how the inputs are transformed by the system, and what flows out of the system. For each of the inputs and outputs, one must identify the number and the range of its values. Appendix B provides a set of UML (Unified Modeling Language) tools and guidelines that can be very useful in formulating a high-level, functional architecture.

6) Define the *linguistic variables* and their values. The linguistic variables allow the system behavior to be expressed in linguistic terms, and the values permit (*linguistic*) quantification of that behavior.
   In selecting values for the variables, it is generally advisable to choose an odd number somewhere between five and nine. The values should overlap so as to ensure stability in the output and to give a smooth transition between regions. The overlap should range from 10 to 50% such that the sum of the overlapped values is less than 1. As the overlap increases beyond 50%, the sum approaches two and the fuzzy subsets approach crisp subsets. The density of the values should be the greatest around the system's critical points.
7) Define and understand the expected behavior of the system. Such a description specifies how the output is expected to change with changes in the system's inputs. Generally, such behavior is expressed as a series of fuzzy implications or *IF-THEN* rules of the form *IF <proposition> THEN <proposition>*.
   The number of such rules is directly related to the number of linguistic state variables. In a control system, these are the control variables. In an information system, such as an expert system, these are the information variables. If a system has two state variables with five values each, there will be a maximum of 25 rules.
8) Select the defuzzification strategy. Several of the more common ones have been discussed in the text. At present, the *center of gravity* method is the most commonly used strategy, although it does not guarantee that it is the best for your system nor the only one to consider.
9) Perform an in-depth failure modes analysis of the system, potentially several times during the design process. Identify possible failure modes, their effects, and estimate the criticality of such a failure. Correct such problems as appropriate, necessary, and predicated upon the criticality of the failure mode.
10) Simulate/test the design. Simulation and/or test is important to ensure that the design behaves/performs as it was specified and as desired. The importance of this point cannot be understated.

8.5 Summary

We opened the chapter with a summary of the growing strengths and major attractions of fuzzy logic. We identified two of the major attractions as its ability to facilitate expressing problems in linguistic terms and supporting applications in a number of general areas both inside and outside of traditional engineering and natural science. These are areas where a designer can encounter vagueness and where the numerical mathematical model of a system may be too complex or impossible to build using conventional techniques.

We then introduced and formulated a formal approach to fuzzy logic design. We pointed out that a fundamental and essential key to any successful design process is knowing and understanding the problem and possible potential hazards. We then introduced and walked through the execution of the major steps in a formal design methodology.

Review Questions

8.1 Identify and describe the major application areas where fuzzy systems have given excellent results.

8.2 What is a *failure modes analysis,* and why is it important?

8.3 Identify and describe the critical aspects of a sound fuzzy logic design methodology.

8.4 Identify and explain the important guidelines that one should follow in executing a fuzzy logic design.

8.5 Identify and explain the major aspects of a system's requirements and a system design specification outlined in Appendix A.

8.6 Identify and explain the major aspects of the UML (Unified Modeling Language) for executing a design as outlined in Appendix B.

9

Moving to Threshold Logic!!!

THINGS TO LOOK FOR...

- What is threshold logic?
- Is threshold logic an entry to fuzzy logic?
- How to execute a threshold logic design?
- How do threshold logic gates compare to classic logic gates?
- How do threshold logic gates compare to fuzzy logic gates?
- What does a threshold logic gate look like?
- Are there any limitations to threshold logic?

9.1 Introduction

As we've learned, a significant challenge with engineering design is transforming our human thoughts, concepts, and language into functioning physical implementations. In an earlier chapter, we reviewed the fundamentals of Boolean algebra as a tool for transforming human thoughts into hardware and software using devices called logic gates. The hardware provided the physical implementation, and the software provided the sequences of events to manage the hardware to produce a desired problem solution.

From our review of classic or crisp logic, we moved to another tool called fuzzy logic that enabled us to expand a variable grade of membership in a set from {0,1} to {±0.0 to ±1.0} and to use what are known as hedges and linguistic variables, both of which thereby enabled us to express and implement our concepts in greater detail.

As we now move forward, we introduce and explore two tools that contribute to the foundation of advanced tools called Neural Networks, Machine Learning, and Artificial Intelligence. These tools are Threshold Logic and Perceptrons.

In our work with classic logic, we learned that the input and output variables to a classic logic design could only be bipolar, that is, they are constrained to be true or false (0 or 1). We then learned that such constraints were relaxed with fuzzy logic such that input variables have an individual input grade of membership in the range of (±0.0 to ±1.0) in an input fuzzy subset and a corresponding output value in the same range.

Introduction to Fuzzy Logic, First Edition. James K. Peckol.

9.2 Threshold Logic

The threshold logic gate, like the crisp and fuzzy logic gates, can have multiple inputs and only a single output. Like the crisp device, the output is constrained to the range of TRUE or FALSE (0 or 1). Unlike the inputs to a crisp logic gate and analogous to the fuzzy logic gate, each of the inputs to the threshold gate has an assigned weight and, as the name suggests, an output threshold. We will learn that such an architecture modification also contributes in the design of the Perceptron.

Functionally, each of the input variable values is multiplied by its associated weight and then those products are summed. The resulting sum is then compared to the specified threshold. If the threshold is met or exceeded, the output of the device is logical 1; otherwise, it is logical 0, hence the name. The values of the weights and the threshold are not constrained; they can be any finite real number and altered as necessary.

We can convert the previous paragraph from a textual description of the threshold gate into its corresponding mathematical description as we see in Eq. (9.1). Inputs are denoted X, outputs Y, the weights W, and the threshold T. The arithmetic sum of the products of the input values and their weights is termed the *weighted sum* as illustrated in Eqs. (9.1a and 9.1b).

$$Y = 0 \text{ if } X_0 \cdot W_0 + X_1 \cdot W_1 + X_2 \cdot W_2 + \ldots + X_{n-1} \cdot W_{n-1} + X_n \cdot W_n < T \tag{9.1a}$$

$$Y = 1 \text{ if } X_0 \cdot W_0 + X_1 \cdot W_1 + X_2 \cdot W_2 + \ldots + X_{n-1} \cdot W_{n-1} + X_n \cdot W_n \geq T \tag{9.1b}$$

The logic symbol for the threshold gate mirrors that for the fuzzy logic gate. The high-level structure is a circle. The inputs appear on the left, and the output appears on the right. The circle is decomposed into two segments. That on the left contains the inputs and weights. Each of the inputs is associated with one of the weights. The portion on the right specifies the output threshold and output. The logical output value will be 0 or 1 depending upon how the weighted sum compares to the threshold value.

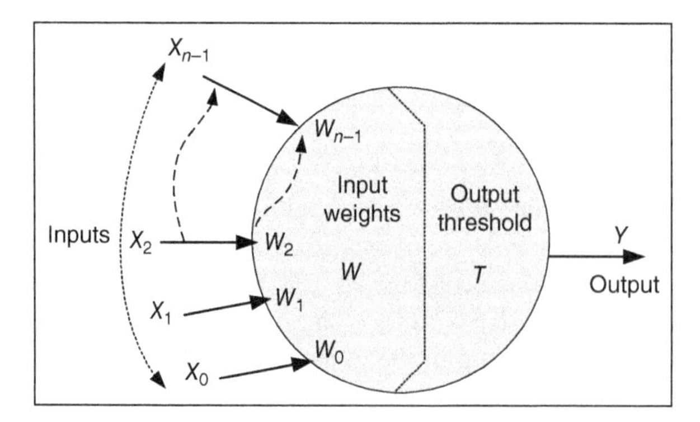

Figure 9.1 Threshold logic gate.

Analogous to what we learned in our work with the grade of membership in a fuzzy logic subset, as illustrated in Eq. (9.1) and Figure 9.1, the output of the threshold logic gate is a function of the weights specified for each of the inputs and the value specified as the device threshold. As is therefore evident, by altering the values of the weights and the output threshold, we can formulate a variety of different Boolean and logic gate functions.

9.3 Executing a Threshold Logic Design

In Boolean algebra, two of the simplest gates are the logical AND and the logical OR. Let's see how to create those functions using a single threshold gate. The truth tables for the two functions are given in Tables 9.1 and 9.2. The weighted sums for the two devices are given in Eqs. (9.2) and (9.3). For both gates, we will set the weights for both inputs to value 1. For the AND gate, we will set the threshold to 1, and for the OR gate, we will set it to 2.

9.3.1 Designing an AND Gate

Computation of the weighted sum for a two input AND gate is given in Eq. (9.2). The weights for both inputs are specified to have a value of binary 1.

$$\begin{aligned} W_{S\,\mathrm{AND}} &= X_0 \cdot W_0 + X_1 \cdot W_1 \\ &= X_0 \cdot 1 + X_1 \cdot 1 \end{aligned} \tag{9.2}$$

Table 9.1 AND gate $T = 2$.

X_1	X_0	W_1	W_0	$X_1 \cdot W_1 + X_0 \cdot W_0 = W_{S\mathrm{AND}}$			Y
0	0	1	1	0	0	0	0
0	1	1	1	0	1	1	0
1	0	1	1	1	0	1	0
1	1	1	1	1	1	2	1

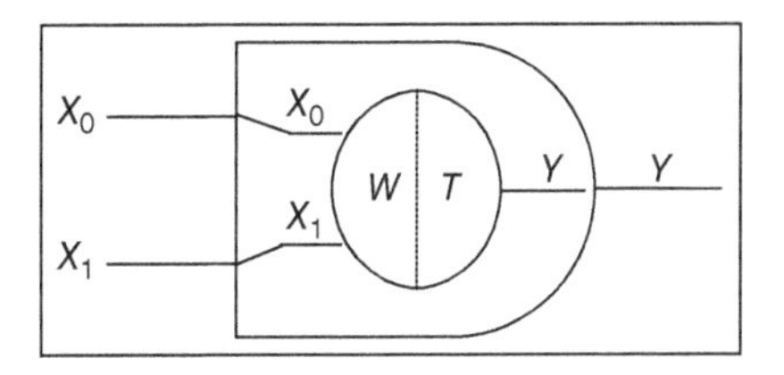

Figure 9.2 Threshold gate disguised as an AND gate.

For the four different input values of X_0 and X_1, for the AND device, the weighted sums will be: 0, 1, 1, 2. The threshold T is specified as 2, so that a weighted sum of 2 or greater is required to produce a value for the output Y of logical 1 (weighted sum $\geq T$).

Only for the input pattern, $[\{X_0, X_1\} = \{1, 1\}]$, will the value of the weighted sum equal or exceed the threshold. The other three input patterns will yield a weighted sum that falls below the threshold and thus will yield a logical 0 for the gate output. Figure 9.2 illustrates the standard logical pattern for a two input AND gate.

9.3.2 Designing an OR Gate

Computing the weighted sum for a two input OR gate is given in Eq. (9.3). The weights for both inputs are specified to have a value of binary 1.

$$W_{SOR} = X_0 \cdot W_0 + X_1 \cdot W_1 \\ = X_0 \cdot 1 + X_1 \cdot 1 \tag{9.3}$$

Table 9.2 OR gate $T = 1$.

X_1	X_0	W_1	W_0	$X_1 \cdot W_1 + X_0 \cdot W_0 = W_{SOR}$			Y
0	0	1	1	0	0	0	0
0	1	1	1	0	1	1	1
1	0	1	1	1	0	1	1
1	1	1	1	1	1	2	1

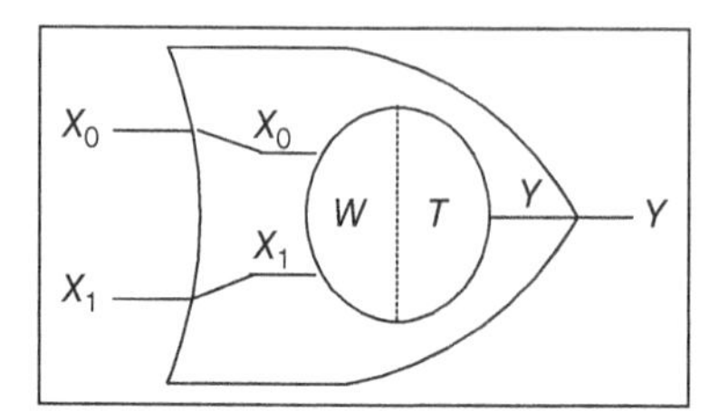

Figure 9.3 Threshold gate disguised as an OR gate.

For the four different input values of X_0 and X_1 for the OR device, the weighted sums will be: 0, 1, 1, 2. The threshold T is specified as 1, so that a weighted sum of 1 or greater is required to produce a value for the output Y of logical 1 (weighted sum $\geq T$).

Three of the input patterns [0 1, 1 0, 1 1] will yield a weighted sum that is equal to or exceeds the threshold and thus will yield a logical 1 for the device output. The weighted sum for the remaining input pattern [0 0] falls below the threshold and thus will yield a logical 0 for the gate output. Figure 9.3 illustrates the standard logical pattern for a two input OR gate.

9.3.3 Designing a Fundamental Boolean Function

In this example, we will begin by executing the design of a basic combinational logical circuit. Such a function can easily be implemented using several standard logic gates. Let's start with that path.

Table 9.3 Truth table.

	X_2	X_1	X_0	Youtput
m_0	0	0	0	1
m_1	0	0	1	0
m_2	0	1	0	1
m_3	0	1	1	1
m_4	1	0	0	1
m_5	1	0	1	0
m_6	1	1	0	1
m_7	1	1	1	0

Table 9.4 K-map.

$X_2 X_1$ / X_0	00	01	11	10
0	1	0	1	1
1	1	0	0	1

As we did earlier in our Boolean algebra review, our first step will be to develop the truth table for our design. From the truth table, we then create the Karnaugh Map. Both of these are given as Tables 9.3 and 9.4. The expression m_i indicates a minterm.

From the K-map in Table 9.5, we extract the following reduced relation in Eq. (9.4).

$$Y = \sim X_1 + \sim X_0 \cdot X_2 \qquad (9.4)$$

~ not
• and
+ or

Table 9.5 K-map.

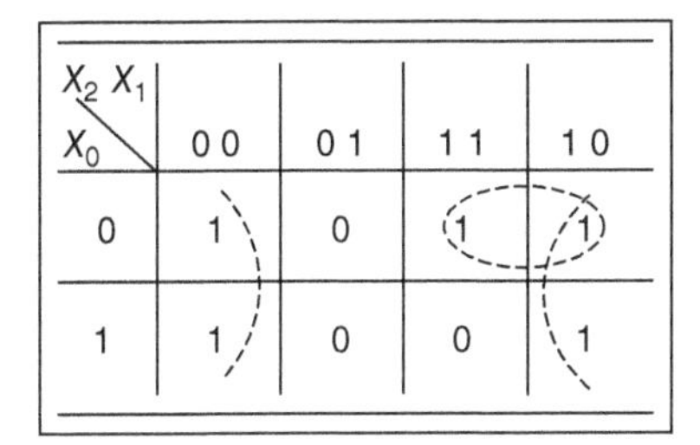

X_2X_1 / X_0	00	01	11	10
0	1	0	1	1
1	1	0	0	1

From Eq. (9.4), we develop the logic diagram in Figure 9.4.

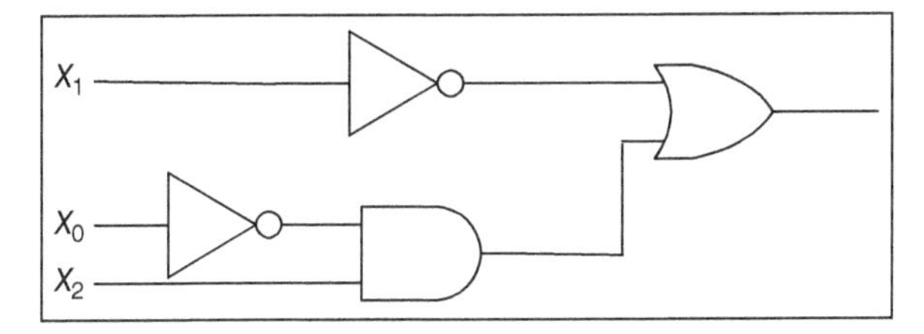

Figure 9.4 Classic logic diagram.

Let's now implement the same logic function using threshold logic.
Such a function can provide one of the logical inputs to a more complex sequential system.

The first step in the design is to decide the number of inputs to the function and the desired output value, based upon the input values. As we did in Figure 9.4, once again we will have three inputs and a single output as illustrated in Figure 9.5.

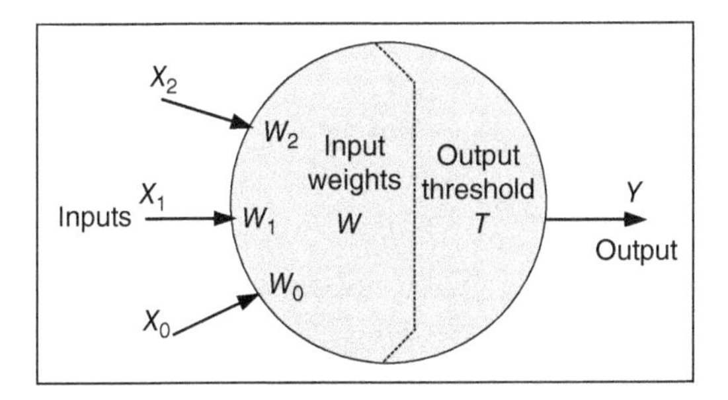

Figure 9.5 Logical function.

The truth table, expressing the minterms (m_i), desired output values, and associated Karnaugh map are given in Tables 9.3 and 9.4. The next step is to specify the threshold T and the weights to achieve the desired output.

We will choose the threshold to equal −3 so that a weighted sum $WS_i \geq 1$ will produce an output Y of 1 and a weighted sum $WS_i < 1$ will produce an output Y of 0.

Columns that indicate the specified weights for the variables comprising each of the minterms, the weighted sum for each row in the truth table, and the threshold output are added to the truth table in Table 9.6.

Table 9.6 Truth table, weights, *WS*, output.

	X_2	X_1	X_0	w_2	w_1	w_0	WS	T = −3	Youtput
m_0	0	0	0	−2	1	−4	0	$0 \geq -3$	1
m_1	0	0	1	−2	1	−4	−4	$-4 < -3$	0
m_2	0	1	0	−2	1	−4	1	$1 \geq -3$	1
m_3	0	1	1	−2	1	−4	−3	$-3 \geq -3$	1
m_4	1	0	0	−2	1	−4	−2	$-2 \geq -3$	1
m_5	1	0	1	−2	1	−4	−6	$-6 < -3$	0
m_6	1	1	0	−2	1	−4	−1	$-1 \geq -3$	1
m_7	1	1	1	−2	1	−4	−5	$-5 < -3$	0

From Table 9.6, we can extract the Boolean function for the output in Eq. (9.5).

$$Y = \Sigma\left(m_0, m_1, m_2, m_3, m_{4,}\ m_5, m_6, m_7\right) \tag{9.5}$$

We can now extract the relationships between the weights and the threshold for each of the minterms.

For each row in the table, we have

m_i: $x_{2i} \cdot w_2 + x_{1i} \cdot w_1 + x_{0i} \cdot w_0 = ws_i$

m_0 $0 \geq T$

m_1 $w_0 < T$

m_2 $w_1 \geq T$

m_3 $w_0 + w_1 \geq T$

m_4 $w_2 \geq T$

m_5 $w_2 + w_0 < T$

m_6 $w_2 + w_1 \geq T$

m_7 $w_2 + w_1 + w_0 < T$

The final threshold gate diagram can now be given in Figure 9.6.

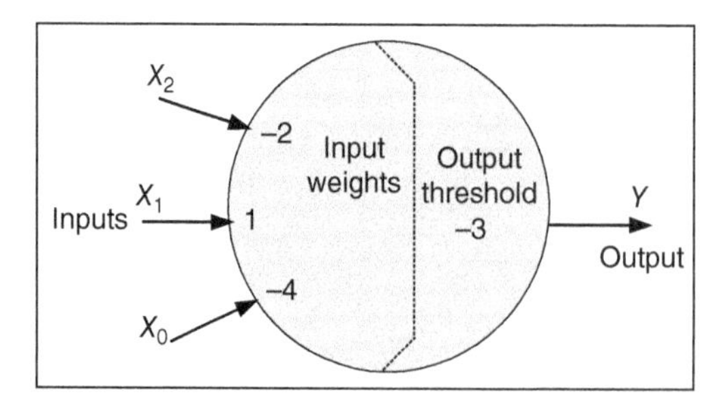

Figure 9.6 Threshold gate final diagram.

9.4 The Downfall of Threshold Logic Design

In this example, we will execute the design of a basic and fundamental combinational logical relationship. Such a function can easily be implemented (and often is) using several standard elementary SSI (small-scale integrated) logic gates or simply a single MSI (medium-scale integrated) gate. Let's see if we can do it using threshold logic.

Our basic SSI logic design implementation and an MSI version are given in Figure 9.7 and the corresponding K-map is given in Table 9.7.

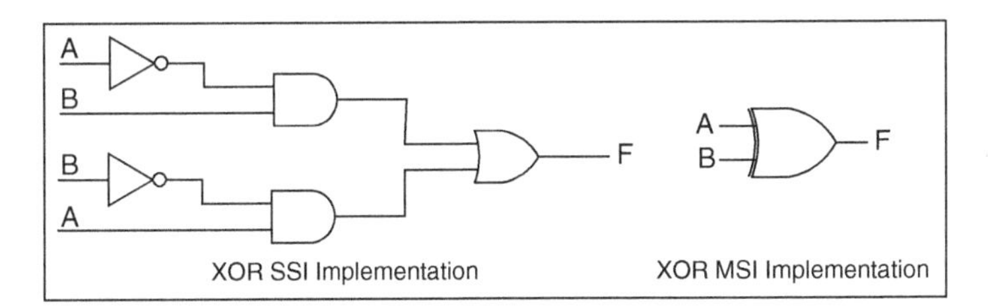

Figure 9.7 Basic combinational logic circuits.

Table 9.7 K-map.

B / A	0	1
0	0	1
1	1	0

Let's now set up a table containing the truth table, weights, WS, and the output for the threshold logic implementation (Table 9.8). Looking at the truth table, there seems to be a problem. Can you spot it?

Table 9.8 Truth table, weights, W_{S},output.

	X_1	X_0	w_1	w_0	W_S	Y_{output}
m_0	0	0				0
m_1	0	1				1
m_2	1	0				1
m_3	1	1				0

Seymour Papert was an early computer scientist, mathematician, and educator who spent most of his career teaching and researching at MIT with threshold logic. He was also one of the early pioneers of artificial intelligence. Marvin Minsky was also an early pioneer at MIT who worked with Seymour Papert.

One day, Minsky raised a question to him. Minsky's question led to the decline of threshold logic and opened the door to what became neural networks.

The logic diagram in Figure 9.5 is known as an exclusive OR or XOR. Its output is a logic 1 only when the inputs are different. The XOR is a very highly used logical operation/device in the field of digital logic design.

Minsky's question to Papert was: "Can threshold logic build an exclusive OR?" Studying Table 9.7, minterm m_3 is creating a problem.

If the weights for X_0 and X_1, individually, are able to produce a weighted sum that is equal to or greater than the specified threshold for minterms m_1 and m_2, then clearly they yield a weighted sum that exceeds the threshold for minterm m_3.

9.5 Summary

We opened the chapter with an introduction to the subfield of digital design known as threshold logic. We learned that such devices are a generalization of the common or classic logic gates and that the input signals to such gates generally can have either of two fixed values. Such values can be logical 0 (typically 0 volts) or logical 1 (typically 5 volts). With the exception of the basic inverter or NOT gate, we know that all traditional devices have at least two inputs and, generally, a single output.

We learned that threshold logic devices follow the same general pattern with exception that each input to such a device has an assigned weight, and the single output has an assigned *threshold* value. The sum of the product of inputs and corresponding weights, called the *weighted sum*, is compared with the specified threshold. If the threshold is met or exceeded, the device will output a logical 1; otherwise, it will output a logical 0. The weights and threshold are assigned at the time that the device is designed.

In contrast to traditional logic, which has different symbols for the different types of gates, a threshold gate is expressed as a circle. Externally, the circle has all inputs coming in on one side, typically the left-hand side, and a single output exiting, typically the right-hand side.

Internally, the circle is divided into two subsections. One subsection, the input side, connects to the input signals and contains the weights. The other subsection, the output side, contains the specified threshold against which the weighted signals are tested and the device output expressing the test result, logical 0 or logical 1.

We learned to design the classic AND and OR gates and a three-input circuit as threshold devices. Finally, we learned that there are certain logical expressions or devices, specifically the exclusive OR (XOR), that cannot be implemented as a threshold device.

Review Questions

9.1 What is threshold logic?

9.2 What is the difference between threshold logic gates compared to classic logic gates?

9.3 What does a threshold logic gate look like?

9.4 What are the limitations to threshold logic?

9.5 How is the threshold determined on the basic device?

9.6 Identify and explain the important guidelines that one should follow in executing a fuzzy logic design.

9.7 What is the purpose of a K-map?

9.8 Can you implement an XOR using a threshold logic gates? If yes, create one and if not, why?

9.9 Why are weights associated with the input signals in a threshold logic device?

9.10 In a threshold logic device, what is the weighted sum?

9.11 What is the purpose of the weighted sum?

9.12 What are the internal portions of the threshold device that affect the output?

9.13 Would it be possible to dynamically control the output of a threshold device? If so, how?

9.14 Why would we want dynamically control the output of a threshold device?

10

Moving to Perceptron Logic ! ! !

THINGS TO LOOK FOR...

- Human thoughts to physical implementations.
- What is a biological neuron?
- What is an artificial neuron?
- What is a perceptron?
- How are artificial neurons and perceptrons related?
- What is perceptron logic?
- What are weights and where are they used?
- Designing basic logic gates.
- How do perceptron logic devices compare to classic, fuzzy, or threshold logic gates?
- What does a perceptron device look like?
- Can a perceptron learn?
- What is the learning rule?
- Are there any limitations to perceptron logic?

10.1 Introduction

In the previous chapters, we noted that a significant challenge with engineering design is transforming our human thoughts, concepts, language, vision, and hearing into functioning physical implementations. We've looked at classic logic, fuzzy logic, and threshold logic. We now raise the question again: "How do we do that?" This remains a very good question. How do we do that?

The first step in addressing and solving contemporary problems such as neural networks, speech recognition, machine learning, IOT, or AI begins with the I in IOT and AI: *intelligence* and *imagination*. We also toss in a C for curiosity, creativity, and cleverness. We now have the first step toward discovering an answer: I C.

If you do C, we have also just found a linguistic variable. Let's start with the human brain and, in particular, its architecture as a model for understanding, designing, and building functioning intelligent machines. Two essential and critical aspects of the brain stand out immediately. The brain has an extremely large, well-organized memory of

Introduction to Fuzzy Logic, First Edition. James K. Peckol.

information supported by parallel, interconnected processing elements. That memory supports access by content and the ability to acquire and store new information and to alter the existing content.

At the foundation of such a process is a biological entity known as a *neuron* or collection of neurons. Such a collection forms what is called a *neural network.*

Now, consider the classic *central processing unit* or CPU. Here, we also have a collection of logic and memory devices interconnected through an artificial *neural network* of conducting paths. Input signals called data and commands are sent to selected logic devices, which then perform operations on or respond to such signals. The results of those operations may then be stored and also sent off to other logic devices where further operations take place and the process is repeated. Alternatively, the results may be sent to the output of the system for display or to another CPU.

Consider now the architecture of the basic threshold logic device that we introduced and studied in Chapter 9 and now will revisit in Figure 10.1. With an interconnected collection of such devices plus other logic, we can begin to form a logical, computational neural network.

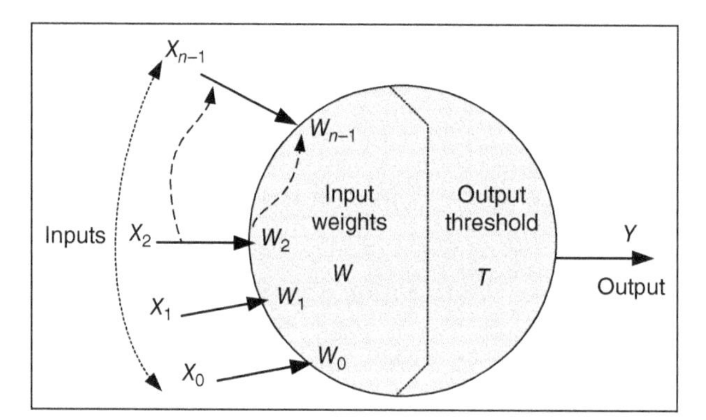

Figure 10.1 Basic threshold device.

Let's start by introducing some vocabulary. Such a device will be the basic building block at the heart of what is called a *perceptron*. A perceptron is also known as an *artificial neuron*. The biological neuron is the fundamental working unit of the brain. It is a specialized cell designed to transmit information to other nerve cells. An artificial neuron is a mathematical function designed to model and replicate the behavior of the biological neuron.

Such artificial neurons become the fundamental components of a *neural network*. They also often contribute in the process called *machine learning*.

10.2 The Biological Neuron

In all vertebrate animals, which include humans, and most invertebrate animals, biological neurons are the core components of the brain. Literally, the human brain contains billions of such neurons. These neurons are interconnected nerve cells involved in the processing, transmission, reception, and interpretation of chemical and electrical signals from and to other neurons. The reception process is affected via *dendrites* from other nerve cells to a neuron's cell body. See the left-hand side of the diagram in Figure 10.2, which gives a high-level schematic representation of the basic biological neuron.

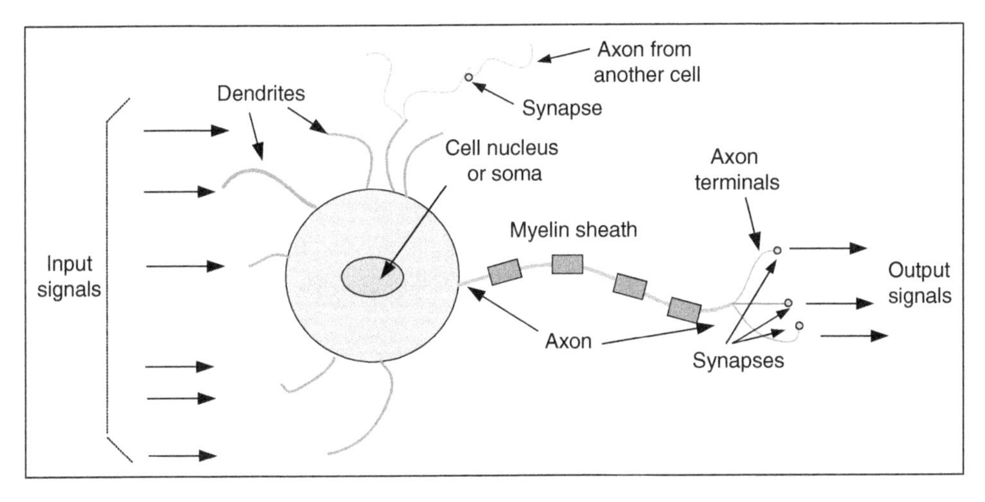

Figure 10.2 High-level basic neuron schematic.

We can now see that the millions of neurons provide the vast memory of information we hypothesized that we would need in formulating an intelligent machine and that dendrites and synapses would provide the necessary interconnections.

Neurons are typically composed of a *dendritic tree* on the input side, a cell body called a *soma*, and an *axon* and *axon terminals* on the output side. Neurons are highly specialized for the fast processing and transmission of cellular signals.

10.2.1 Dissecting the Biological Neuron

10.2.1.1 Dendrites

Starting on the left-hand side of Figure 10.2 we have the *dendrites*. These protrusions form a branched extension of or from the nerve cell and thereby provide a path for electrochemical signals from over 1000 other nerve cells into the cell body of the *soma*. Such signals act as inhibiting or exciting signals, that is, negative or positive input signals (ions) to the neuron.

We can view these as analogous to the input circuitry on a digital logic block or microprocessor input subsystem.

10.2.1.2 Cell Body – Soma

In the neuron, the *soma* (cell body) inherently performs as a summation function on the arriving signals (ions). Such an operation occurs as a result of inherent co-mixing of the inhibiting (negative) or exciting (positive) incoming ions.

10.2.1.3 Axon – Myelin Sheath

The *axon* is a conducting path from the soma (cell body) to the neuron's output tree. The opening to the axon enables a sampling of the electrical potential (the result of a summation process) inside the soma. When the soma reaches a certain potential, the axon will transmit a signal down its length to the *axon terminals*, which then provide the neuron's output signals to other neurons. We can call the collection of *axon terminals* an *output tree*.

The *myelin sheath* is a lipid-rich substance that surrounds the axons to insulate and protect them and also thereby increase the rate at which electrical signals propagate along the axon path to the neuron's output tree and ultimately to other neurons. Such signals are designated *action potentials*.

One can easily view the axon and myelin sheath combination as analogous to insulated wires conducting data and control signals inside of an embedded system. Notice, however, in Figure 10.2, that the myelin sheath is not contiguous along the axon path. Rather, each axon comprises multiple myelin sheath segments separated by short gaps. Such gaps are called *nodes of Ranvier*.

Because the nodes are spaced out, they allow the signal(s) to rapidly jump from node to node. The rate at which the axon fires becomes the rate at which neighboring cells receive signaling ions.

The myelin sheath was discovered and named by the German anatomist Rudolf Virchow in 1854. Later, the French pathologist and anatomist Louis-Antoine Ranvier discovered the gaps in the myelin sheath that were later named after him.

10.2.1.4 Synapse

A synapse is an interconnection or junction between two nerve cells. Such a connection enables an initiating neuron to pass or convey chemical or electrical signals and information to a receiving neuron or to what is called a *target effector cell*. The target cell actively responds to the stimulus and affects or initiates some change.

10.3 The Artificial Neuron – a First Step

The artificial neuron, as the name suggests, is an attempt to model the behavior of an actual biological neuron. Such a model can be implemented mathematically, physically using logic devices of one form or another, or a combination of both.

Some of the first successful modeling of the brain cell grew out of the research of Walter Pitts and Warren McCulloch and was presented in their paper in 1943 describing their simplified model. Their artificial neuron was referred to as the *McCulloch–Pitts* or *MCP* neuron.

Pitts and McCulloch's early artificial neuron closely follows the threshold logic device we introduced in Chapter 9. Drawing from the digital world, they suggested that the nerve cell was simply a basic logic gate with binary outputs. The outputs were binary in the sense that they either fired or they didn't.

The artificial neuron, like the crisp and fuzzy logic gates, can have multiple inputs and only a single output. For the MCP neuron, the inputs (Xi) have the value (1 or 0), and, like the crisp device, the output (Y) is constrained to the values of True or False, 1 or 0. An early schematic model of the artificial neuron is illustrated in Figure 10.3 .

During operation, the input values (Xi) are summed and the aggregated result is passed to the function Θ(), called a *threshold parameter*, which makes a decision and passes the result, true or false, to the output Y.

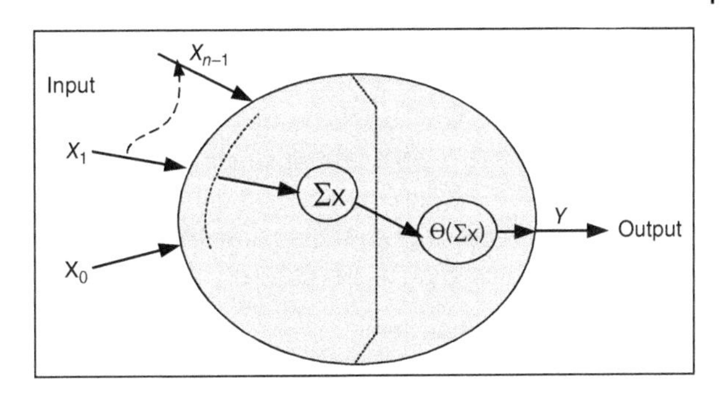

Figure 10.3 Basic artificial neuron schematic.

We can compare the application of the device to the process of implementing several standard basic Boolean functions. It can be shown that a variety of much more complex functions can be implemented using such basic functions as we'll see shortly.

Example 10.1 The Basic Logic Gates

The basic artificial neuron configuration is illustrated in Figure 10.4a and is accompanied by three other possible configurations. Specifying the input polarity and the threshold parameter for the device determines its logical function. The small circle at the tip of an arrow on three of the configurations indicates a logical inversion: 1 to 0 or 0 to 1. A logical 1 coming in is inverted to a 0 and vice versa.

Let's now examine the architecture, properties, behavior, and application of these devices starting with the logical AND and the logical OR configurations. The logical implementation methodology will follow that we used in Chapter 9. The accompanying diagrams present the standard symbol for each of the logical functions.

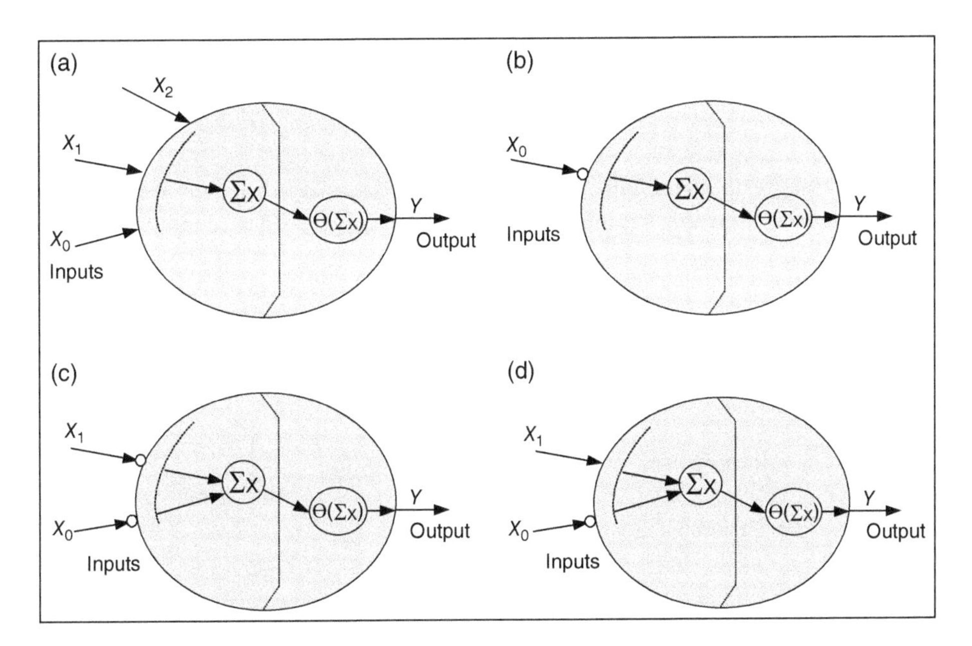

Figure 10.4 AND, OR, NOT, NOR, inhibit gate configurations.

AND Gate – Figure 10.4a

For a three-input logical AND gate configuration, the inputs are X_0, X_1, and X_2. We have the same thing for a logical AND function as shown in the accompanying figure. Borrowing from the threshold logic device, each input X_i has a value of 1, and we specify that the threshold parameter for the artificial neuron is 3, which gives us Eq. (10.1a).

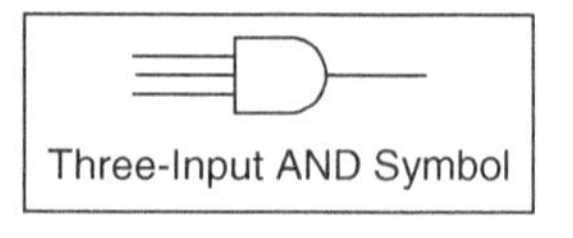
Three-Input AND Symbol

Σ indicates summation

$\Theta(\ )$ is the *threshold parameter*

$$\begin{aligned}&(\Theta(\Sigma X_i) => 3)\text{ neuron fires}\\&(\Theta(\Sigma X_i) < 3)\text{ neuron does not fire}\end{aligned} \tag{10.1a}$$

As a result, if all inputs X_i are ON, the output will be a logical 1 or ON; otherwise, if any or all of the inputs are OFF, the output will be a logical 0 or OFF,

OR Gate – Figure 10.4a

The inputs are X_0, X_1, and X_2, and for a logical OR configuration, we specify that the threshold parameter is 1, which gives us Eq. (10.1b).

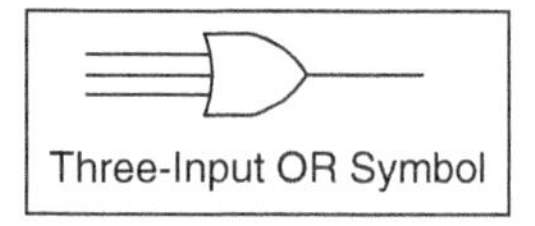
Three-Input OR Symbol

$$\begin{aligned}&(\Theta(\Sigma X_i) => 1)\text{ neuron fires}\\&(\Theta(\Sigma X_i) < 1)\text{ neuron does not fire}\end{aligned} \tag{10.1b}$$

As a result, if 1, 2, or 3 of the inputs X_i are ON, the output will be a logical 1 or ON; otherwise, if all of the inputs are OFF, the output will be a logical 0 or OFF.

We see that the logical AND and the logical OR devices are implemented using the same configuration with the logic and output managed by the *threshold parameter.*

Inhibitory Inputs – Figure 10.4b–d

As the name suggests, an inhibitory input "inhibits or alters" the behavior of a signal by inverting its state. The inversion is indicated by a small circle on the device input where the signal enters.

NOT Gate – Figure 10.4b

For the logical NOT device, the input is X_0 and for a NOT function, we specify that the threshold parameter is <1, which gives us Eq. (10.1c).

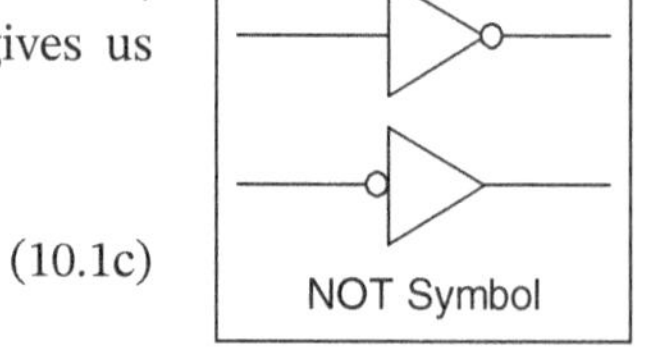
NOT Symbol

$$\begin{aligned}&(\Theta(\Sigma X_i) < 1)\text{ neuron fires}\\&(\Theta(\Sigma X_i) => 1)\text{ neuron does not fire}\end{aligned} \tag{10.1c}$$

As a result, if the input is ON, logical 1, the output will be a logical 0 or OFF; otherwise, if the input is OFF, the output will be a logical 1 or ON.

As the name suggests, an inhibitory input blocks or inhibits a signal. Such an input is indicated by a small circle on the output of an input signal.

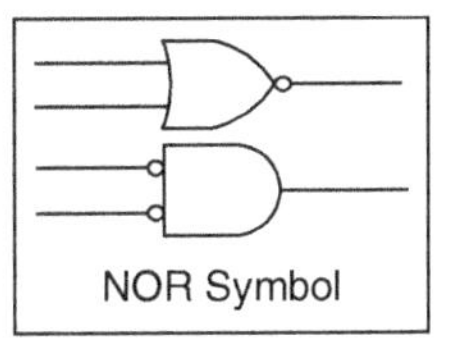

NOR Gate – Figure 10.4c
The inputs are X_0 and X_1, and for a logical NOT configuration (either one or both inputs are logical 1), we specify that the threshold parameter Θ is 0, which gives us Eq. (10.1d).

$$\left(\Theta\left(\Sigma X_i\right) => 1\right) \text{neuron does not fire}$$
$$\left(\Theta\left(\Sigma X_i\right) < 1\right) \text{neuron fires} \tag{10.1d}$$

As a result, if both of the inputs are OFF (both inputs are 0), the output will be a logical 1 or ON; otherwise, if neither or both of the inputs are ON, the output will be a logical 0 or OFF.

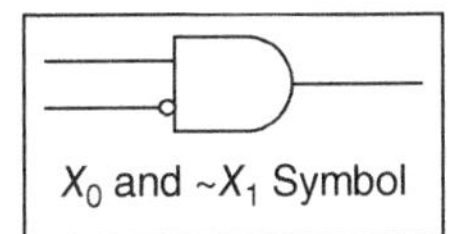

Mixed Logic – Figure 10.4d
The inputs are X_0 and X_1. For a mixed configuration, we specify that $X_0 = 1, X_1 = 0$ (the 0 will be inverted to 1), and the threshold parameter is 2, which gives us Eq. (10.1e).

$$\left(\Theta\left(\Sigma X_i\right) = 2\right) 0 + 1 \text{or} 1 + 0 \text{neuron fires}$$
$$\left(\Theta\left(\Sigma X_i\right) <= 1\right) \text{neuron does not fire} \tag{10.1e}$$

Example 10.2 Adding Weights
Unlike the inputs to a crisp logic gate and analogous to the threshold device and the fuzzy logic gate, the basic device configuration on the artificial neuron can be altered by assigning and adding a weight to each of the inputs. The application and benefits of such a technique are described and discussed in the next section.

The devices in Figure 10.4 are now modified by adding weights (w_i) in series with each of the inputs and are illustrated in Figure 10.5. With the weights added, specifying the

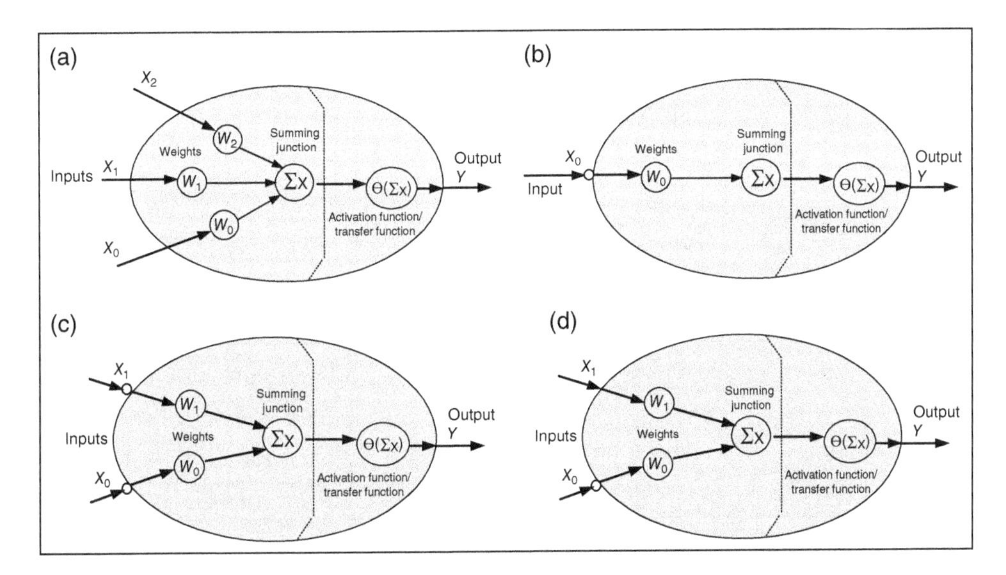

Figure 10.5 AND, OR, NOT, NOR, INHIBIT weighted gate configurations.

input's polarity, the weights' values, and the *activation* and *transfer* functions, in combination, determine the logical operation of the devices.

Functionally, within each device, the value of each of the inputs is multiplied by its associated weight and then the resulting products are summed. The sum is then passed through a nonlinear function that is known as an *activation function* or *transfer function* and compared to the specified threshold. If the threshold is met or exceeded, the output of the device is *true* (logical 1 or another value); otherwise, it is false (logical 0 or another value), hence the name.

The values of the weights and the threshold are not constrained; they can be any finite real positive or negative numbers. The choice is based on or dependent upon the intended application as specified by the designer.

We can convert the previous paragraph from a textual description of the threshold gate into a corresponding mathematical description as we see in Eq. (10.2). Inputs are denoted X, outputs Y, the weights W, and the threshold T.

The arithmetic sum of the products of the inputs and the weights is termed the *weighted sum* as we saw with the threshold logic device earlier.

$$\begin{aligned} &\text{if}\left(X_0 \cdot W_0 + X_1 \cdot W_1 + X_2 \cdot W_2 + \ldots. + X_{n-1} \cdot W_{n-1} + X_n \cdot W_n\right) < T\, Y = 0 \\ &\text{if}\left(X_0 \cdot W_0 + X_1 \cdot W_1 + X_2 \cdot W_2 + \ldots. + X_{n-1} \cdot W_{n-1} + X_n \cdot W_n\right) >= T\, Y = 1 \end{aligned} \tag{10.2a}$$

The equation can be expressed in shorthand notation as shown in Eq. (10.2b).

$$\begin{aligned} &\text{if}\left(\Sigma X_i \cdot W_i\right) < T\, Y = 0 \\ &\text{if}\left(\Sigma X_i \cdot W_i\right) >= T\, Y = 1 \end{aligned} \tag{10.2b}$$

In Chapter 9, using basic threshold logic, we also illustrated the design and implementation of the elementary AND and OR gates. As we've noted, such a logic closely parallels that used in the basic *McCulloch–Pitts* artificial neuron. The *MCP* neuron could also be used to implement the inversion or complement (NAND, NOR, NOT) of such functions. They are given in Tables 10.1–10.3.

Table 10.1 NAND Gate $T > -2$

X_1	X_0	NAND	W_1	W_0	$X_1 \cdot W_1 + X_0 \cdot W_0 = W_{NAND}$			Y
0	0	1	−1	−1	0	0	0	1
0	1	1	−1	−1	0	−1	−1	1
1	0	1	−1	−1	−1	0	−1	1
1	1	0	−1	−1	−1	−1	−2	0

Table 10.2 NOR Gate $T > -1$

X_1	X_0	NOR	W_1	W_0	$X_1 \cdot W_1 + X_0 \cdot W_0 = W_{NOR}$			Y
0	0	1	−1	−1	0	0	0	1
0	1	0	−1	−1	0	−1	−1	0
1	0	0	−1	−1	−1	0	−1	0
1	1	0	−1	−1	−1	−1	−2	0

Table 10.3 NOT Gate $T > 0$

X_0	$\overline{X_0}$	W_0	$X_0 \cdot W_0 = W_{NX}$		Y
0	1	−1	0	1	
1	0	−1	−1	0	0

10.4 The Perceptron – The Second Step

Several years after McCulloch and Pitts formulated the first artificial neuron, Donald Hebb (1949), a Canadian psychologist, sought to understand how the functions of biological neurons were able to contribute to psychological processes such as learning and memory. Hebb's research led to his theory that was later called *Hebbian learning*. He is also called the father of *neural networks*.

Based on Hebb's work, a basic McCulloch–Pitts neuron, previously illustrated in Figure 10.3, was modified to assign a weight to each input (W_i). We initially saw this in Figure 10.5. Such weights are now also added to the device designated Neuron B and illustrated in Figure 10.6. Neuron B accepts signals or an input from Neuron A as well as synapses or connecting links from other neurons to inputs ($X_0 - X_{n-1}$).

An input synapse is characterized by the letter X with an associated subscript indicating the source and destination neurons for that synapse. Each input has an associated specified weight, synapse identifier, and the neuron identifier. Thus, the weight on synapse a in neuron b would be expressed as W_{ab}.

The synapses indicated in Figure 10.6 by X_0 to X_{n-1} are merely illustrating the possibility of additional inputs to Neuron B. Those connections, potentially from other neurons, are given a temporary working weight label in the range of $W_0 - W_{n-1}$.

The weights associated with each of the neuron inputs are collectively passed to what is termed a *summing junction* where they are summed. The resulting sum, called the *weighted sum*, is passed to an *activation function* where it is assessed against a threshold to produce the appropriate output. The neuronal output then connects to other neurons.

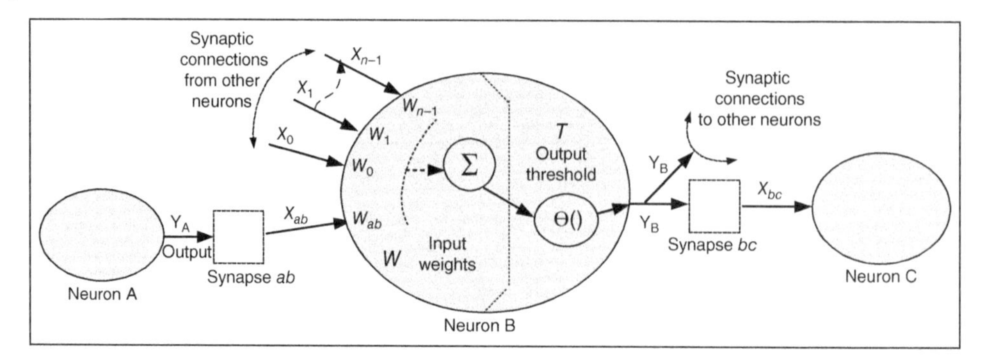

Figure 10.6 Simple McCulloch–Pitts neuron network.

A simple artificial neuron network can be formed as illustrated in Figure 10.6. The output from Neuron A, (Y_A), becomes an input to Neuron B, (X_{AB}), where it is assigned the weight W_{AB} and is summed with the other weighted inputs. Similarly, the output from Neuron B, (Y_B), becomes an input to Neuron C, (X_{BC}), where it is assigned the weight W_{BC} and summed with the other incoming weighted inputs. Thus, we have:

$$Y_A \rightarrow X_{AB} \cdot W_{AB} \tag{10.3}$$

$$Y_B \rightarrow X_{BC} \cdot W_{BC} \tag{10.4}$$

For each neuron, the resulting sum is then compared against the specified threshold value for success or failure.

As expressed in Eq. (10.4), an input to Neuron C is the output value Y_B from Neuron B. That value is then multiplied in Neuron C by the specified weight W_{BC} and similarly summed with the any other weighted input values. The resulting sum is then compared against the specified threshold for Neuron C and the result passed on.

At this stage, we have assembled most of the physical pieces for the basic artificial neuron and the rudiments of a neural network.

10.4.1 The Basic Perceptron

We now introduce the basic perceptron. As the early research continued, Frank Rosenblatt (1962), a research psychologist at the Cornell Aeronautical Laboratory, combined the work of McCulloch and Pitts with that of Hebb to develop the first *perceptron* in 1957.

His device, following Hebb, had the ability to learn by the appropriate weighting of the device inputs. His contributions were later influential in the development of neural networks. Sadly, Rosenblatt was lost to science at the age of 43 in a boating accident on his birthday in 1971.

Rosenblatt's fundamental Perceptron device is illustrated as Neuron B in the schematic diagram in Figure 10.7 with:

- Inputs denoted as X_i and the weights as W_{Bi}
- A potentially negative or positive input X_0
- A number of input signals X_1, X_2, and X_{n-1}

- A summing junction Σ
- A potential bias input
- A threshold input Θ_B
- An activation function α( )
- One output Y_B

The B subscript is arbitrary. It just indicates that we are working with Neuron B.

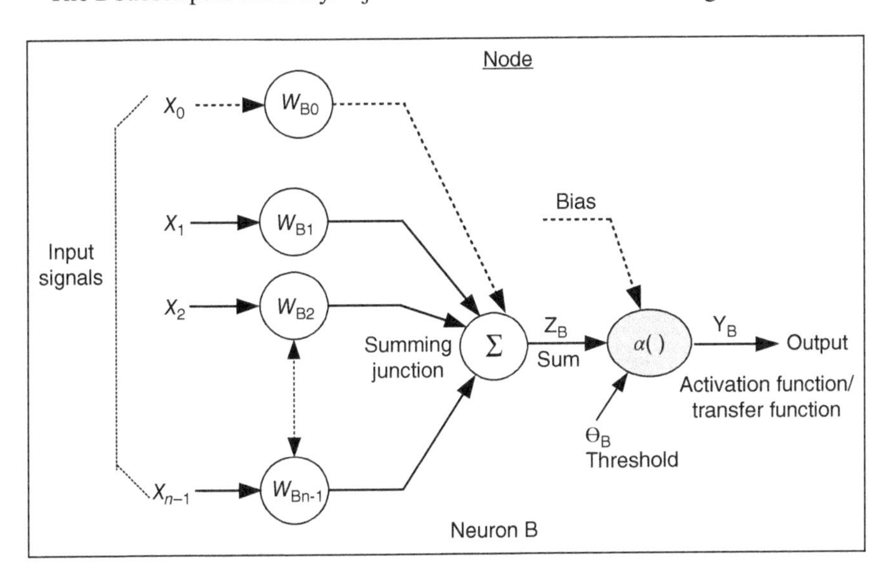

Figure 10.7 Basic perceptron schematic.

The artificial neuron, like the standard crisp, fuzzy, and threshold logic gates, can have multiple inputs and only a single output. The number of inputs is not limited to a particular number, and the single output Y_B feeds to other neurons.

The input X_0 can potentially be used to alter the *weighted sum*. Each of the inputs, X_1 to X_{n-1} to the perceptron has an assigned weight. The weights are not required to all be of the same value and can also be of different polarities.

The *Node* accepts the individually weighted inputs, which are then summed at the *summing junction*. The arithmetic *sum* of the products of the inputs and weights is termed the *weighted sum* and is expressed as Eq. (10.5).

$$Z_B = X_0 \cdot W_0 + X_1 \cdot W_1 + X_2 \cdot W_2 + \ldots. + X_{n-1} \cdot W_{n-1} \tag{10.5}$$

$$n-1$$

$$= \Sigma X_i \cdot W_i$$

$$i = 0$$

Figure 10.7 shows three inputs to and one output from the *activation function* block. The inputs are the *weighted sum* (Z_B), the *bias* is (B), the *threshold* is (Θ_B), and the output is (Y_B). The *bias* (B) and the threshold (Θ_B) are optional and are external to the *activation function* block.

Typically, the *threshold* will be negative. Leaving the *bias* (B) out, the perceptron output will be given by Eq. (10.6), indicating that the output is a function of Z_B, the weighted sum and Θ_B, the threshold.

$$Y_B = f\left(Z_B - \Theta_B\right) \tag{10.6}$$

Bringing in the threshold (Θ_B) has the effect of applying an offset to the perceptron output signal. Introducing the bias (B) could be used to counter the effect of the threshold.

As shown in Eq. (10.7), the weighted sum is then compared against the *threshold* (Θ_B). The result (Y_B) is ultimately sent to the output.

$$\begin{aligned} Y_B &= 0 \text{ if } X_0 \cdot W_0 + X_1 \cdot W_1 + X_2 \cdot W_2 + \ldots. + X_{n\ 1} \cdot W_{n-1} < \Theta_B \\ Y_B &= 1 \text{ if } X_0 \cdot W_0 + X_1 \cdot W_1 + X_2 \cdot W_2 + \ldots. + X_{n\ 1} \cdot W_{n-1} >= \Theta_B \end{aligned} \tag{10.7}$$

As illustrated in Figure 10.7, the result could possibly be affected by the bias as well.

10.4.2 Single- and Multilayer Perceptron

In Figure 10.8, we introduce two fundamental types of perceptrons: *single-layer* and *multilayer*. They are both specified as *feed-forward* architectures, that is, they do not support feedback capability. The descriptive terminology indicates that signals enter the perceptron through an input layer, then propagate through the network in a single direction, and ultimately exit directly as the output.

The *single-layer* implementation has, as illustrated in Figure 10.8, a single input layer and a single output layer for processing. Figure 10.8 further illustrates that the two *multilayer* implementations also have a single input layer and a single output layer. However, in addition, they also have a single, hidden processing layer connected in sequence between the input and output layers.

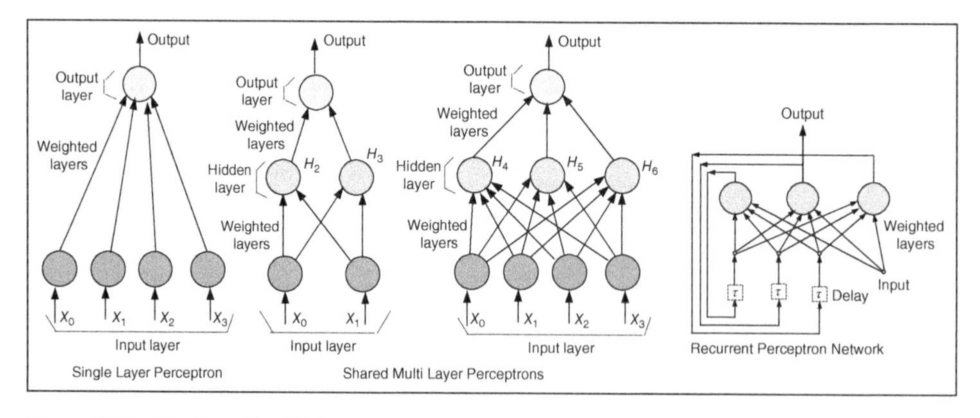

Figure 10.8 Single and multi-layer perceptrons.

The diagram in Figure 10.7 illustrates that for any of the perceptron models in Figure 10.8, the output of the *summing junction* is passed on to the *activation function*, which is indicated by the symbol $\alpha()$. If the specified threshold is reached, the *activation function* (*transfer function*) is activated, and an output Y is generated.

In addition to the single-layer and multilayer architectures, the perceptron also supports a *Recurrent* configuration that supports at least one feedback loop. Such a network is also illustrated in Figure 10.8. The illustrated architecture incorporates three feedback paths and a single input. Observe that each of the feedback paths has a path delay incorporated, which is shown as the symbol τ.

10.4.3 Bias and Activation Function

As we noted earlier, the input X_0, weight W_0, threshold $\Theta(\,)$, and the *bias* input all play roles in establishing and managing the activation threshold (*transfer function*) on the perceptron and that the threshold $\Theta(\,)$, bias input, and input X_0 are potentially optional.

Working with single-layer devices, let's now examine each of these in greater detail starting with the *activation* or *transfer function*. Figure 10.9 illustrates four different models of the basic perceptron, two that include the bias and two that don't.

The diagram further illustrates that in each of the models, the output of the *summing junction* is passed on to the *activation function*, which is indicated by the symbol $\alpha()$. If the specified threshold is reached, the *activation function* (*transfer function*) is activated, and an output Y is generated.

One version of the output relationship can be expressed mathematically as was given earlier in Eqs. (10.6) and (10.7).

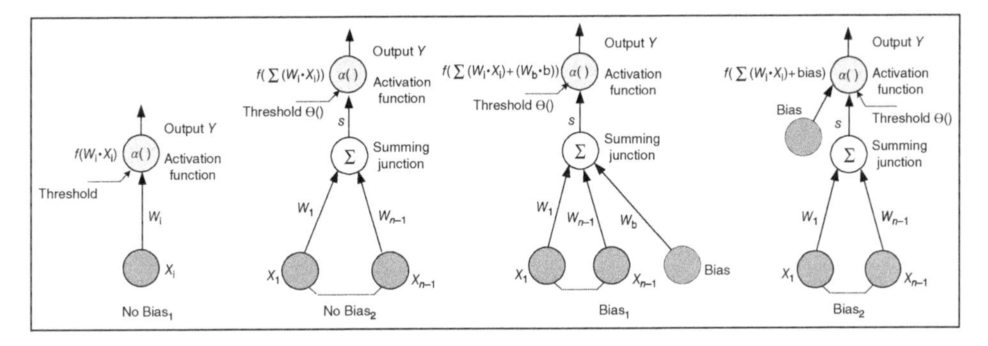

Figure 10.9 Basic perceptron models with and without bias.

The symbol Θ identifies the *threshold*, and the symbol $\alpha(\,)$ indicates the *activation function*. As we've seen, typically, the output will be 0 if the sum is below a specified threshold and 1 if the sum equals or exceeds the threshold. Based on the design, the output can also be expressed as -1 or $+1$. Other output values are also possible as we will see shortly.

As we see from Eq. (10.6), the *activation function* takes the weighted sum as an input and potentially a *threshold* then determines the neuron output. Adding the bias input enables the *activation function* to shift the computed output value Y. The shift is accomplished by adding a constant (the bias) to the *weighted sum* or the bias could be an additional input to the output logic as illustrated in the models Bias_1 and Bias_2 in Figure 10.9. The bias can also be incorporated as indicated in Eq. (10.8a) or (10.8b). Note also that the bias can be specified as positive or negative.

$$\text{Output Y} = \Sigma_{\text{i}}\left(W_{\text{i}} \cdot X_{\text{i}}\right) + \left(W_{\text{b}} \cdot \text{bias}\right) \tag{10.8a}$$

$$\text{Output Y} = \text{bias} + \Sigma_{\text{i}}\left(W_{\text{i}} \cdot X_{\text{i}}\right) \tag{10.8b}$$

The purpose of the addition of a bias is to alter the behavior of the *activation function*. In the case of Bias_1, in addition to the *weighted sum*, the effect on the activation function will be to shift its value by a constant amount specified by the product of the connection weight and the bias value thus: (W_b·b). The added weight increases the steepness of the *activation function*. The intended resulting effect is to manage how quickly the function will trigger.

In contrast to the weights, the intent of the bias can be to delay the triggering of the output signal. Thus, in the case of the Bias_2 design, the bias is not an integral part of the *weighted sum* yet is typically used in conjunction as an independent tool to affect the neural output.

The inclusion of an externally applied positive or negative *threshold* can have a similar effect by mathematically raising or lowering the net input to the *activation function* as shown in Eq. (10.9) and Figure 10.10.

$$\text{Output Y} = \alpha\left(\Sigma_{\text{i}}\left(W_{\text{i}} \cdot X_{\text{i}}\right) - \Theta(\tau)\right) \tag{10.9}$$

Let's now look at a number of different *activation functions*. If we go back to the discussions of membership functions under fuzzy logic, we will see that the graphic shapes of *activation functions* and membership functions have a lot of similarity.

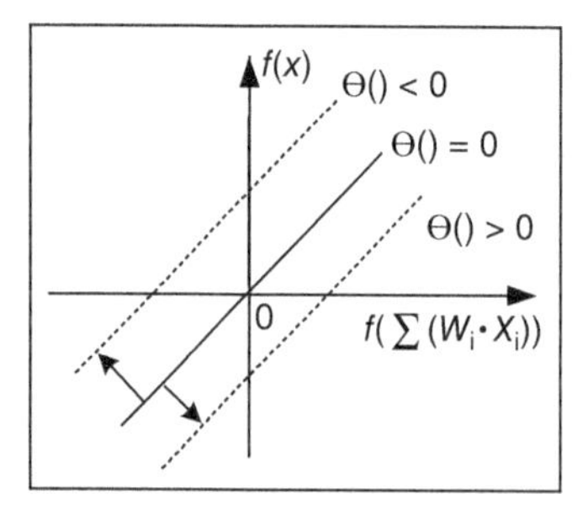

Figure 10.10 Linear activation function effect of a threshold signal.

Our primary goal with the activation/transfer functions is to enable the management of the transmission of signals from one neuron to others. In doing so, we are concerned about the timing, rate, and value of the output signal(s).

Example 10.3 Common Activation Transfer Functions
The following Figures (10.11–10.15) illustrate several of the common activation/transfer functions.

Linear $f(x) = x$

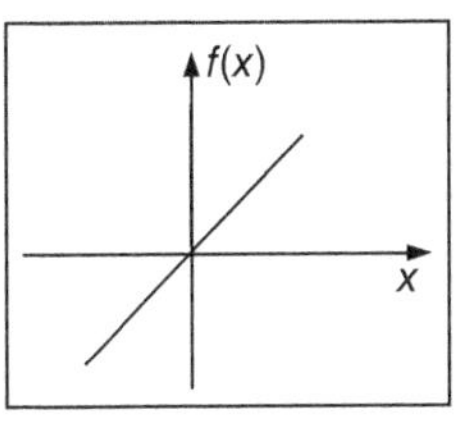

Figure 10.11 Linear.

Threshold

$$f(x) = 0 \qquad x < 0$$
$$f(x) = 1 \qquad x >= 0$$
$$(\text{binary})$$

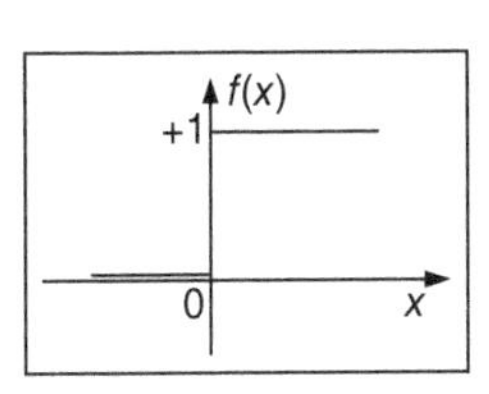

Figure 10.12 Threshold.

Unit step

$$f(x) = 0 \qquad x < 0$$
$$f(x) = 0.5 \qquad x = 0$$
$$f(x) = 1 \qquad x > 0$$

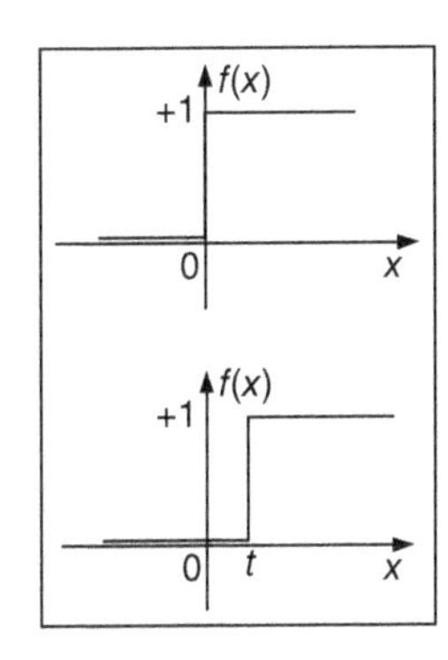

Figure 10.13 Unit step.

Unit step with delay

$$f(x) = 0 \qquad x <= t$$
$$f(x) = 1 \qquad x >= t$$

Sign

$$f(x) = -1 \qquad x < 0$$
$$f(x) = 0 \qquad x = 0$$
$$f(x) = 1 \qquad x > 0$$

Piecewise linear

$$f(x) = 0 \qquad x <= -½$$
$$f(x) = x + ½ \qquad -½ <= x <= ½$$
$$f(x) = 1 \qquad x => -½$$

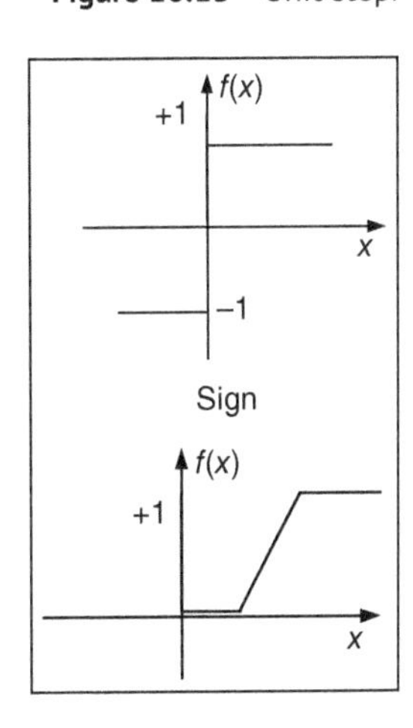

Figure 10.14 Piecewise linear.

Sigmoid $f(x) = \frac{1}{1+e^{-x}}$

Hyperbolic tangent $f(x) = \frac{e^{x} - e^{-x}}{e^{x} + e^{-x}}$

ReLU $f(x) = 0 \quad x < 0$

$f(x) = x \quad x > 0$

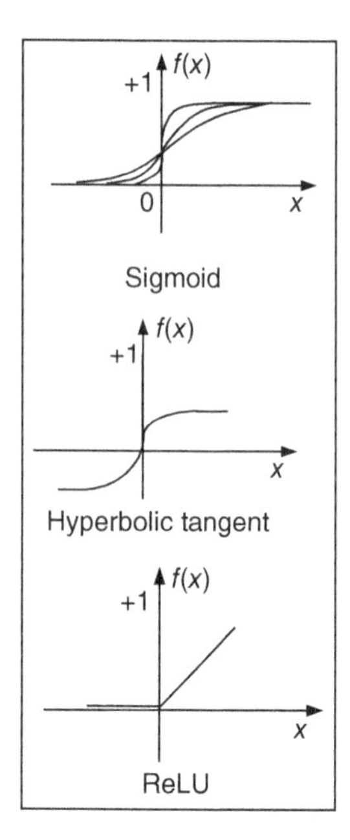

Figure 10.15 ReLU.

10.5 Learning with Perceptrons – First Step

We've looked at the high-level structure, basic functionality, and several possible architectural designs for Rosenblatt's perceptron. Let's now examine how those pieces work together to accomplish portions of his early goals.

We opened this book with a brief introduction and discussion of human learning and reasoning. We noted Edward Feigenbaum proposed a five-phase learning process. We further noted that the learning situation is considered to consist of two parties, the *learner* and the *teacher* or *environment*, and a body of knowledge to be transferred from the environment to the learner. At that time, our assumption was that the learner and the teacher were humans. As we now move to learning with perceptrons, the learner will typically be an electronic logic device, and the teacher may be either a human or a similar logic device.

Working with Feigenbaum's learning process, the requirements and information will be entered into storage on the perceptron by a designer or operator. When the device is operational, input signals will be entered into the perceptron, processed, and a result sent to the output subsystem where it is analyzed, evaluated, and the appropriate output generated. If the output is as desired, the process continues. On the other hand, if incorrect or not what was expected, the operational logic can self-modify, and the process can then continue or potentially be shut down.

This is similar to the process that we often use when we first try to learn something new. We either self-learn or someone teaches and guides us. We then try it, possibly failing at first, but practicing and ultimately succeeding. As the old saying goes: "We learn to fail successfully."

Now moving to the perceptron. We have introduced and examined the McCulloch–Pitts neuron and its evolution to the device called the perceptron. The challenge, however, is that the ability of a single neuron or perceptron on its own is limited. Generally, to design and implement a more complex system, we often require numerous neurons or perceptrons in cascade. In such a system, we will typically have a large number of inputs with activation flows of different strengths flowing and interconnecting the system devices.

Let's start with a high-level schematic of a basic single artificial neuron system illustrated in Figure 10.16.

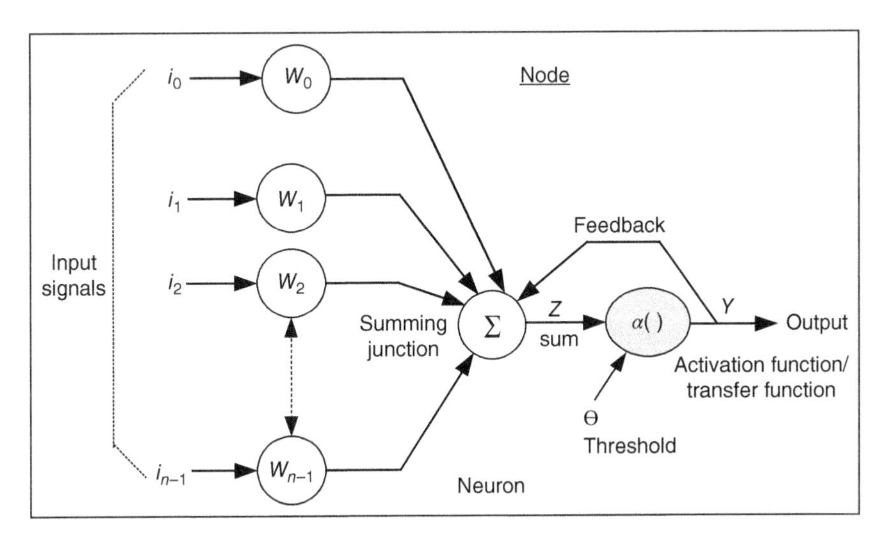

Figure 10.16 Single-neuron system schematic.

As a first step, we will introduce some of the descriptive terminology with which we will be working. The symbols and terms appear in the schematic in Figure 10.16.

Terminology

- Inputs $x = (i_0, i_1, \ldots i_{n-1})$ // (i_i) system input
- Weights = $(w_{B0}, w_1, \ldots w_{n-1})$, // ($w_{B0}$) system bias unit // ($w_i$) system weights
- Feedback // Adjust weights and threshold
- Summation = Z // $Z = \Sigma\,(w_i \cdot i_i)$
- Output = y // y = Perceptron output
- Correct output = O // Specified correct output
- Activation function = $\alpha()$ // Computes and enables output
- Threshold = Θ // Limit on computed output value
- Learning rate = C // Also referred to as step size // Amount weights are updated // during training. (typically 0.0–1.0)

10.5.1 Learning with Perceptrons – The Learning Rule

As we move from humans to the systems that we have created, the essential first step that we must take is that we need to step back and identify the problem. What do we know? What are we trying to teach? What do we want the system to learn? What does system learning really mean?

Looking at the problem, we could teach the values for the weights and thresholds by showing the system the proper or correct answers that we want to generate and when it will be necessary to do so. Although this sounds complex, and to some extent, it is. However, this is also something that is routine when we design and program an embedded system to execute a certain task, following certain specified constraints, on command or in response to certain events.

Let's first look at the input and output sides of the system:

The input signals will be those comprising the vector: $x = (i_0, i_1, \ldots i_{n-1})$.
The output signal will be: $y = 0$ or 1.

We also know the following:

- Given the input vector, the activation function will produce the output signal y.
- We are told the expected correct output: O which may be 0 or 1.
- If the activation function produces the correct answer(s), we do nothing.
- If the correct output is $O = 1$ and the activation function produces $y = 0$, then the computed value for y is too small. Thus, the weights, W_i, for the active input lines (non 0) need to be increased and the threshold needs to be reduced. The combination will increase the output.
- If the correct output is $O = 0$ and the activation function produces $y = 1$, then the computed value for y is too large. Thus, the weights, W_i, need to be reduced, and the threshold needs to be increased. The combination will decrease the output.
- If the correct output is $O = 0$ and the activation function produces $y = 0$ or if the correct output is $O = 1$ and the activation function produces $y = 1$, then we have success, and there is no need to change anything.

Working from our knowledge and understanding of the system, we can formulate a high-level learning algorithm that can run continuously while the system is operational. Observe that both the values for the weights and the threshold are to be learned.

- C is the learning rate or step size. It specifies the amount that the weights or the threshold are updated during training (typically, 0.0–1.0).
- Given the input vector, $(i_0, i_1, \ldots, i_{n-1})$, the activation function will produce the output signal y.
- The correct output $= O$ is specified.
- For all values of i, the value of a weight, w_i is defined as: $w_i + C\,(O - y)\, i_i$.
- The threshold Θ is defined as: $\Theta - C\,(O - y)$.

Example 10.4 Simple Learning Example 1
Let's build the basic perceptrons that implement the functions of the three input OR and AND logic gates. As noted earlier, such devices are commonly found in sophisticated embedded systems.

Logical OR
System specifications:

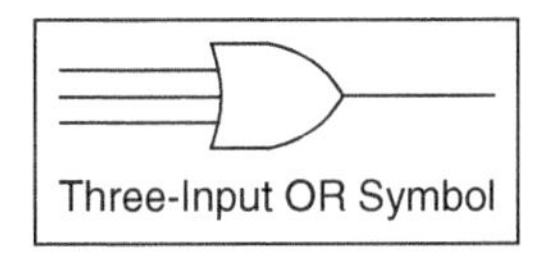

- Perceptron input vector, (i_0, i_1, i_2)
- Weights (w_0, w_1, w_2), (1, 0, 0) (0, 1, 0) (0, 0, 1),
  (1, 1, 0) (1, 0, 1) (0, 1, 1),
  (1, 1, 1)
- Threshold, $\Theta = 1$
- Correct output $O = 1$
- Summation $Z = \Sigma\,(w_i.i_i)$
- Perceptron output y

If any 1 or more of the three inputs has a value of 1, Z will have a value => 1, equaling or exceeding the threshold, and thus the output y will fire. If all of the inputs are 0, Z will have a value = 0, and y will not fire. Changing a weight will not help.

Logical AND

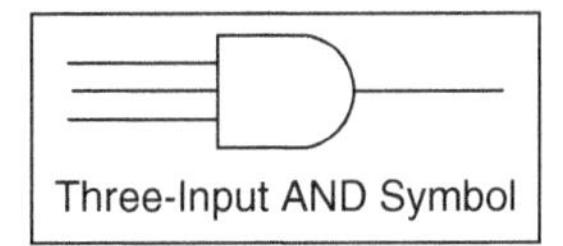

- Perceptron input vector, (i_0, i_1, i_2)
- Weights (w_0, w_1, w_2), (1, 1, 1)
- Threshold, $\Theta = 3$
- Correct output $O = 1$
- Summation $Z = \Sigma\,(w_i.i_i)$
- Perceptron output y

If the three inputs all have a value of 1, Z will have a value of 3, equaling the threshold, and the output y will fire. If any (or all) of the inputs = 0, then Z will have a value <3, and thus, y will not fire. Changing a weight may get y to fire; however, the logic would be invalid.

Example 10.5 Simple Learning Example 2
Let's now start to explore, examine, and begin to understand how such a system operates and can learn and self-correct. We first encountered similar systems in Chapter 7, Figures 7.17 and 7.18. We'll now start by looking at the behavior of a simple similar such system in Figure 10.17. Such a system, which can be either analog or digital, is called a *closed-loop feedback control system*. The high-level behavior of the system is our key focus, not the specific detailed design.

The goal with such a design is to accurately and tightly control the behavior (or output) of a system to a specific and potentially precise value. The system output is continually monitored, sampled, and compared against a specified input value and routinely adjusted to ensure the desired behavior. On the input side of the diagram, we have the relation:

✓ Input – Feedback Signal = Error

We want Error to be (0).

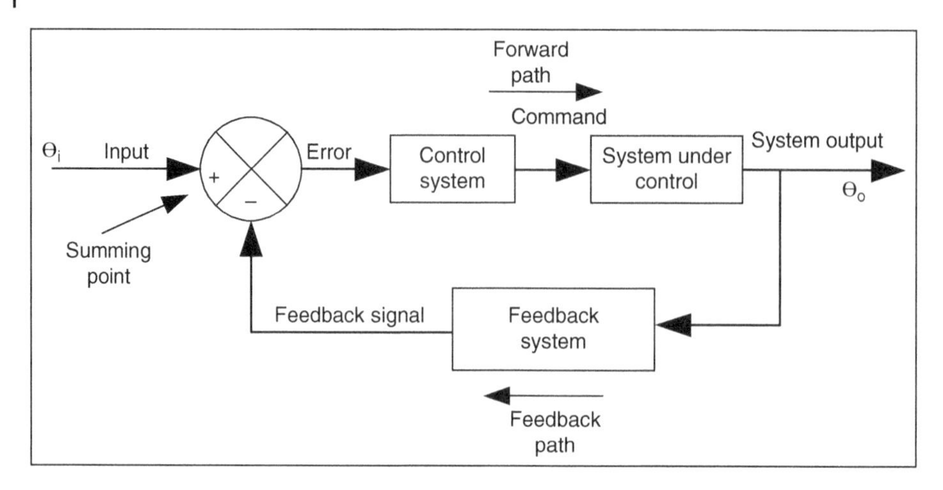

Figure 10.17 Closed-loop feedback control system.

To achieve the goal, the intended operation of the *Control System* commences when the *Input signal* is applied. The *Feedback Signal* (FS) and the *Input signal* are applied to the *Summing Point* to produce an *Error signal*. Error becomes:

- Error = Input – (FS × Θo)

Initially, the Feedback Signal is (0). As a result, when summed with the Input signal, the Error signal equals the Input signal.

The Error signal is an input to the Control System block, which produces the commands that are applied to the System Under Control to produce the desired System Output signal.

The Output signal Θo is sampled and the sampled value then propagates back through the Feedback System (FS), (possibly modified) to the summing point to repeat the whole process. If the Input and Feedback signals are equal, the Error is "0" indicating that the proper output behavior has been achieved.

10.6 Learning with Perceptrons – Second Step

Using the behavior of the feedback control system as a model, let's now move back to the basic perceptron. As we saw in Figure 10.15, the general architecture of the perceptron device is a sequence of functional layers, analogous to what we saw with the feedback control system.

Also, as we learned earlier, there are two types of perceptrons: single-layer and multi-layer. We will work with the Single layer system with a binary output. Repeating the fundamental architecture will be a contiguous series of functional blocks comparable to our control system.

10.6.1 Path of the Perceptron Inputs

Following Figure 10.15, the first step into the perceptron is the path through which signals from other perceptrons or other sources enter. That path is therefore naturally termed the *input subsystem*. The input subsystem on the device can support multiple input nodes.

However, typically, in practice and based on the application, that number may be increased or reduced to two to four. The input set is expressed as the following vector:

$$\text{input } x = (i_0, i_1, i_2, \ldots i_{n-1}). \text{ or}$$
$$= (i_1, i_2, i_3, \ldots i_n)$$

Typically, the inputs originate as outputs from other external devices. The input i_0, however, may be reserved as an internal bias input local to the device.

When an input signal i_i enters the system, it encounters the *weight layer* where a weight w_i is then associated with it as Figure 10.17 illustrates and as we saw with the threshold logic devices.

Let the inputs $x = (i_1, i_2, i_3, i_4)$ have the specified values: (0.2, 3.1, 1.4, 4). The weights will also have specific values as illustrated in the example in Figure 10.18.

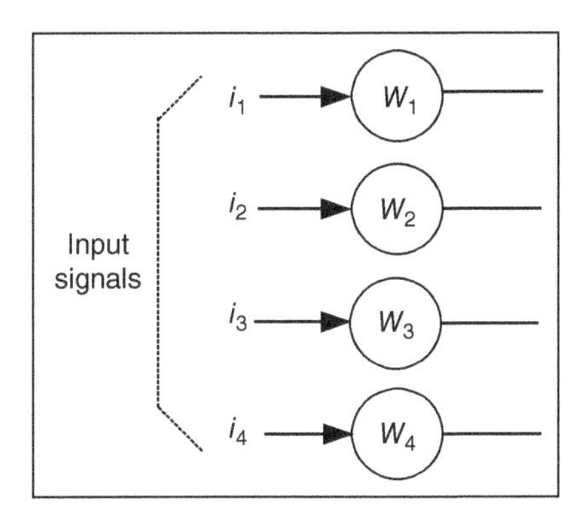

Figure 10.18 Perceptron weight layer.

At the output of the weight layer, we now have: $x = (0.2{\cdot}W_1, 3.1{\cdot}W_2, 1.4{\cdot}W_3, 4{\cdot}W_4)$.

The weighted signals now arrive at the *summing junction* where the sum of the weighted inputs is computed as illustrated in Figure 10.19.

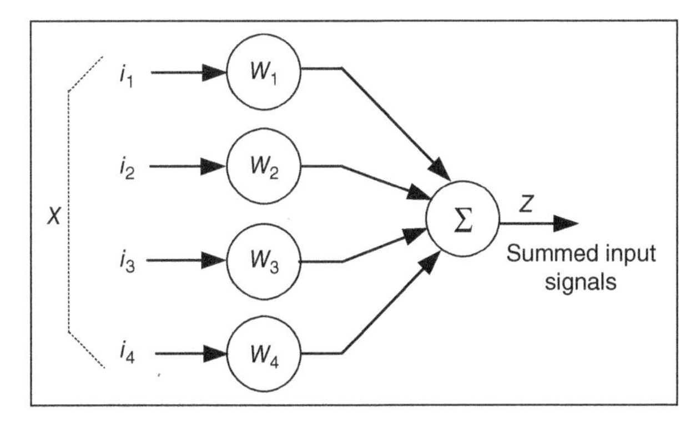

Figure 10.19 Summed input signals.

The output of the *summing junction* is now given in Eq. (10.10):

$$\Sigma(w \cdot i_i) = Z = (0.2 \cdot W_1 + 3.1 \cdot W_2 + 1.4 \cdot W_3 + 4 \cdot W_4) \tag{10.10}$$

The sum is over the range $i = (1...4)$, and the weights will have been assigned values.

From the *summing junction*, the resulting summation signal (z) of the weighted signals goes to the *activation function* layer. At this stage, the signal (z) is compared against the threshold Θ as illustrated in Figure 10.20 and Eq. (10.11).

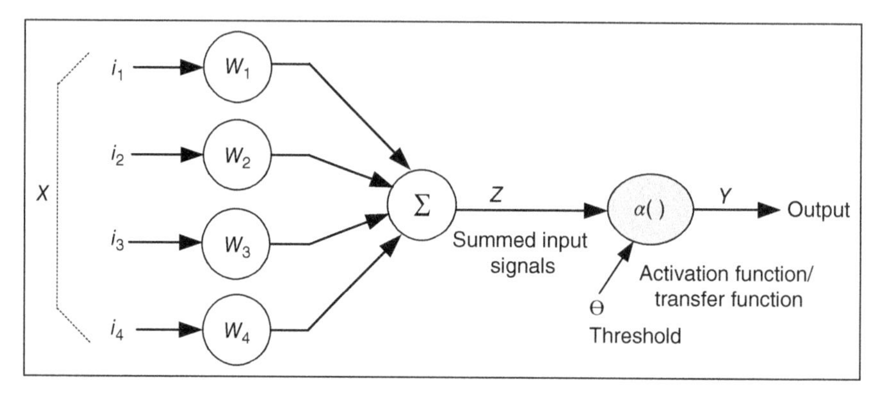

Figure 10.20 Activation function.

The weighted sum of the inputs now arrives at the activation function where the output is determined.

$$\begin{aligned}
&\text{if } \Sigma(w_i \cdot i_i) >= \Theta \text{ then } y = 1 \text{ the output node fired} \\
&Z >= \Theta \\
&\text{else } \Sigma(w_i \cdot i_i) < \Theta \text{ then } y = 0 \text{ the output node did not fire} \\
&Z < \Theta
\end{aligned} \tag{10.11}$$

At this stage, we have a simple function that takes a set of real inputs from a set of outside devices, assigns a weight to each, processes them, and ultimately produces a binary output that potentially becomes an input to a number of successive devices.

Reflecting back to the design and function of the closed-loop feedback system, we may need to have similar control over the magnitude and polarity of the perceptron's weights. Consider the following potential concerns:

10.6.1.1 Implementation/Execution Concerns

- A large input that has been assigned a negative weight could potentially result in a reduced sum that would cause the node not to fire.
- Some inputs may be positive and others negative of equal value thereby canceling each other out.
- The need to alter a weight based upon an incorrect output value.
- The need to remove an input from the sum: $W_i = 0$.

The need to accommodate such issues indicates a potential need for feedback support. A version of such support is illustrated in Figure 10.21.

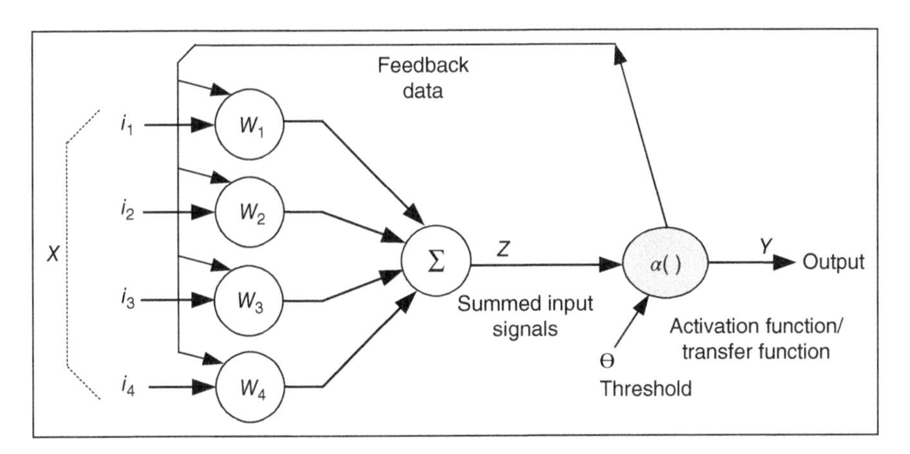

Figure 10.21 Feedback support.

The underlying implementation, however, depends upon the requirements of a specific design. Here, the feedback data source originates at the activation function and terminates at each of the weight blocks. Logic within the blocks responds to and incorporates the command; it learns. Alternatively, based on the design, the feedback could terminate at and be incorporated at the summation junction.

10.7 Testing of the Perceptron

In the two appendices, we have outlined and discussed the need to create *System Requirements* and *System Design Specifications* as a prelude to beginning a formal design process. The same requirements hold true when designing networks comprising perceptrons. Because of the devices with which we are working, we will have a number of additional requirements that we will outline here.

1) At the start of any design, the initial fundamental requirements are to fully understand the problem and accurately specify the necessary inputs and outputs.
2) Keep things simple. Start with the simplest design that should be able to solve a basic version or sub-portion of the problem.
3) Determine what needs to be tested, the testing methodology, the testing signals and their characteristics, and the expected results.
4) Determine the necessary and appropriate connection weights and required neuron thresholds to ensure the proper responses for any training data.
5) Verify and confirm the fundamental system performance with the original test data and then with additional test data for any additional performance deemed necessary.
6) Identify and correct any incorrect or inappropriate operations or errors and retest.

7) Repeat steps 4 and 5 until compliance with the original specifications is confirmed.
8) Perceptrons have the ability to learn. Confirm that the learning ability is compliant with the original specifications, and repeat step 6 as appropriate and necessary.
9) On systems like the perceptron, working through what is called a FMECA analysis is strongly recommended. The acronym stands for *Failure Modes Effects* and *Criticality Analysis*. The analysis examines potential failures, their impact, causes of failure, and the severity or criticality of failure.
10) If the design is changed or modified for any reason, repeat the test suite, and confirm that everything is still operating properly.

10.8 Summary

We opened the chapter with the question: *How do we transform human thoughts and concepts into functioning physical implementations*? This is, indeed, a challenging question. Our response began with the introduction of the biological entity known as a *biological neuron* or collection of neurons. We learned that with such a collection we can form what is called a *neural network*. We reflected that such a configuration sounded very much like the classic Central Processing Unit (CPU) and examined some of the typical CPU features.

We then introduced a high-level basic schematic of the biological neuron and learned that the human brain contains billions of neurons that are interconnected nerve cells involved in the processing, transmission, and reception of chemical and electrical signals from other neurons. We then dissected a basic biological neuron and learned about the fundamental components that comprise such a device.

From the biological neuron, we moved to the fundamental artificial version based on the work of Walter Pitts and Warren McCulloch. With a little background, we then moved to the next step and what is called the perceptron, which was based upon the work of Donald Hebb and Frank Rosenblatt. We learned about and explored the basic version and how it worked. Starting with the basic version of the device, we learned how to use the device to implement four classic logic devices: AND, OR, NOT, and NOR gates. We also learned that weights could be added to the device inputs to alter or eliminate the input signals.

From the ground floor, we moved to the multilayer version and what are called a bias, a threshold, and an activation function (transfer function). We then introduced and examined a variety of transfer functions and their graphical representations.

From the basics, we moved to an important and essential feature of perceptrons: learning. The topic opened with a brief review of the book's introduction and discussion of human learning, then moved to the presentation and analysis of the critical features in a single neuron system. Those features were presented in a basic high-level perceptron schematic accompanied by an explanation of the key associated terminology.

We then introduced and worked through what is called the perceptron learning rule. We opened by summarizing what we knew about the available features in the system. Before proceeding with perceptron learning, we digressed to examine a similar system called a *closed-loop feedback control system* to see how that one operated.

With what we knew and what we learned from the feedback control system, we walked through each of the major functional blocks from inputs to output in the basic perceptron and also pointed out potential implementation and operational concerns and possible solutions of which to be aware.

We concluded with a short overview of a prelude to beginning a formal design process contained in the book's two accompanying appendices. We also incorporated a number of additional important requirements in the test area.

Review Questions

10.1 Identify and describe how we transform human thoughts into physical implementations.

10.2 What is a biological neuron?

10.3 What is an artificial neuron?

10.4 Identify and describe the primary aspects and components of a biological neuron.

10.5 Identify and describe the critical aspects and components of an artificial neuron.

10.6 How many inputs does a perceptron have? How many outputs?

10.7 What types of logic gates can you implement with an artificial neuron?

10.8 What is the purpose of adding weights on neuron inputs?

10.9 What is a perceptron and for what might it be used?

10.10 Identify and explain the major components of a perceptron.

10.11 What is a threshold function? How many versions did we introduce?

10.12 What are the differences among the various threshold functions, and why are there different versions?

10.13 What is the difference between single-layer and multilayer perceptrons, and what is the purpose for different layers?

10.14 What is the purpose of the path delay in the *Recurrent* perceptron?

10.15 Assume a perceptron output that is plotted as a straight line at a 45° angle through the center of a standard X–Y graph. Now introduce and plot the perceptron output with a positive, then a negative-incorporated threshold. What will the resulting plots look like?

10.16 How did Rosenblatt alter the basic McCulloch–Pitts neuron and neural network?

10.17 What is a bias, and what is its purpose?

10.18 When training a perceptron and the correct output is $O = 1$ yet, the output is $y = 0$, why don't we increase the weights for the $W_\mathrm{i} = 0$?

10.19 What is a perceptron's learning rate?

A

Requirements and Design Specification

THINGS TO LOOK FOR…

- Items to consider when formulating requirements documentation.
- Factors to consider when formulating design documentation.
- The traditional product development life cycle.
- The need to understand the environment and the system being designed.
- The difference between requirements specification and design specification.
- Motivation for and timing of static and dynamic analysis of design.
- Requirements traceability.

A.1 Introduction

As we study and practice the design and development of any serious large-scale systems, we must consider both the *system* to be designed and the *environment* in which it must operate. The abstract view given in Figure A.1 captures this. We will refine the model as our discussion evolves and we iteratively develop ideas.

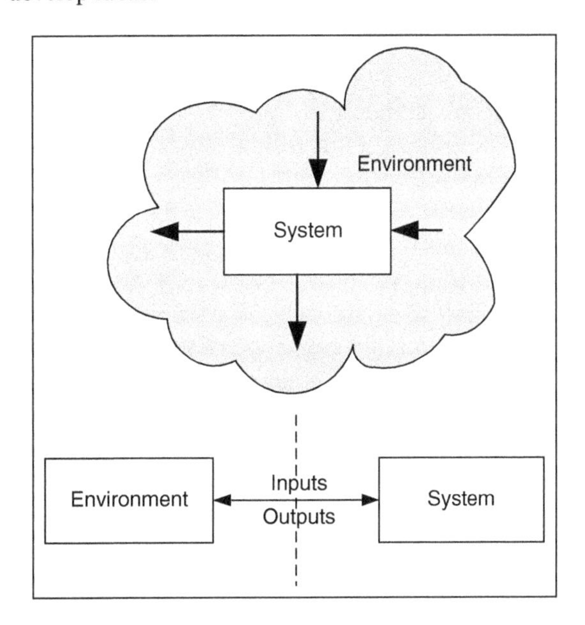

Figure A.1 The system and its environment.

Introduction to Fuzzy Logic, First Edition. James K. Peckol.

As our design progresses, we decompose the top-level system block into modules and subsystems. Some of those modules will be hardware, some will be software, and some may be combinations of both.

Components that must be either hardware or software are generally clear. We can say: this part must be hardware or this part must be software. For example, the power supply, display, and communication port are necessarily hardware; the operating system and associated communication drivers are necessarily software.

A major task, once we start to move inside the system, will be that of decomposing and refining the design from a nebulous entity that someone needs into a product which meets that need. We will first decompose (organize) the collection of the customer's wishes into functional blocks that are then partitioned and mapped into an architecture. That architecture provides the aggregate of hardware and software modules that will make up the ultimate system. The final step in the design cycle is that of bringing the design together into a prototype, testing, and finally into production.

Because there is not one right answer, the problem represents a challenge and an opportunity to be creative. A colleague who worked on numerous designs of a particular piece of measurement technology once said, "although each design performs exactly the same function, each also represents an opportunity to explore a new approach that is better than the old." That colleague built a career around doing what everyone else said could not be done, including some of the top names in the industry. One of the best ways to learn how to do something is simply to do it; so,

Let us get started.

Stating the Problem

As a senior development engineer at *Your Time Is Our Frequency, Ltd.com*, you've just finished one project and are now getting ready to head off to the next. As part of the early planning of that project, you and one of the marketing folks are traveling around the country, talking with people from a number of different engineering firms. You are trying to determine what features your customers would like to see in the next-generation product.

You've been on the road with this guy for a couple of weeks now and are anxious to get home. All the cities are beginning to look exactly alike. Tuesday, this must be Cleveland ... hmmm, looks just like the last three cities. Oh, well. This is the last customer of this trip. This morning, you're talking with *High Flying Avionics, Inc.* They're interested in a new counter that can be used on several of their avionics production lines.

Following several hours of discussion with one of the manufacturing managers, you identify most of their requirements. Your discussion with them follows.

Business is a little slow right now and money is tight, so we don't have a large budget to purchase a lot of different new instruments. In fact, ideally, we'd like to be able to use the same instrument on several of our lines.

Today, we have our technicians running most of the tests manually, but, in future, we'd like to be able to automate as many of these tests as we can. As we upgrade our systems, we'd like to be able to operate several of these counters remotely from a single PC. Here are some of the other things that we'd like to be able to do.

Example A.1
Designing a Counter
As part of our ongoing efforts to improve production and flow through our lines, we monitor the rate at which units arrive into each of the major assembly areas. To do that, we need to be able to track how many of our navigation radios come down a production line each hour. Because we support small-quantity builds of different kinds of radios, the rate at which the units come past the monitoring points is not constant. As each radio arrives at an entry point, it breaks an IR beam. On most of the lines, breaking the beam generates a 1-μs-wide, negative-going 5.0 V pulse. However, we do have several older lines that we must still support. On these, the pulse is positive going.

On several of the newer lines, we have to measure frequency up to 150 000 MHz. We also have several tests for which we must measure frequencies in the range of 50 KHz ± 0.001 KHz and 100 Hz with 0.001 Hz resolution.

On another line, we have several instruments with output signals that have a duration of up to 1.0000 ± 0.0001 ms and others that have a duration of up to 9.999–10.000 ms and up to 1.000 ± 0.001 s. These signals are not periodic. Finally, we have several periodic signals on those same units that we must be able to measure with the same accuracy and resolution.

A.2 Identifying the Requirements

The development of a well-conceived and well-designed system must begin with a requirements definition. Such a need holds, independent of the life-cycle model, crisp or fuzzy that one chooses to work with.

The goal of the *requirements identification* process is to capture a formal description of the complete system from the customer's point of view and then to document these needs as written definitions and descriptions. Such documentation forms the subsequent basis for the formal *design specification.*

Very often, we use the natural language of the customer and of the application context. We do so because such a formal expression of the requirements forces the early discussion and resolution of many complex problems, involving a variety of people with expertise in many different areas, particularly those who are knowledgeable in the application domain. We express the role that the requirements definition plays between the customer and those who execute the design with the accompanying simple graphic in Figure A.2.

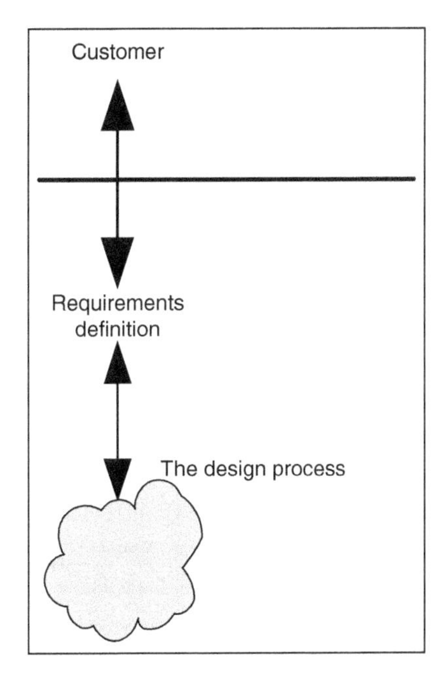

Figure A.2 The interface between the customer and the design process.

The requirements definition provides the interface between the customer and the engineering process. It is the first step in transforming the customer's wishes into the final product. One can see, then, that the requirements definition is a description of something that is wanted or needed. It identifies and captures a set of required capabilities or operations. As one begins to identify all the requirements, as noted earlier, it is important to consider *both* the system to be designed *and* the environment in which it is to operate.

At this early stage in the product life cycle, the goal is to capture and express purely external views of the environment, the system, and their interaction. With respect to the system, one refers to such a view as its public interface. One tries to identify *what* needs to be done (and *how well* it needs to be done) starting with the user's needs and requirements.

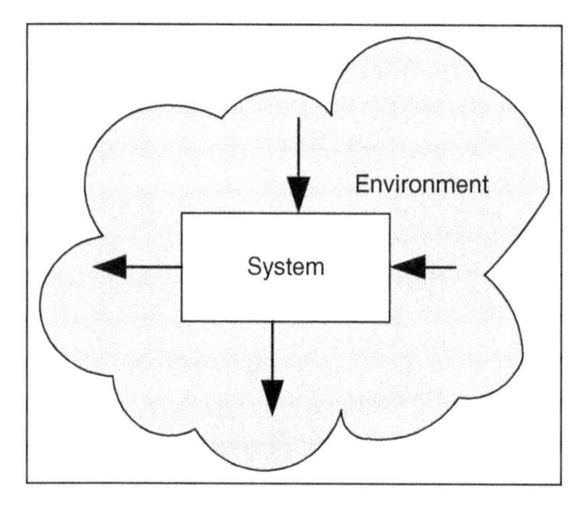

Figure A.3 The system and its environment – Step 0.

The first abstract model of the environment and of the system given earlier is repeated in Figure A.3 and captures this. From the diagram, it is evident that the environment surrounds the system. The inputs to and outputs from the system can come from or go to anywhere in the environment. As one begins, one should make no assumptions about the extent of either.

The first step is to abstract and consolidate that view so that both appear as the black boxes seen in Figure A.4. The initial focus must be on the world or environment (the application context) in which the system is to operate. Next, one follows with an increasingly detailed description of the role played by the system in that environment, and at each step one adds to and refines the requirements.

From the perspective of the environment, one can see that the requirements definition must include a specification for the containing environment, a description/definition of the inputs and outputs to and from that environment, a description of necessary behavior of the system, and a description of how the system is to be used.

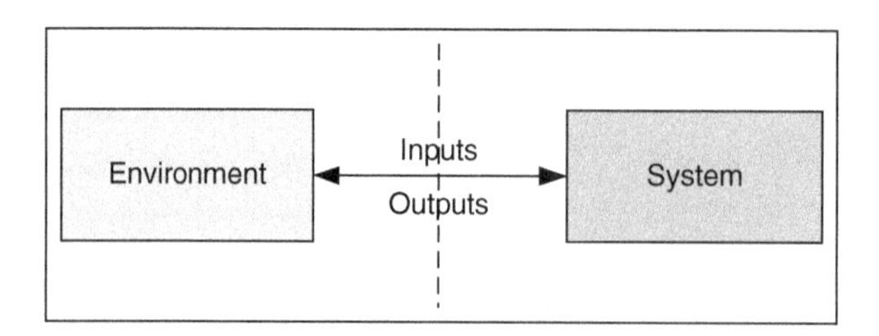

Figure A.4 The system and its environment – Step 1.

From the system's point of view, one starts at a high level of abstraction with an *outside view*. One develops the definition(s) that are appropriate for that level. As was done when specifying the environment, through progressive refinement, one moves to lower levels of abstraction and a more detailed understanding and definition.

At this stage in the development life cycle, as the definition of the requirements solidifies and is ultimately formalized into a specification, one should be unencumbered by plans for implementation. The focus should be on the high-level behavior of the system. The complete, accurate, and internally consistent specification must be available before one can start formal design. Ideally, it should be executable and thereby able to work in conjunction with a modeling tool suite. Such an executable specification ultimately serves as the basis for validation of the system.

Although an executable specification is a laudable goal, achieving that goal can become difficult when one must include support for nonfunctional constraints, integrate legacy components into an abstract model, and potentially combine different domain-specific languages and semantics.

A.3 Formulating the Requirements Specification

The objective of specification process is to capture the description of both the complete system and the environment. Such a description should be structured, understandable, and verifiable. Our primary focus must be on the world or environment in which system is to operate. We follow with an increasingly detailed description of the role played by the system in the application. At each step, we add to the specification.

From Figures A.3 and A.4, and as stated earlier, our design must include a model of the containing environment, a description / definition of the inputs and outputs, a description of necessary behavior, and a description of how system is to be used. We start at a high level of abstraction with an outside view of the system. We then develop the model(s) appropriate for that level. Through progressive refinement, we move to lower levels of abstraction, to a more detailed model.

Let's now examine some of the things that one should think about when starting to identify and capture the requirements and when trying to define them in a formal specification. The form, extent, and formality of such a specification depend on the project on which one is working, the target audience, and the company for which one is working. Remember, too, that it is a product that is being delivered, not a pile of paper. As a rule of thumb, the specification should be the absolute minimum necessary to capture and clearly identify all of the necessary requirements.

In capturing *requirements*, one strives to be very specific about the details from the user's point of view. Bear in mind that one is identifying and formalizing the *requirements*. One still cannot begin to design until the *specification* has been completed and the customer has agreed to it. Remember, too, that one should not be discussing microprocessors, memory, peripheral chips, or software modules at this point in the development process.

As one begins the designs, one usually has some general ideas, casual discussion, and thoughts but nothing firm. One can use these as a guide in directing the steps, but one cannot design from them. It is important to be careful, however, not to rely too heavily on preconceived ideas; one should always be open to alternative approaches. Starting to code or draw logic diagrams at this point is inviting major problems as the project proceeds. In all likelihood, the project will fail.

For the environment component of the specification, one must identify and establish a detailed picture that includes all inputs, outputs, and characterization of the functional behavior for each of the relevant entities that make up the target environment. We must know and understand how the environment is interacting with and affecting our system as well as the effect(s) on the environment of the system's output(s). For the system, we require a description of all inputs and outputs as well as a complete description of the functional and operational behaviors and the technological constraints.

At this juncture we can naturally ask: how can one get such information about (let alone model) the system and the environment without describing or knowing implementation of the system? The internals are inherently unknown at this point. How does one capture the desired behaviors?

A.3.1 The Environment

A reasonable first step begins with defining and describing the environment, the world in which the system must operate. Consider the following small sampling:

- On Earth
- Under an ocean
- At some level above Earth and in the Earth's atmosphere
- In outer space or on another planet
- In or on a human body

The environment is a temporal world; it is a heterogeneous collection of entities of one form or another. It comprises the collection of physical devices to which the system is interconnected as well as any physical world attributes that the system intends to measure or control or that can have an effect on the system. The initial goals in understanding the environment are to identify all relevant entities, then characterize their effects on the system, and vice versa. When the requirements specification has been completed, one should have all the necessary information about such entities, with sufficient details to support moving forward to begin developing the solution.

A.3.1.1 Characterizing External Entities

Each entity that makes up the environment is characterized by a name and an abstracted public interface. That interface consists of the entity's inputs and outputs as well as its functional behavior. The specification of the external environment should contain the following for each entity.

- *Name and Description of the Entity*

  The name should be suggestive of what the entity is or does. The description should present the nature of the entity. Is it data, an event, a state variable, a message? An entity may be something that is to be controlled, for example, the rudder on an aircraft or the clear air turbulence that must be accounted for in such a control system.

- *Responsibilities – Activities*
  What activities or actions is the environment expected to perform? The hydraulic system moving the rudder is part of the environment. Its action or responsibility is to move the rudder in response to the signal coming from the system being designed.
- *Relationships*
  What are the relationships between the entity and its responsibilities or activities? Is that relationship causal or responding? Is it a producer or a consumer?
- *Safety and Reliability*
  Safety and reliability issues must be included early in the specification process. With respect to the environment, at the requirements stage, the focus is primarily on safety. The goal is to identify all safety critical issues and hazards so that they can be addressed in detail in the system design specification. One should also identify any regulatory agencies under those auspices the system will operate.

A.3.2 The System

The focus next shifts to the system's point of view. The same questions posed for the environment are now asked about the system. As with the characterization of the environment, the initial goals are to identify all the aspects of the public interface of the system and then characterize their effects on the environment and vice versa. For the system component, we must have a description of all inputs and outputs, a description of the functional and operational behavior, and an identification of all technological constraints.

Let us use the requirements description and definition as a starting point. Such a definition describes the customer's need; it is something that is desired. We talk with the customer. The analysis of the *system* leads to a synthesis of reality in the form of a model.

For the *system,* we formulate the design from three perspectives:

- *Functional view* – defines the system's internal functions and the relationships between and among those functions.
- *Operational view* – captures and expresses the behavior of those functions.
- *Technological view* – formulates a hardware and software solution to the problem, identifies the components comprising the solution, and implements the functional behavior on the hardware that is consistent with the identified constraints.

At this stage, for our system and based upon these perspectives, we develop three types of specification: *functional specification*, *operational specification* and *technical specification.*

1) **Functional Specification**
   The functional specification enumerates and describes the functions or operations to be performed by the system on or in the environment. These are the external functions in contrast to internal implementations. They describe the behavior of the environment under the operation of the system for these functions. The specification poses and answers the question: "*How does the system affect the environment?*"

   We view the system in terms of the user's needs and requirements. We must observe or hypothesize what the system must do in its environment, which can be done in UML through use cases. We must observe how the system interacts with objects in its environment; such a view is purely external. Knowing the environment means modeling the objects without the system and understanding and describing the relationships between them. The functional specification gives us the bulk of our high-order requirements.

2) **Operational Specification**
   We are now capturing the detailed requirements and constraints. The operational specification focuses on the behavior, performance, information details, methods, and approaches to be used in system. It leads ultimately to the *Design Specification.*
3) **Technological Specification**
   The technological specification includes high-level timing and timing constraints, geographic distribution constraints, characteristics of the interface, and implementation constraints.

Our high-level model now takes on the form of the diagram in Figure A.5.

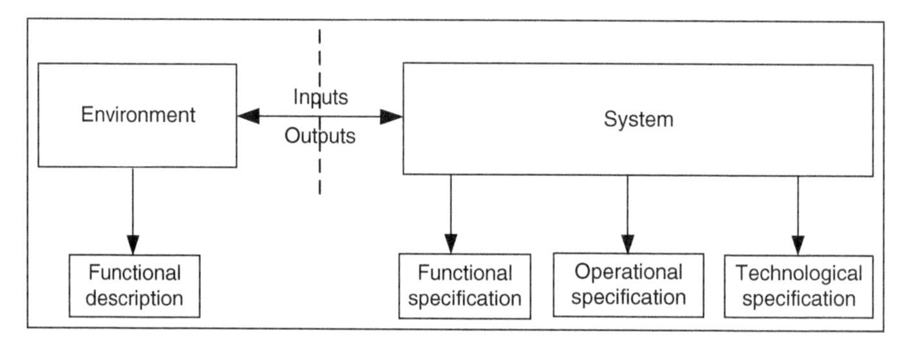

Figure A.5 Refining the system specification

A.3.2.1 Characterizing the System

Characterization of the system begins with identifying inputs and outputs.

A.3.2.1.1 System Inputs and Outputs The system interacts with the real world through the entities described and defined in the environmental characterization. The inputs to the system are the outputs from the environmental entities, and the outputs from the system are the inputs to the environmental entities. One can easily see that the system I/O has already been characterized in environmental entity specification.

For each such I/O variable, the following information is already available:

- The *name* of the signal
- The *use* of the signal as an input or output
- The *nature* of the signal as an event, data, state variable, and so on

Working with the environment specification, one can write the structure, domain of validity, and physical characteristics of each signal. To these, one can add any technical or technological constraints that are identified.

A.3.2.1.2 Functional View As was done with the environment specification, focus now turns to the function that the system is intended to perform. Before it is designed, the system appears as a black box. It can only be viewed from an external point of view. A section on functional behavior is now included in the specification.

The functional description defines the system's external behavior. It characterizes the effects of the system outputs on the environmental entities and the system's intended response to inputs from the environmental entities. It elaborates on how the system is used and to be used by the user. Such a specification is equivalent to developing a model of the system.

The functional description approach is to use the UML tools discussed earlier. One can construct one such view through *use case* and *class diagrams*. Another view can be gained through high-level *state charts* and *activity diagrams*; *data and control flow diagrams* commonly used in structured design methodologies give a third view.

As one formulates these diagrams and the specification, care must be taken to ensure the following:

- The specified (and ultimately modeled) *states* are appropriate to the application.
- The *actions* associated with system I/O that are captured in the specification are necessary to express functional specification and accurately reflect the desired (external) behavior of the system as perceived and intended by the customer.
- The *conditions* or *constraints* on its behavior are only a function of the system inputs, the specified states, the internal events, and the appropriate time demarcation (relative or absolute).
- The possible *exceptions* associated with each of the use cases.

A.3.2.1.3 Operational View With respect to the operational specification, operational means the manner in which a function must operate, what conditions are imposed on the operation, and the range of operation. This specification is dynamic.

Such a specification must consider concrete numbers (precisions and tolerances) quantifying all variables in the functional specification, all operating conditions, and all ordinary and extraordinary operating modes. The known information may contribute to producing and evaluating a design and may include domain-specific knowledge and proprietary or heuristically known to the customer.

A.3.2.1.4 Technological View The technological specification includes all specifications relevant to the hardware and software design.

We can easily identify six areas that should be considered:

1) Geographic constraints
   For distributed applications, we must consider items such as topographies and communication methods. We must also identify restrictions on usage and environmental contamination.
2) Electrical considerations for interface signals
   Such considerations include characteristics and constraints on any electrical I/O signals. These are driven by the external environment and may be beyond the control of the designer.

3) User interface requirements
   A system such as a medical or instrumentation device may have an interface to the external world. We must consider presentation method(s) and protocols.
4) Temporal considerations
   The system may have hard or soft real-time constraints imposed. The constraints may specify delays on signals originating from external entities, responses to system outputs by external entities, and internal system delays.
5) Maintenance, reliability, safety, security
   The system may have requirements for diagnostic tests, remote maintenance, or remote upgrade. We must have concrete numbers for MTTF and MTBF (Meantime To or Before Failure) and environmental and safety issues. We must address performance under partial or full failure. We must identify security vulnerabilities.
6) Electrical considerations
   Electrical characterization of internal signals and behavior include power consumption, supplies, tolerance to degraded power, characterization of signal levels, times, frequencies, etc.

A.3.2.1.5 Safety, Security, and Reliability In formulating the safety, security, and reliability requirements for the system, the focus is on the high-level objectives of each and on the strategy for achieving those goals. Relevant information can be taken from the exceptions component of the UML Use Cases or a preliminary Failure Modes and Effects analysis.

The safety considerations should address the following:
  Safety guidelines, rules, or regulations under the governing agencies identified under the environment portion of the specification.
The security considerations should address the following:
  Potential vulnerabilities
  Potential pre-, during, and post-attack defense strategies
With respect to reliability, one can specify the following aspects:
  The system uptime goals
  Potential risks, failures, and failure

Example A.2
Identifying the Requirements
Starting from the trip report from *High Flying Avionics, Inc.*, which discussed their needs for a new counter, let's put the requirements specification together.

As a first step in the thought process, one extracts and summarizes the essential information from the trip report. By doing so, one can begin to focus on what should be included in the requirements specification. From the discussions with the customer, a high-level sketch of the system and the environment captures the essential parts of the problem. The next step is to begin to formalize the model of the system and the environment as illustrated in Figure A.6. Let's put the Requirements Specification together.

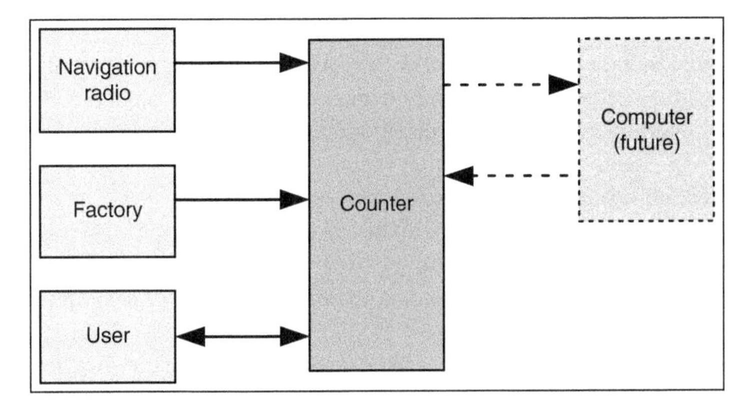

Figure A.6 The system and Its environment – Step 2.

In its initial configuration, the environment contains the following:

- A set of navigation radios that are to be tested
- The user who is doing the testing
- The factory

Signals flow from the navigation radio to the counter, but not the reverse. The factory has inputs to the counter as well; these include the power system and the ambient environment in the factory. The user's interaction is bidirectional. The user must select and configure the measurement to be made and then view the results once the measurement is complete. For the computer, the signal interchange with the counter similarly occurs in both directions.

In the developing model, the factory can be viewed as an aggregation of test lines and the radios to be tested. Later, the remote computer is to be added. The system to be designed, that is, the counter, interacts with all three entities. Such an interaction is reflected in Figure A.7.

Now let's move to the next level of detail.

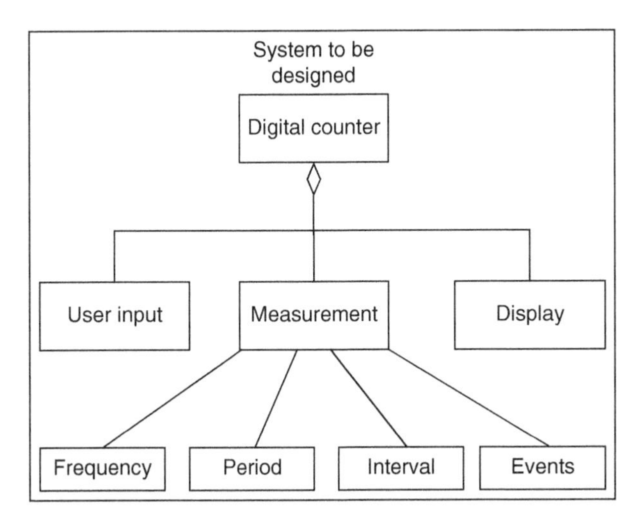

Figure A.7 The system as an aggregation of components

The Environment

- The customer has stated that the counter is to operate in a factory environment on any of several productions lines. Based on such an understanding, one can make certain assumptions about temperature, power, and ambient lighting.
- Time intervals and frequencies on the navigation radios and events from equipment monitoring the production line are to be measured.
- The time intervals may be either periodic or aperiodic but cannot be both.
- The polarity of the event signal to be counted can be either positive or negative going.
- The data display and the annunciation for mode and range are the only outputs expected from the counter.
- The assumption is made that the signals to be measured are independent of one another.
- In future, commands will be sent from a computer to the counter to direct its operation. Data will be sent from the counter to the computer.

The Counter

- The counter must have the ability to measure time intervals and frequencies and to count events.
- The frequencies are fixed but span a range of values.
- The time intervals span a range of values and may be either periodic or aperiodic, but they cannot be both.
- The counter will support the user's ability to manually select mode and measurement range for all input signals.
- The counter will continue to make and display the selected attribute of the signal until power to the system is turned off or until the user makes another selection.
- The counter will measure only one signal at a time.
- An event can be modeled as an aperiodic time signal.
- The design will be sufficiently flexible to allow future inclusion of the ability to send commands from a computer to the counter to direct its operation.
- The response of the counter to remote commands will be the same as its response to front panel selections, with the exception that measured data will be sent from the counter to the computer as well as to the front panel display.
- The next step is to formalize, in a specification, what is known about the system to be designed. The document, the *System Requirements Specification*, opens with a summary of the design.

A.4 The System Design Specification

We have just looked at the design specification for a digital counter. Let's examine such a specification more closely.

The *System Design Specification* formalizes the qualitative view of the system given by the *System Requirements Specification* to present a more quantitative view. Thus, the purpose of the *Design Specification* step, in part, is to capture, express, and formalize the purely external view of the system identified during requirements definition. We have identified *WHAT* needs to be done starting from needs and the user's requirements; we now quantify those *WHATs*. The step requires a solid understanding of the system behavior, the environment, and the system in the environment. At the end of the day, it continues to focus on the *what* and *how well* of the design, rather than the *how*.

The formal design specification must be written in precise language stating specific requirements of the system. It can include the following:

- Tables
- Equations or algorithms
- State or flow diagrams
- Formal design language
- A pseudo language

Unless there are exceptional and limited circumstances, it does not include:

- Schematics
- Code
- Parts lists

Nonfunctional specifications have to be added; we use these to explain constraints such as

- Performance and timing constraints
- Dependability constraints
- Cost, implementation, and manufacturing constraints
  While the *Requirements Specification* provides a view from the outside of the system looking in, the *Design Specification* provides a view from the inside looking out as well. Notice also that the *Design Specification* has two masters:
- It must specify the system's public interface from inside the system.
- It must specify *how well* the requirements defined for and by the public interface are to be met by the internal functions of the system.

The specification is written in the designer's language and from the designer's point of view. It serves as a bridge between the customer and the designer, as we see in Figure A.8.

We have seen that the *Requirements Specification* is written in less formal terms with the intent of capturing the customer's view of the product. The *Design Specification* must formalize those requirements in precise, unambiguous language.

Putting the inevitable changes that occur during the lifetime of any project aside for the moment, we find that the design specification should be sufficiently clear, robust, and complete that a group of engineers could develop the product without ever talking to the author of the specification.

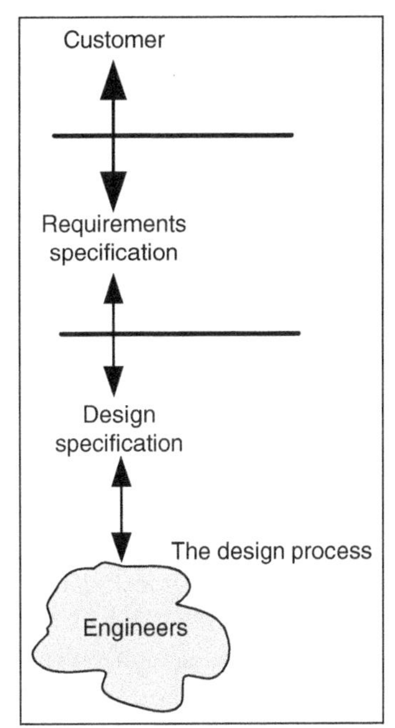

Figure A.8 The customer, the requirements, the design, and the engineer.

Design Note

A good litmus test of the viability of a design specification is the question "If I send this to my colleague (who is working for one of our subcontractors in another country), will he or she understand this?" If the answer is no, the specification should be reexamined.

A.4.1 The System

As part of formalizing and quantifying the system's requirements, one must attach concrete numbers, tolerances, and constraints to all of the system's input and output signals. All timing relationships must be defined. The system's functional and operational behaviors are described in detail.

A.4.2 Quantifying the System

The quantification of the system's characteristics begins with the inputs and outputs, based on the specified requirements. The necessary technical details are added to enable the engineer to accurately and faithfully execute the actual design.

System Inputs and Outputs

For each I/O variable, the following are specified.

- The name of the signal.
- The use of the signal as an input or output.
- The nature of the signal as an event, data, state variable, and so on.
  Starting with the requirements specification, we provide detailed descriptions as necessary and incorporate any additional technical or technological constraints that may be needed.
- The complete specification of the signal, including nominal value, range, level tolerances, timing, and timing tolerances.
- The interrelationships with other signals, including any constraints on those relationships.
  Responsibilities – Activities
- Functional and Operational Specifications
  The functional and operational specifications that will quantify the dynamic behavior of the system are now formulated. The functional requirements specification identifies the major functions that the system must perform from a high-level view. The operational specification endeavors to capture specific details of how those functions behave within the context of the operating environment.
  The manner in which a particular function must operate, the conditions imposed on the operation, and the range of that operation are now captured. The specification must consider concrete numbers – *precisions* and *tolerances*.
  All variables in the functional specification, all operating conditions, and all ordinary and extraordinary operating modes must be quantified. The specification may

include domain-specific knowledge that is proprietary or heuristically known to the customer. Such knowledge can be very important to the design.

In stating the specific design requirements for the system, one can use tables, equations, or algorithms, formal design language, or pseudo code, flow diagrams, or detailed UML diagrams such as state charts, sequence diagrams, and timelines. Schematics, codes, or parts lists are not included, except in limited circumstances.

- Technological (and Other) Specifications

The technological portion includes all detailed and concrete specifications that are relevant to the design of the system hardware and software. Five areas that should be considered can easily be identified.

1) Geographical constraints

   Distributed applications can span a single room, can expand to include a complete factory, or can encompass several countries. Consequently, one must address both the technical items such as interconnection topologies, communication methods, restrictions on usage, and environmental contamination and nontechnical matters such as costs associated with the physical medium and its installation.

2) Characterization of and constraints on interface signals

   The assumption is made that signals between the system and the external world are electrical, optical, or wireless or that they can be converted into or from such a form. The necessary physical characterization of each is obviously going to depend on the type of signal. That is, an electrical signal is specified differently from an optical signal. Since many of the interface signals may be driven by the external environment, potentially they are beyond the designer's control. Therefore, it is important to gain as much information about them as possible.

3) User interface requirements

   If the system interfaces to external world devices such as medical or instrumentation equipment, how information is presented and whether any relevant and associated protocols exist must be considered. There may also be standards that govern how such information must be presented.

   Consider the significant risk that would arise if each avionics vendor presented critical flight information and controls to the aircraft pilot in a different way. The near disaster at Three Mile Island in 1979 arose, in part, because of the confusion caused by too much information.

4) Temporal constraints

   The system may have to perform under hard or soft real-time constraints. Such constraints may specify delays on signals originating from external entities, responses to system outputs by external entities, and/or internal system delays.

5) Electrical infrastructure considerations

   There must be a specification for the electrical characteristics of any electrical infrastructure. Included in this portion of the specification are power consumption, necessary power supplies, tolerances, and capacities of such supplies, tolerance to degraded power, and power management schemes.

- *Safety and Reliability*
  In formulating the design requirements for the safety and reliability of the system, the focus shifts to the detailed objectives of each and to the strategy for achieving those goals.

Safety considerations should address

- Understanding and specifying any environmental and safety issues
  The reliability specification should include
- Requirements for diagnostic tests, remote maintenance, remote upgrade, and their details
- Concrete numbers for MTTF and MTBF of any built-in self-test circuitry
- Concrete numbers for MTTF and MTBF of the system itself
- Consideration of system performance under partial or full failure

Let's now bring everything together.

Example A.3
Quantifying the Specification
We will now continue with the development of the counter. The system *Design Specification* will follow, but extend, what has been captured in the *Requirements Specification*. The focus will now be on providing specific numbers, ranges, and tolerances for signals that are within the system.

Once again, we will put together any thoughts about the environment and the system prior to writing the specification.

Environment
Specifications related to the environment have been discussed earlier. There are no changes here.

Counter

- When specifying measurement and stimulus equipment, the specifications for that equipment are generally 10 times (one order of magnitude) better than those for the signals that must be measured or generated.
- That margin is provided when specifying the range and tolerances on the counter's measurement capabilities.
- Specifications on counting events are based on the granularity of the timing of the interval during which the events are counted.
- The values to be displayed at the measurement boundaries are now defined.
- The next step is to provide any additional detail that may be needed and to fully quantify the counter specifications.

System Design Specification for a Digital Counter

System Description

This specification describes and defines the basic requirements for a digital counter. The counter is to be able to measure frequency, period, time interval, and events. The system supports three measurement ranges for each signal and two for events. The counter is to be manually operated with the ability to support remote operation in future. The counter is to be low-cost and flexible, so that it may be utilized in a variety of applications.

Specification of the External Environment

The counter is to operate in an industrial environment in a commercial-grade temperature and lighting environment.

The unit will support either line power or battery operation.

.***System Input and Output Specification***

System Inputs

The system shall be able to measure the following signals.

Frequency in three ranges

- High range up to 150.000 MHz
- Midrange up to 50.000 KHz
- Low range up to 100.000 Hz

Period in three ranges

- High resolution up to 1.0000 ms
- Mid-resolution up to 10.000 ms
- Low resolution up to 1.000 sec

Time interval in three ranges

- High resolution up to 1.0000 ms
- Mid-resolution up to 10.00 ms
- Low resolution up to 1.000 s
- Voltage range 0.0–4.5 VDC

Events

- Events up to 99 events in 1 minute
- Signal level 0–4.0 V }0.5 V
- Transition time

Voltage sensitivity

- 50 mV RMS to ·} 5.0 V ac signal + dc signal

All signal inputs will be

- Digital data
- Voltage range 0.0–4.5 VDC

User Interface

The user shall be able to select the following using buttons and switches on the front panel of the instrument.

Mode

Frequency, Period, Time Interval, Events

Range

Frequency, Period, Time Interval - —High, Mid, Low

Events - —Fast, Slow

All signal inputs will be

- Digital data
- Voltage range 0.0–4.5 VDC

Trigger Edge

Frequency, Period, and Events

Rising or falling edge

Time Interval

Rising to rising edge

Falling to falling edge

Rising to falling edge

Falling to rising edge

Reset

The reset button will clear the display to all 0"s and reset the internal timing/counting chain.

System Outputs

The system shall measure and display the following signals using a six-digit display.

Frequency in three ranges

- High range up to 200.000 ± 0.001 MHz
- Midrange up to 200.000 ± 0.001 KHz
- Low range up to 200.000 ± 0.001 Hz

Period in three ranges

- High resolution up to 2.000 ± 0.0001 ms
- Mid-resolution up to 20.00 ± 0.01 ms
- Low resolution up to 2.000 ± 0.001 sec

Time interval in three ranges

- High resolution up to 2.0000 ± 0.0001 ms

  Measure: 0

  2.00000 ± 0.00001 ms

  Display: 0- - 2.0000 ± 0.0001 ms
- Mid-resolution up to 20.00 ± 0.01 ms

  Measure: 0

  2.00000 ± 0.0001 ms

  Display: 0–20.0000 ± 0.001 ms
- Low resolution up to 2.000 ± 0.001 s

  Measure: 0

  2.0000 ± 0.0001 s

  Display: 0 – 2.000 ± 0.001 s

(Continued)

Events in two ranges

- Fast up to 200 events in 1 minute
  Measure: 0–200 ± 1 event
  Display: 0–200 ± 1 event
- Slow up to 2000 events in 1 hour
  Measure: 0–2000 ± 1 event
  Display: 0–2000 ± 1 event

User Interface

The user shall be able to select the following using buttons and switches on the front panel of the instrument.

Mode

Frequency, Period, Time Interval, Events

Range Frequency, Period, Time Interval

High, Mid, Low

Events – Fast, Slow

Trigger Edge

Frequency, Period, and Events

Rising or falling edge

Time Interval

Rising to rising edge

Falling to falling edge

Rising to falling edge

Falling to rising edge

Reset

Power ON/OFF

The reset button will clear the display to all 0's and reset the internal timing/counting chain.

The measurement results shall be presented on a six-digit display; leading zeros will be suppressed. The display shall be readable in direct sunlight and from any angle.

The front panel will appear as follows

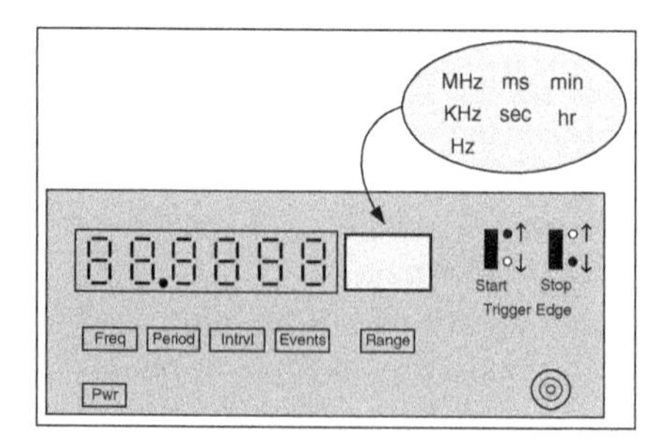

Use Cases

The use cases for the counter are given in the following two diagrams.

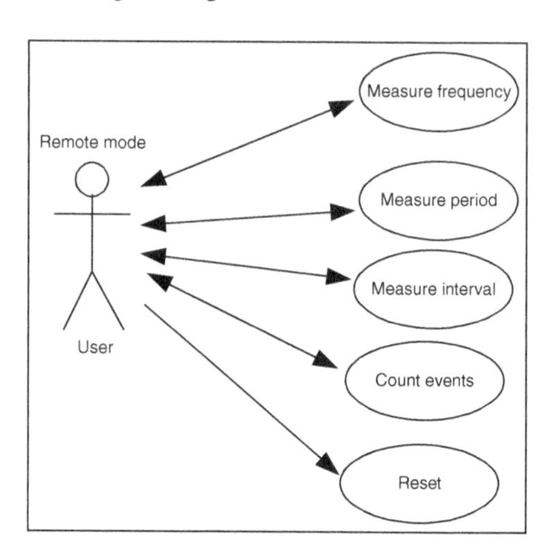

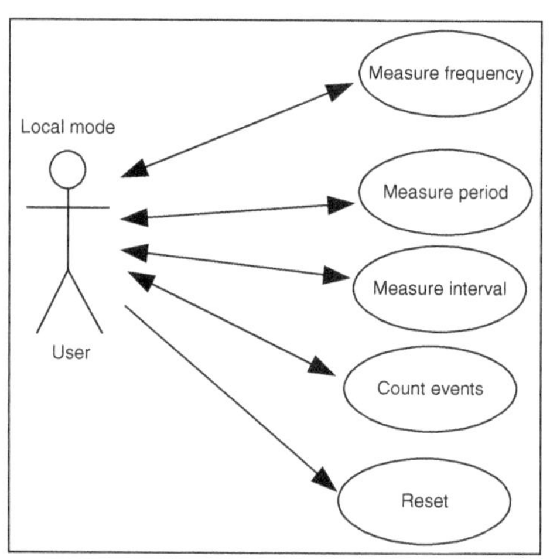

The first indicates manual operation through the front panel, and the second through a remote connection to a computer.

The remote option will not be included in the initial model, but will be incorporated in a later release. The time of that release is to be determined.

Execution of the selected measurement function will not depend on how (local or remote) that function was selected.

At power ON, the default mode is to measure frequency.

All ranges will default to their highest value.

Measure Frequency The counter will continuously measure and display the frequency of the input signal on the currently selected range as long as the *Frequency* mode is selected. The following use cases are defined for the *Frequency* mode.

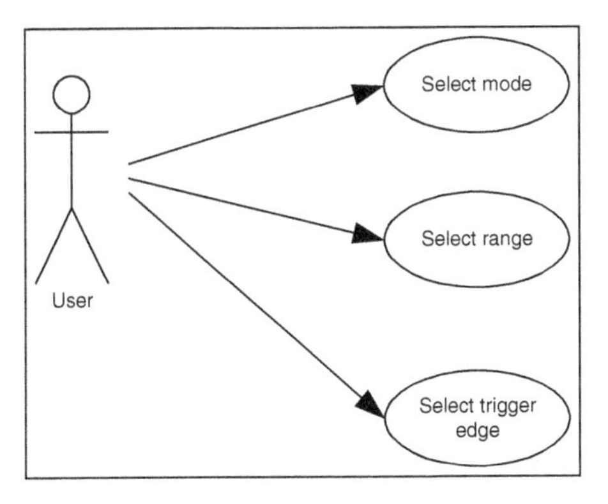

If the frequency of the input signal exceeds the maximum allowable value on the selected range, the display will present the full-scale reading and will flash and will present one of the following values based on the selected range,

- 200.000 MHz
- 200.000 KHz
- 200.000 Hz

If the frequency of the input signal is below the minimum allowable value on the selected range, the display will present a zero reading.

If the input signal returns to a value within the bounds of the range, the value of the frequency will be displayed.

The range may be changed at any time by depressing the *range select* pushbutton.

The user may elect to measure frequency starting on the positive or negative edge of the signal by depressing the *start trigger edge* pushbutton.

Measure Period The counter will continuously measure and display the period of the input signal on the currently selected range as long as the *Period* mode is selected. The following use cases are defined for the *Period* mode.

If the period of the input signal exceeds the maximum allowable value on the selected range, the display will present the full-scale reading and will flash. If the period of the input signal is below the minimum allowable value on the selected range, the display will present a zero reading.

If the input signal returns to a value within the bounds of the range, the value of the period will be displayed.

The range may be changed at any time by depressing the *range select* pushbutton.

The user may elect to measure the period starting on the positive or negative edge of the signal by depressing the *start trigger edge* pushbutton.

Measure Interval The counter will continuously measure and display the duration of the selected portion of the input signal on the currently selected range as long as the *Interval* mode is selected. The following use cases are defined for the *Interval* mode.

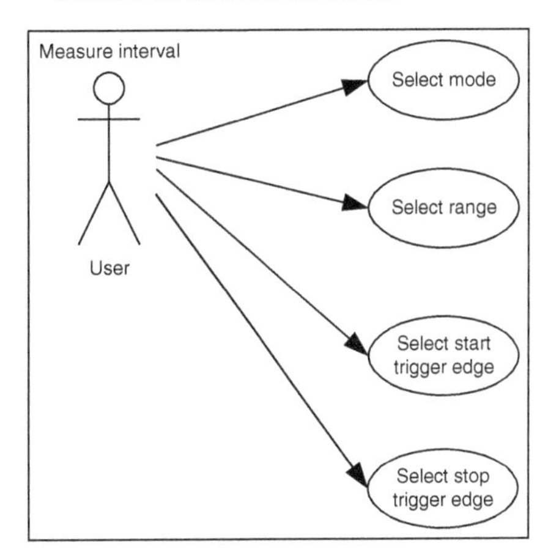

If the duration of the selected portion of the input signal exceeds the maximum allowable value on the selected range, the display will flash and will present one of the following values based on the selected range.

- 2.0000 ms
- 20.000 ms
- 2.000 s

If the duration of the selected portion of the input signal is below the minimum allowable value on the selected range, the display will display zero.

If the input signal returns to a value within the bounds of the range, the value of the duration of the selected portion of the input signal will be displayed.

(Continued)

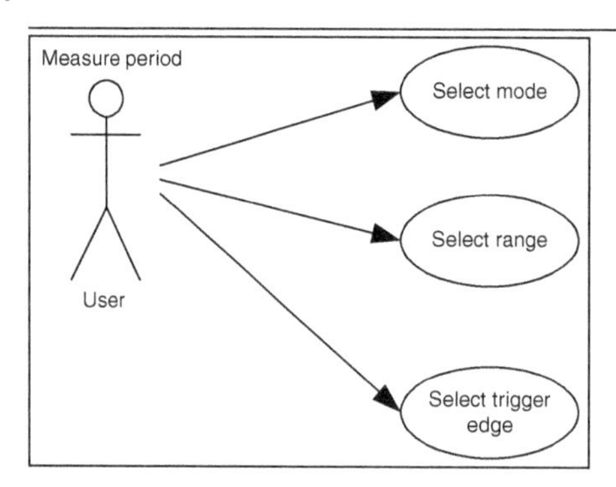

If the period of the input signal exceeds the maximum allowable value on the selected range, the display will flash and will present one of the following values based on the selected range:

- 2.0000 ms
- 20.000 ms
- 2.000 s

The user may elect to terminate the measurement interval on the positive or negative edge of the signal by depressing the *stop trigger edge* pushbutton.

The user may elect to terminate the measurement interval on the positive or negative edge of the signal by depressing the *stop trigger edge* pushbutton.

Note that the signal duration from the positive edge to positive edge or negative edge to negative edge is the same as the period of the signal.

Events The counter will continuously count and display the number of occurrences of the input signal on the currently selected range. The accumulated count will be reset to 0 at the end of the select count duration. The following use cases are defined for the *Events* mode.

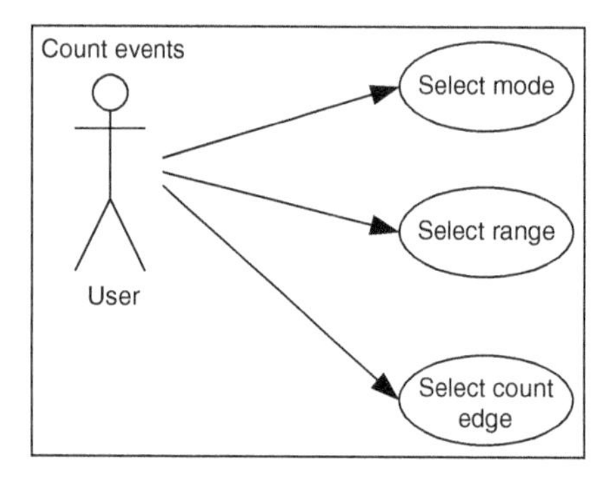

The range may be changed at any time by depressing the *range select* pushbutton.

Note that the signal duration from the positive edge to the positive edge or the negative edge to the negative edge is the same as the period of the signal.

The user may elect to commence measuring the interval on the positive or negative edge of the signal by depressing the *start trigger edge* pushbutton.

System Functional Specification

The system is intended to make four different kinds of digital measurement in the time and frequency domains comprising frequency, period, time interval, and events. The activities associated with the *measure frequency* mode are shown in the following diagram.

The time and frequency measurements will be implemented to provide three user selectable resolution ranges: high frequency range/shorter duration signals, a second for midrange frequency/midrange duration signals, and a third for low frequency/longer duration signals. The events measurement capability will support two selectable counting durations, shorter and longer.

For frequency, period, and events measurements, the user will be able to select either a positive or negative edge trigger. For interval measurements, the user will be able to select the polarity of the start and stop signals independently.

The system will be designed so as not to preclude the incorporation of a remote access option in future.

The system comprises six major blocks as given in the following block diagram.

Operating Specifications

The system shall operate in a standard commercial/industrial environment

Temperature range 0–85 C

Humidity up to 90% RH noncondensing

Power automatic line voltage selection

- 100–120 VAC $\pm$ 10% 50, 60, 400 Hz $\pm$ 10%
- 220–240 VAC $\pm$ 10% 50, 60 Hz $\pm$ 10%

The system shall operate for a minimum of eight hours on a fully charged battery.

The system time base shall meet the following specifications.

Temperature stability 0–50 C

$<6 \times 10^{-6}$

The range may be changed at any time by depressing the *range select* pushbutton.

The user may elect to increment the count on the positive or negative edge of the input signal by depressing the *start trigger edge* pushbutton.

If the number of accrued counts exceeds the maximum allowable value on the selected range, the display will present the full-scale reading and will flash.

If the number of accrued counts exceeds the maximum allowable value on the selected range, the display will flash and will present one of the following values based on the selected range,

- 200 minutes
- 2000 hours

The range may be changed at any time by depressing the *range select* pusbutton. The user may elect to increment the count on the positive or negative edge of the input signal by depressing the *start trigger edge* pushbutton.

Aging rate

90 days $<310^{-8}$

6 months $<610^{-7}$

1 year $<2510^{-6}$

Reliability and Safety Specification

The counter shall comply with the appropriate standards

Safety: UL-3111-1, IEC-1010, CSA 1010.1

EMC: CISPR-11, IEC 801-2, -3, -4, EN50082-1

MTBF: Minimum of 10 000 hours

(*Continued*)

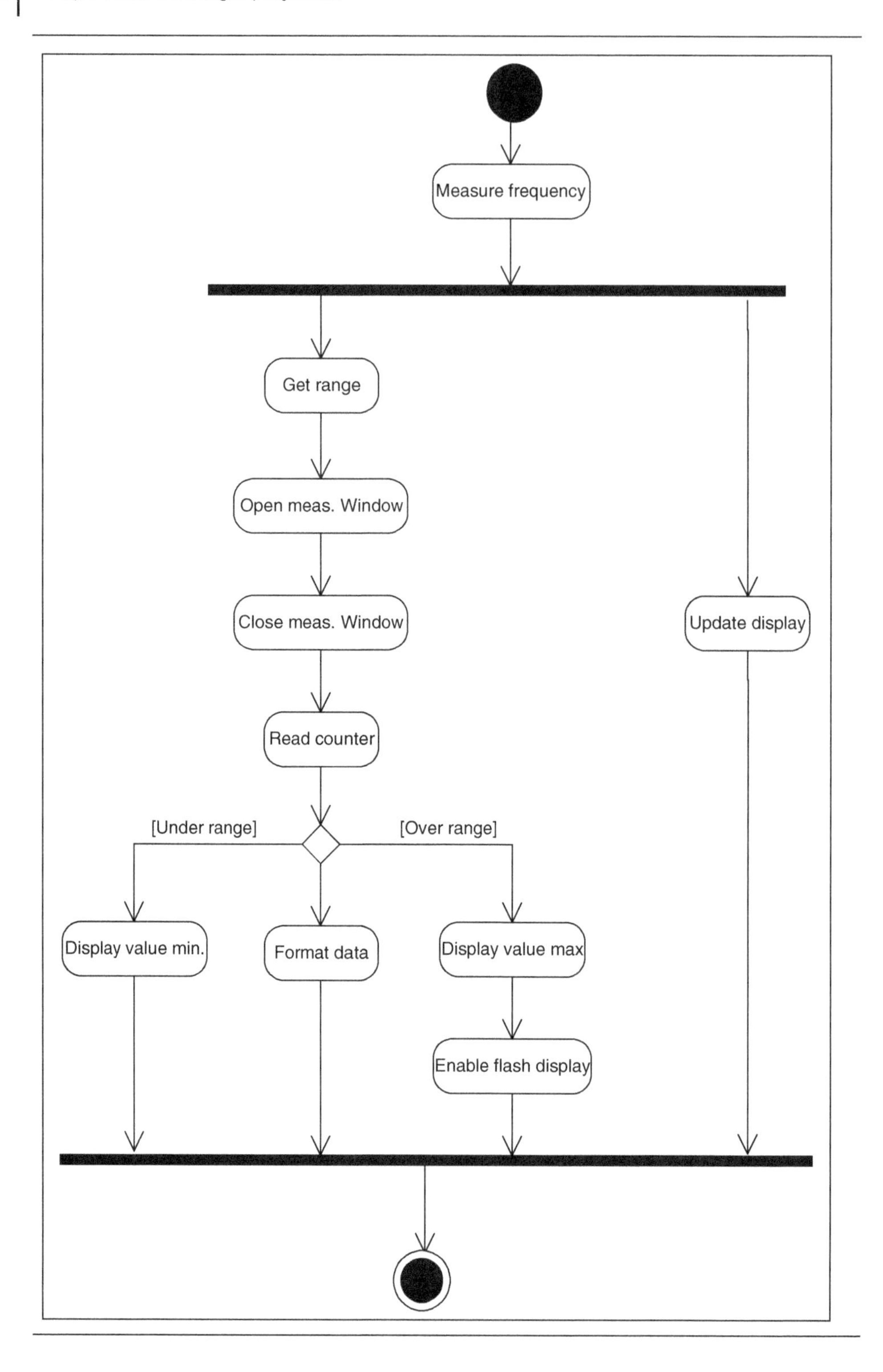
Measure frequency
Get range
Open meas. Window
Close meas. Window
Read counter
[Under range]
[Over range]
Display value min.
Format data
Display value max
Enable flash display
Update display

A.5 System Requirements Versus System Design Specifications

Examining the different steps that have been outlined up to this point, we find a lot of duplication. It would seem that the *System Design Specification* and *System Requirements Specification* are just different names for the same thing. But they are not; requirements and specifications are fundamentally different types of descriptions.

> *Requirements:* Give a description of something wanted or needed. They are a set of needed properties.

Generally, requirements come from the marketing, product planning, or sales department, and they represent the customer's needs. The requirements definition and specification is not concerned with the internal organization of the system. Rather, it is intended to describe *what* a system must do and *how well* it has to do it, not *how* it does it.

The *System Design Specification* is generated by engineering as an answer to and a description of how to implement the requirements. Then the two groups negotiate and iterate until the requirements and specifications are consistent.

> *Specification* is a description of some entity that has or implements those properties.

The system design specification is a means of translating the description of needs into a more formal structure and model.

Nonetheless, every part of the design needs another specification. Specifications can and do exist at various levels as the design is refined and elaborated. Different things must be quantified and at different levels of detail during different phases of the product development. The *System Design Specification* may require that an intersystem communication channel transfer data at the rate of 10 000 bytes per second at a specific bit error rate. The detailed *Hardware* and *Software Specifications* establish the requirements and constraints on their respective components to be able to meet those specifications.

A specification is a precise description of the system that meets stated requirements. Ideally, a specification document should be

- Complete
- Consistent
- Comprehensible
- Traceable to the requirements
- Unambiguous
- Modifiable
- Able to be written

The *Design Specification* should be expressed in as formal a language or notation as possible, yet readable. Ideally, it should also be executable. It should focus precisely on the system itself and should provide a complete description of its externally visible characteristics, that is, its public interface. External visibility clearly separates those aspects that are functionally visible to the environment in which the system operates from those aspects of the system that reflect its internal structure.

While the *Requirements Specification* nonfunctional specifications have to be added, we use these to explain constraints such as

- Performance and timing constraints
- Dependability constraints
- Cost and implementation provide a view from the outside of the system looking in; the *Design Specification* provides a view from the inside looking out as well. Notice also that the *Design Specification* has two masters:
- It must specify the system's public interface from inside the system.
- It must specify *how well* the requirements defined for and by the public interface are to be met by the internal functions of the system.

The specification is written in the designer's language and from the designer's point of view. It serves as a bridge between the customer and the designer,

We have seen that the *Requirements Specification* is written in less formal terms with the intent of capturing the customer's view of the product. The *Design Specification* must formalize those requirements in precise, unambiguous language.

Putting the inevitable changes that occur during the lifetime of any project aside for the moment, we find that the design specification should be sufficiently clear, robust, and complete that a group of engineers could develop the product without ever talking to the author of the specification.

Once again, using the space program as an example of the importance of ensuring that all team members know, understand, and agree on all specifications. The Mars Climate Orbiter was launched in 1999. Four months into the mission, the ground computers were being used to adjust the spacecraft's trajectory. Almost immediately, it was recognized that the something wasn't functioning correctly. The data from the ground computers and the spacecraft didn't agree. It turned out that the spacecraft was using metric units and the ground computer was using imperial units. The spacecraft made it to Mars; however, the unit mismatch led to loss of the craft during a landing attempt.

B

Introduction to UML and Thinking Test

THINGS TO LOOK FOR...

- Unified modeling language overview and diagrams.
- Use case diagrams.
- The major static modeling diagrams in UML and the utility of each.
- UML class diagrams and use cases.
- The need for dynamic modeling.
- Preparing for system test.

B.1 Introduction

Unified modeling language (UML) uses diagrams and models as a first step toward expressing static and dynamic relationships among objects. While an important part of the standard, the authors do not see such diagrams as the main thrust of the approach. Rather, a philosophy of a model-driven architecture (MDA) in which UML is used as a programming language is more common. The high-level goal is to create an environment in which tool vendors can develop models that can work with a wide variety of other MDA tools. On the user side, designers who work with UML range from those who are putting together a "back of the envelop" sketch to those who utilize it as a formal (high-level) design and programming language. UML provides a very good mechanism for quickly exchanging ideas with other designers and for capturing the critical elements of a design.

In addition, the current standard recognizes 13 different classes of drawings. As a design evolves, these different perspectives offer a rich set of tools whereby we can formulate and analyze potential solutions. Such tools enable one to model several different aspects of a design. It's rare that all of the types are used in a single design.

The different diagram types are presented in Table B.1.

As is suggested by their names, diagrams in the first four categories provide the means for a developing a static or structural view. The next four add dynamic analysis; the final five bring the pieces together.

We'll begin with several of the static components and relationships. These will be sufficient to get us started. We will spend only limited time on the diagram types in the last category.

Introduction to Fuzzy Logic, First Edition. James K. Peckol.

Table B.1 Common UML diagrams.

• Class
• Use case
• Component
• Communication
• State chart
• Timing
• Sequence
• Activity
• Object
• Package
• Composite structure
• Interaction
• Deployment

B.2 Use Cases

The first diagram that we'll look at is the use case illustrated in Figure B.1. The use case gives us an outside view of the system. It describes the public interface for the module or system. It answers the questions, "What is the behavior that the user sees?" "What is the behavior the user expects?" The use case repeatedly poses the question, "What?" until the external view of the system has been satisfactorily captured.

The use case diagram is intended to present the main components of the system and how the user interacts with those components. Like many of the diagrams we'll work with, the use case diagram can be hierarchical in nature. From the top level drawing, one can expand each use case into sub use cases as necessary.

The use case diagram comprises three components: the system, the actor(s), and the use case(s). The meaning of system is self-evident; after all, that's what is being designed. It's expressed in the diagram as a box – we'll often leave this off the diagram. The actor(s), drawn as simple stick figures, represent anyone or anything that might be using the system. They are viewed as being out-side of the system. The use cases, represented as a solid oval, identify the various behaviors of the system or ways that it might be used. They encapsulate the events or actions that must occur to implement the intended behavior of the system and are stated or expressed from the point of view of the user. Accompanying each use case is a textual component fully describing it. Use case diagrams can be a very powerful tool during the early stages of a project when one is trying to identify, define, and capture the requirements for the system.

As we construct the diagram, we place the actor that executes the use case on the left-hand side. Supporting actors appear on the right-hand side. Supporting actors are not restricted to human users; an actor can be a computer or other system as well. The set of use cases appears in the center of the drawing with arrows indicating the actors involved in the use case.

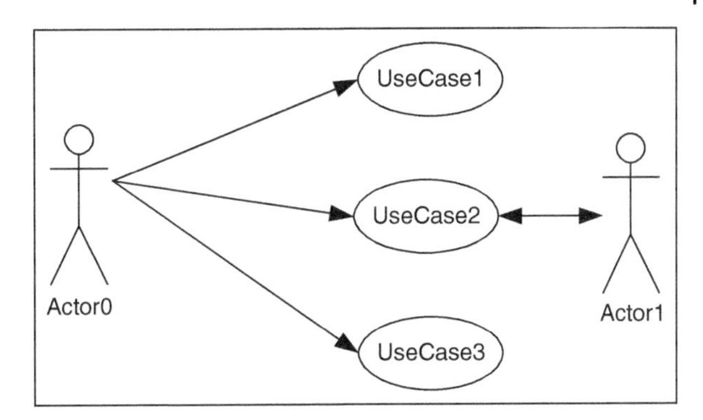

Figure B.1 The UML use case diagram.

We see that the system comprises three use cases. Actor0 is using the system and appears on the left-hand side. Actor1 is supporting UseCase2 and is placed on the right-hand side.

It's important to remember to keep things simple when putting the use case diagram together. If a system being designed is showing 25–50 use cases on the top level drawing, then it's time to rethink the design. In this next example, we are working on a simple data acquisition system.

A basic data acquisition system that has the ability to measure voltage and temperature is to be designed. The use case diagram for the system begins with the user shown as Actor0. After the data has been collected, it is to be analyzed for trends, alarm conditions, or other specific patterns. In addition, because the temperature sensor is a nonlinear device, a linearization operation must be performed.

The data is to be collected at very high speed from a number of measurement points; as a result, hardware co-processing capability is probably going to be necessary. That entity is included as a second actor and labeled as Data Processor. A possible use case diagram for the data acquisition system is given in Figure B.2.

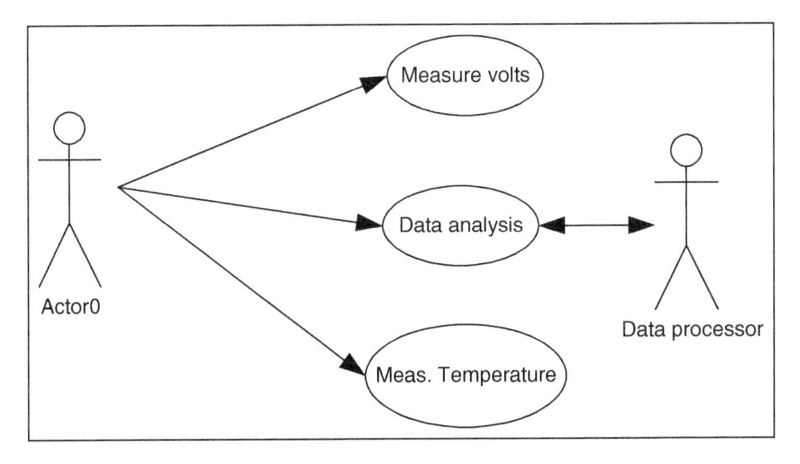

Figure B.2 Use case diagram for a simple data acquisition system.

B.2.1 Writing a Use Case

The use case diagram captures a graphical representation of the public interface to the module or system. Associated with each use case is a textual description of what actions the actor is to perform and how the system is expected to respond. Such a description can be decomposed into two activities: the normal activity of the use case and how exceptional conditions are to be handled.

Let's examine the *measure volts* use case for the data acquisition system. We specify how the user is to select the task, any options associated with the task, and how exceptions are handled as is done in Figure B.3. Do not forget, a use case description is not intended to be War and Peace. Keep things simple.

```
User
  Select measure volts mode
  Select measurement range or auto range
System
  If range specified
    Configure to specified gain
    Make measurement
      If in range – display results
      If exceed range – display largest value for range and flash the display
  If auto range
    Configure to midrange gain
    Make measurement
      If in range – display results
      If above or below range adjust gain to next range and repeat the measurement
      If exceed range – display largest value for range and flash the display
```

Figure B.3 Writing a use case description.

B.3 Class Diagrams

Once we have identified how the user intends or expects to interact with the system, the next step is to begin to identify and to formulate to modules that give rise to that external behavior. That process begins with the class diagram. This diagram gives a description of the objects in a system coupled with the relationships that exist amongst them. Such a description is frequently found among the foundation elements of most modeling tools. The class diagram enables one to specify the public interface to the object; the interface expressed in the use cases. Such a description includes the properties and the operations that instances of the object can perform and identifies any constraints the application imposes on those operations. The public interface should always be one of the earlier views that one takes of any design. We want to see the design from our user's point of view.

The class diagram presents the various kinds of objects in the system and identifies the relationships called associations amongst them. Object Name Objects are expressed as a rectangle, subdivided into three areas as illustrated in the diagram in Figure B.4. The top area gives the name of the class or object. The middle section identifies all of the proper

UML class diagram ties of the object. These will generally be declared inside the module implementation and thereby hidden from the casual user. The third pane identifies the operations that the object is intended to perform. These operations establish the external behavior of the object; they provide the public interface to the object.

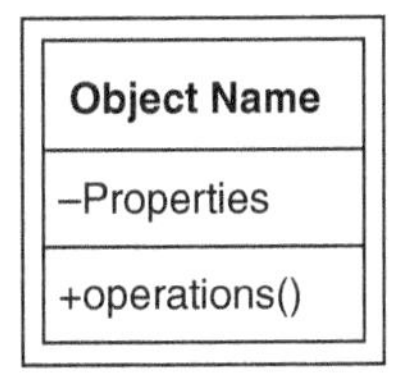

Figure B.4 Class diagram.

The properties of an object provide a mechanism to capture the structural features of that object. A property may be further elaborated as attributes or associations. The former describe a particular characteristic of a property such as the address of an output port. The latter capture how the object relates to other objects within the system. A property can be quantified by a multiplicity attribute identifying how many objects may fill the property. The wheel on an automobile has a multiplicity of four.

B.3.1 Class Relationships

We can define number of different relationships among classes. These include the following:

- Parent–child or inheritance or generalization
- Containment or aggregation

B.3.1.1 Inheritance or Generalization

We express inheritance using a solid line terminating in a hollow arrow. The diagram in Figure B.5 presents a portion of the design of an external world communications interface in an embedded system. Therein, we represent the relationship between the parent – Driver and two children – Serial and Parallel. We say that Serial and Parallel are a kind of (AKO) Driver.

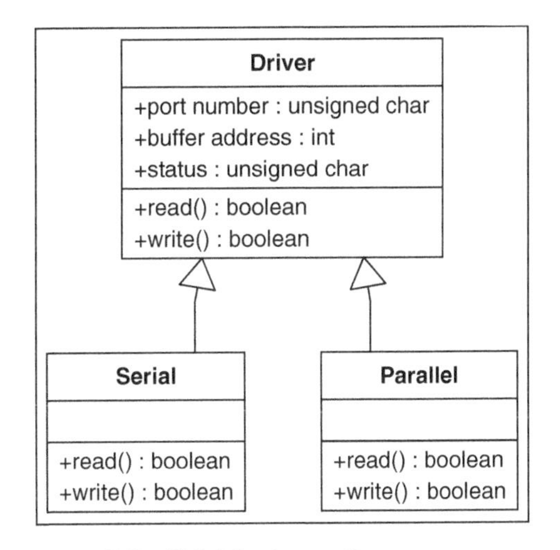

Figure B.5 UML inheritance diagram.

The diagram captures the requirement (through the parent Figure B.5 UML Inheritance Diagram interface) that each of the different types of interface must support a common subset of capabilities. Specifically, there must be a port number associated with each interface, the driver must provide the address to an I/O buffer, it must manage a status flag, and it must implement the read() and write() functions to execute the transfer. The "+" appearing in the diagram indicates that each of the corresponding elements is publicly visible.

While we generally think of inheritance as supported by the Java or C++ languages; the concept naturally applies as we begin the design of an I/O interface and its associated drivers. It seems reasonable that there should be a common way of communicating with each driver. The C language does not formally support inheritance; however, such a limitation should not preclude using the concept as a hardware or software design tool.

B.3.1.2 Interface

An interface is a wrapper around one piece of functionality that allows us to present a different set of capabilities as a public view. We express an interface in a manner that is similar to that which we use for inheritance. We use a dashed line terminating in hollow arrow.

Here, in Figure B.6, we illustrate the concept of an interface with a standard laboratory instrument. We'll apply the same concept shortly when we are working with several different data structures.

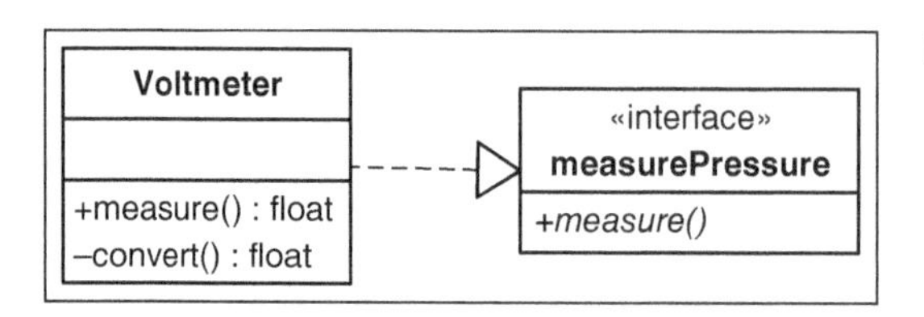

Figure B.6 Representing an interface.

In the diagram, the interface measurePressure, gives the underlying voltmeter the public appearance of a pressure meter. The hidden operation, convert(), performs the necessary math to transform the raw voltage reading from a transducer into the corresponding and proper pressure reading.

B.3.1.3 Containment

Containment conveys the idea that one object is made up of several others, that is, a whole–part relationship. Under UML, we can express two different forms of containment, aggregation and composition.

B.3.1.4 Aggregation

Let's look first at aggregation which expresses a whole–part relationship in which one object or module contains another module. A key characteristic of an aggregation is that the owned module may be shared with other modules outside of the aggregation. Under such conditions, rules must be established to ensure proper management of the shared module.

Continuing with the voltmeter system, let's assume one of the design requirements specifies that, in addition to executing pressure measurements, it must also perform several different kinds of analyses on the data that is collects. In partial support of such analysis, we design an algorithm that performs a series of statistical computations such as trend, mean, limits test, or rate of change, on the collected data.

To perform the necessary computations, the algorithm utilizes a number of different library functions. While the individual functions may be collected under the umbrella of the analysis package in the design, they can exist without that module and certainly could be used by other modules within the system as well.

The statistical analysis algorithm is an aggregation of many specific algorithms. As shown in Figure B.7, to perform the necessary computations, the algorithm utilizes a number of different library functions. Figure B.7, presents both the whole and the parts connected via a solid line that originates at an open diamond on the end associated with the whole and terminates on the end associated with the part.

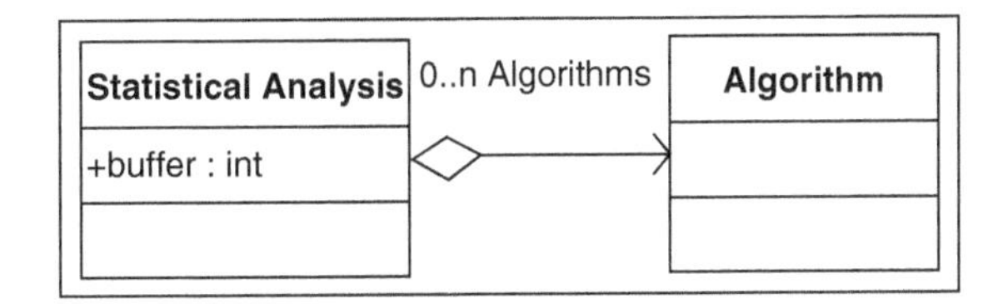

Figure B.7 Representing the aggregation relationship.

B.3.1.5 Composition

The composition relationship is similar to aggregation. However, the notion of ownership of the parts by the whole is much stronger. The elements of the composition cannot be part of another object and, unlike the aggregation relationship, they cannot exist outside of the whole object. While this may sound a little strange, at the core of the issue is the proper management of memory. The idea is loosely analogous to local variables in a function. Once one leaves the scope of the function, the local variables disappear.

In an embedded system, we often build an application as a collection of tasks. Each of these tasks executes according to a designated schedule. The schedule is made up of a number of intervals. Without the schedule, the intervals have no meaning. We express such a relationship in a composition diagram as shown in Figure B.8.

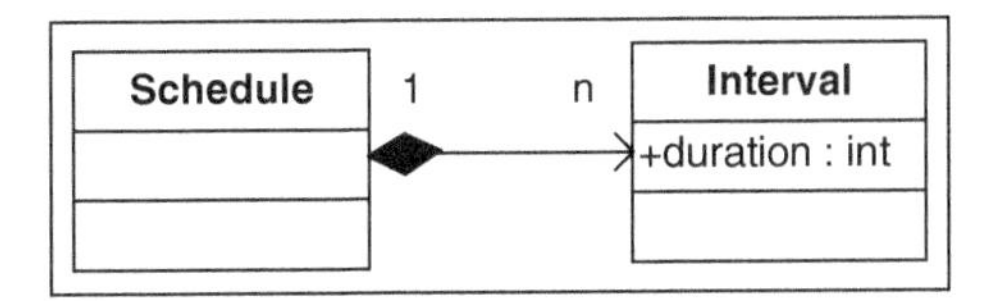

Figure B.8 Representing the composition relationship.

The schedule is composed of 1 to n intervals. Observe that the diagram is similar to that for the aggregation. The connecting line now originates in a solid rather than open diamond. We annotate the relationship as a 1 to n composition.

B.4 Dynamic Modeling with UML

Dynamic modeling provides the means to capture, to understand, and to design the intended behavior of a system. The static structure gives an architecture; the dynamic aspects of the design get the real work done. Important elements of a dynamic model include the following:

- Inter module interaction and communication
- Ensuring the proper order of task execution
- Understanding what activities can be done in parallel
- Selecting alternate paths of execution
- Identifying which tasks are active and when they are not

In the next several sections, we will study UML diagrams that will enable us to explore, to express, and to make trade-offs on these elements of a design.

B.5 Interaction Diagrams

The first diagram that we will study is the *interaction diagram*. For embedded design, understanding and modeling the dynamic behavior of the system is essential. Dynamic behavior gives information about the lifetime of a task, identifies when that task is active or inactive and models interactions amongst tasks. Such interaction often takes the form of messages. A message is a means of communication between two or more tasks. It can take several forms as follows:

- Event
- Rendezvous
- Message

Generally the receipt of a message results in the initiation of one or more actions. Such actions are executable functions within the task and result in a change in the values of one or more attributes associated with the task.

UML explicitly supports five kinds of actions.

- Call and Return
  The call action invokes a method on an object and the return action returns a value in response to call.
- Create and Destroy
  The create action creates an object; the destroy action does the opposite.
- Send
  The send action sends a signal to an object.

Each of these actions is directly applicable to later work with tasks. These actions are shown in the following diagrams. The dashed line emanating from each object or class is called a lifeline. The lifeline captures the notion of the persistence of the object.

B.5.1 Call and Return

A call action is expressed by a solid arrow from the calling object to the receiving object and the return action by a dashed, open arrow from the receiving object to the calling object. Such an interaction is shown in Figure B.9.

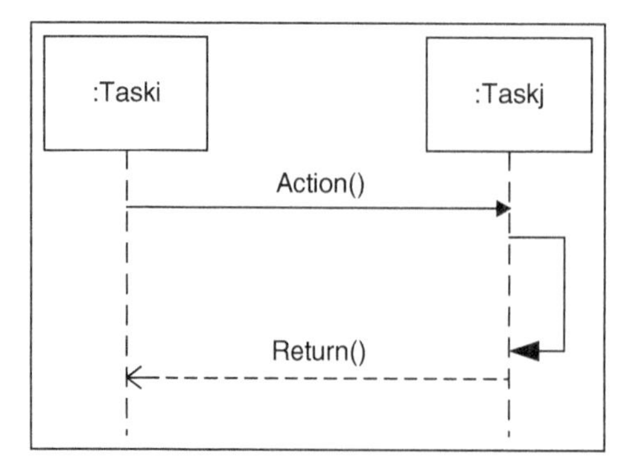

Figure B.9 The call and return interaction diagram.

B.5.2 Create and Destroy

The create action is represented by a solid arrow from the creating object to the created class instance and the destroy action by a solid arrow from the destroying object to the destroyed class instance. This relationship is presented in Figure B.10.

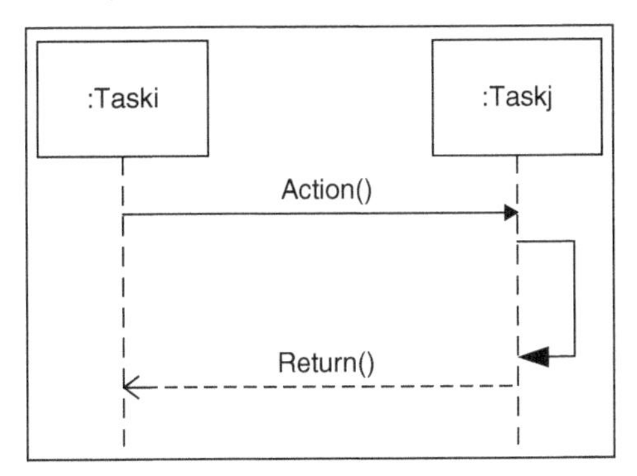

Figure B.10 The create and destroy interaction diagram.

B.5.2.1 Send

The send action is captured by a solid arrow with an open half arrow head from the sending task to the receiving task as shown in Figure B.11.

The sender does not expect a response.

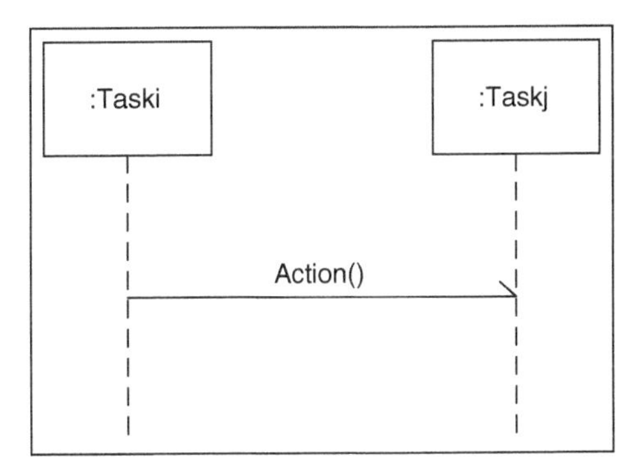

Figure B.11 The send interaction diagram.

B.6 Sequence Diagrams

Figure B.12 identifies the principal components of the UML sequence diagram. The drawing in Figure B.13 illustrates an application of those components in a *sequence diagram* for making, converting, and displaying a time interval measurement in a simple counter design. The initial selection of the specific function spawns the *measure* task.

The measure task retrieves the range and measurement edge information from an internal buffer and sends these to the *execute measurement* task. The execute task returns the raw reading to the measure task which spawns the convert task to process the raw reading into a format that can be displayed. The convert task will also perform the bounds check on the reading and return the bounds exceeded value if necessary.

- Objects
  Objects appear along the top margin of the diagram as they did in the interaction diagrams. In our designs, these will be the tasks.
- Lifeline
  The lifeline, drawn as a dashed line leaving the object, captures the notion of object persistence.
- Focus of Control
  The focus of control reflects the durations in the object's life during which it is considered to be active. It's expressed as a thin rectangular box that straddles the object's lifeline and indicates the time during which object is in control of the flow, that is, executing a method or creating another task. This is the time when a task has the CPU.
- Messages
  The messages show the actions that objects perform either on themselves or on each other.

Figure B.12 Principal components of the UML sequence diagram.

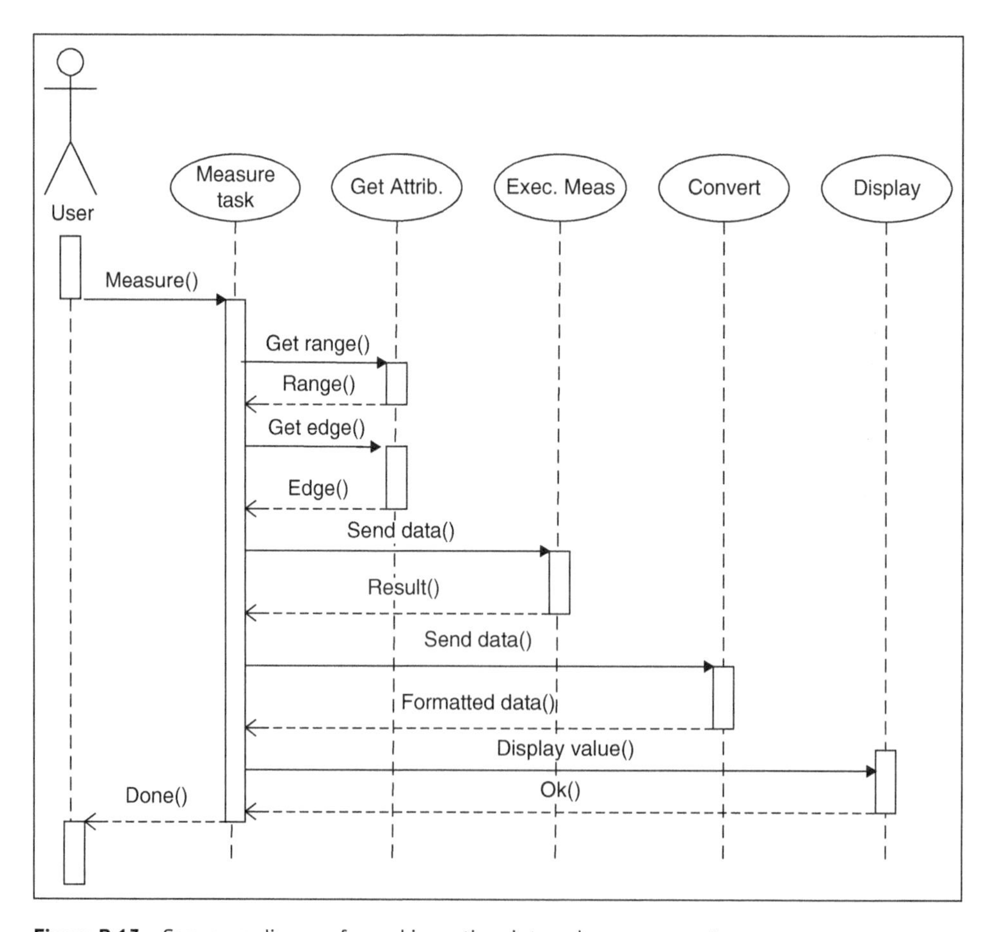

Figure B.13 Sequence diagram for making a time interval measurement.

Finally, the measure task sends the measurement to the *display* task which presents it to the user via the front panel display.

B.7 Fork and Join

When working with a multitasking embedded system, a common sequence of operations is for a parent process to start then spawn several child tasks to do the real work. The child tasks complete their jobs and terminate then the parent class follows. The process of splitting the control flow into two or more flows of control or subtasks is called a fork. Each subtask represents a separate, independent thread of control. When the subtasks are brought back together or resynchronized, it is called a join.

Such behavior of tasks and subtasks is modeled using a fork and join diagram as reflected in Figure B.14

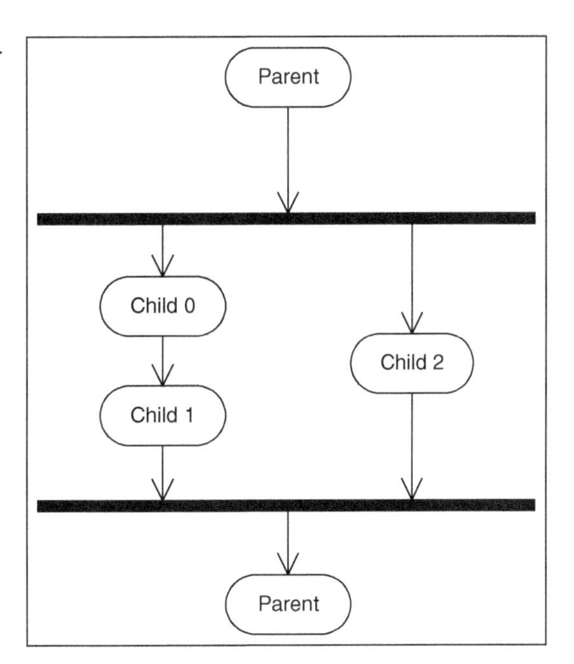

Figure B.14 UML fork and join diagram..

The tasks are represented by a cartouche or rounded rectangle. Sequential flow is given by a solid arrow. Forks and joins are represented by thick bar or rectangle called a synchronization bar. The fork occurs after the first parent activity or action completes. Following such an action, we see that the task spawns subtasks then suspends itself until subtasks complete. Once all subtasks have completed, the join occurs, and the parent task resumes its activities.

In the diagram above, the parent spawns two child tasks (Figure B.14). One child performs its task and completes; the second similarly finishes its task, then, spawns a second. When all activities are completed, the child tasks terminate and the parent continues.

B.8 Branch and Merge

Another form of flow of control is the branch in which the thread of execution is determined by the value of some control variable. Such a structure permits one to model alternate threads of execution. A merge brings the flow back together again. Each is represented by the diamond symbol that is commonly found in the familiar flow chart. Sequential flow is shown by a solid arrow and individual tasks or activities are shown using a rounded rectangle.

A simple diagram with two alternate paths of execution for a portion of the overall task is given in the diagram shown in Figure B.15.

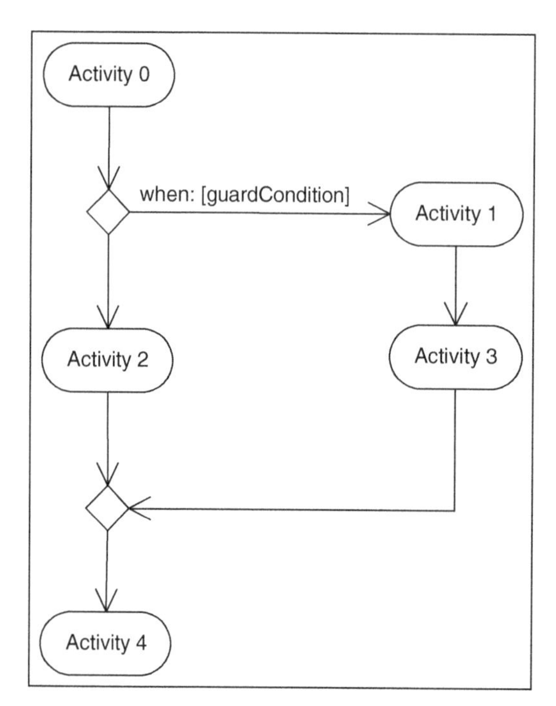

Figure B.15 UML branch and merge diagram.

Following the completion of the activities in the right-hand path, the flow of control merges back to a single path. At each branch point one can associate a guard condition to stipulate under what conditions the branch is to be taken. The guard condition is shown in square brackets on the transition arrow.

B.9 Activity Diagram

An activity diagram permits the capture of all of the procedural actions or flows of control within a task. Such actions may be a branch and merge, a fork and join, or a simple transition from state to state.

The initial node in the diagram is given by a solid black circle; the final node is a solid black circle surrounded by a second circle. The diagram in Figure B.16 shows how we might combine our earlier activities into a larger task. Conversely, one can show how a larger task is decomposed into its components.

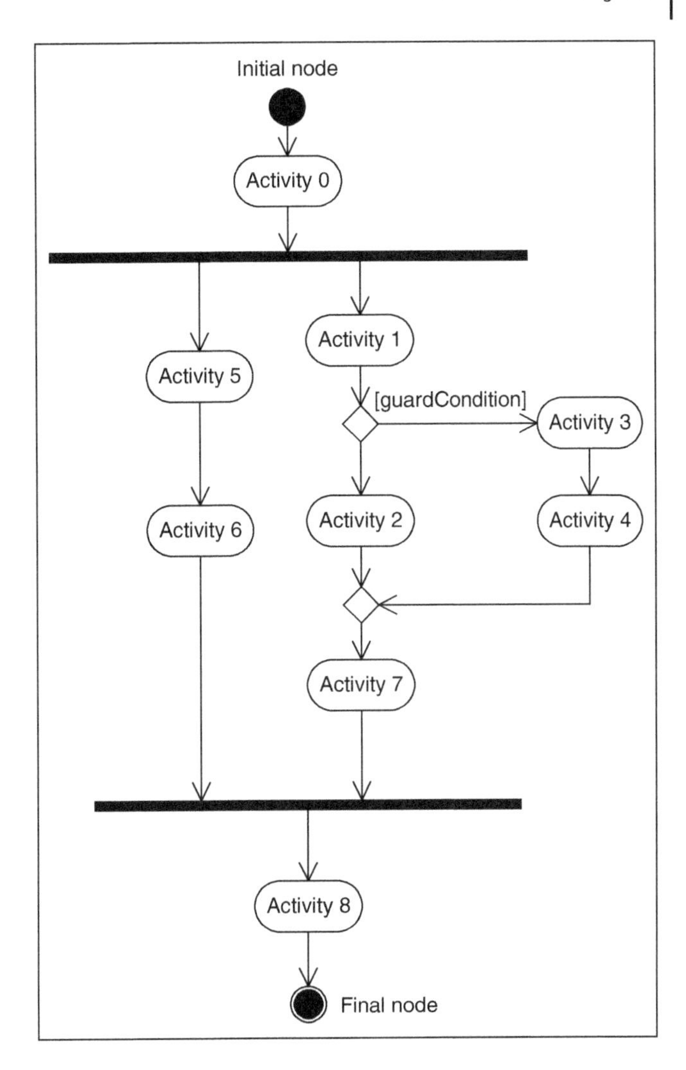

Figure B.16 UML activity diagram.

B.10 State Chart Diagrams

The state chart diagram, like the familiar state diagram, finds its roots in the mathematics of graph theory. Using the diagram, we can begin to capture and to model the state behavior of the (software) system as well as the myriad external and internal events that are affecting that behavior.

B.10.1 Events

Any embedded application must interact with world around it. The system will accept inputs and produce outputs. Inputs generally result in some associated action and the actions may or may not lead to an output. Such inputs, outputs, and actions are referred to by various names. Under the UML umbrella, they are collected under name events. An event is any occurrence of interest to the system; more specifically and typically to one of the tasks in the system.

UML supports the four kinds of events given in Figure B.17.

- Signal
  A signal is an asynchronous exchange between tasks.
- Call Event
  A call event is a synchronous communication that involves sending message to another task or sending a message to self.
- Time Event
  A time event occurs after a specified time duration has elapsed following another event.
- Change Event
  A change event occurs after some designated condition has been satisfied

Figure B.17 UML events.

B.10.2 State Machines and State Chart Diagrams

We have studied and used the finite state machines (FSM) to model and to implement a system's behavior in time. The term state machine is used to describe the following:

- The states that a system can enter during its life time
- Events to which the system can respond
- Possible responses the system can make to an event
- Transitions between possible states

Because of its simplicity, the FSM gives a good first order model of a system's behavior. UML supports and extends the traditional notion of state machines.

B.10.2.1 UML State Chart Diagrams

A state chart diagram is nothing more than the familiar state diagram with some extensions/modifications under UML. The diagram begins with the notion of a state. A state is written as a cartouche – a rectangle with rounded corners. Transitions between states reflect a change in system from one state to another and are expressed as an arrow directed from the source state to the destination state.

Mathematically, the UML state chart is a directed graph. Because cycles are permitted, it's a cyclic directed graph.

B.10.2.2 Transitions

A transition between states occurs under the following conditions: an event of interest to the system occurs or the system has completed some action and is ready to move to next state. The latter transition is called a triggerless transition. One may associate an action with the transition and a transition to self is permitted. All four types of transition are illustrated in Figure B.18.

B.10.2.3 Guard Conditions

A guard condition can be associated with a transition. A guard condition is a Boolean expression that must evaluate to true before the transition can fire. As was done in the branch and merge diagram, a guard condition is shown in square brackets on the transition arrow. UML supports several different kinds of guard.

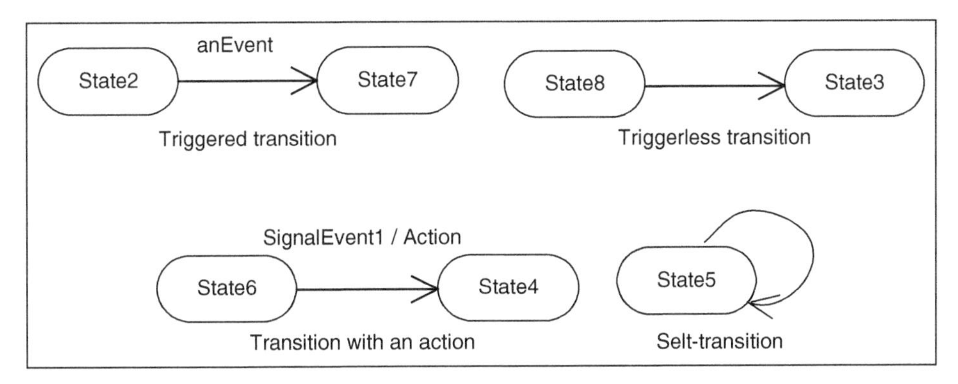

Figure B.18 Possible transitions in an UML state chart.

- An event and a guard condition are written as,
  EventName [guardCondition]
  on the state transition edge. If the guardCondition evaluates to false, the transition will not be taken.
- An event, guard condition, and action triple, appear as
  EventName [guardCondition] / Action
  on the state transition arrow. If the guardCondition evaluates to false, the action is not executed and the transition not taken.
- A guard condition by itself, is described as
  [guardCondition]
  Under such a condition, there is a repeated transition to self until the guard condition is met. Through such a mechanism, one can model the polling operation or blocking on an event or variable's state.

In the diagram shown in Figure B.19, a solid black circle represents the initial state and a solid circle with a surrounding open circle represents the final state. Illustrated in Figure B.19 is a transition with an action and a guarded event.

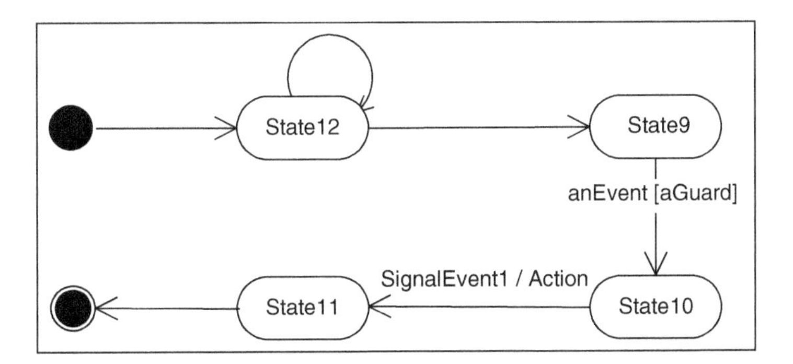

Figure B.19 Transitions with guard conditions in a UML state chart.

UML also makes the following definitions:

- An entry action is an action that the system always performs immediately upon entering a state. The requirement appears as entry/actionName within the state symbol.
- An exit action is an action the system always performs immediately before leaving the state. The constraint appears as exit/actionName within the state symbol.
- A deferred event is an event that is of interest to the system. Handling the event is deferred until system reaches another state. The deferred event appears as eventName / defer within the state symbol. Such events are entered into a queue that is checked when the system changes to the new state.

B.10.2.4 Composite States

The states that we've looked at so far are called simple states. UML extends the notion of a simple state to include multiple nested states – called composite states. These come in several different varieties.

B.10.2.5 Sequential States

If the system exists in a composite state and in only one of the state's substates at a time such substates are called sequential substates. Transitions between such substates are permitted as expected. Using sequential substates, the behavior of a state can be decomposed into smaller components as shown in Figure B.20.

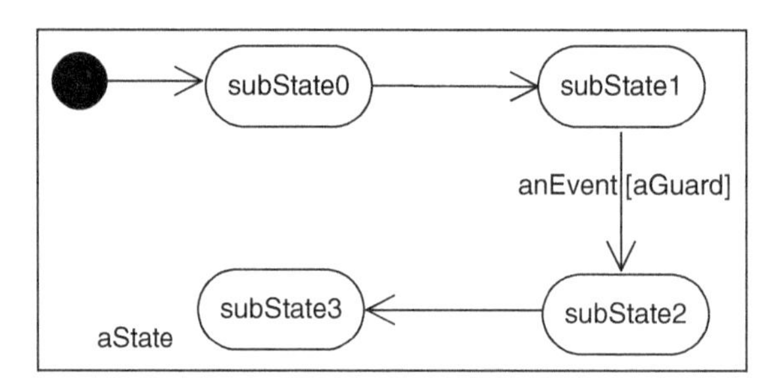

Figure B.20 Composite states in a UML state chart

B.10.2.6 History States

When system makes a transition into a composite state, typically the flow of control will start in the initial substate. However, it may be desirable to or necessary to begin in some other state. UML includes the concept of a history substate to support such capability. The history substate, shown in the state chart in Figure B.21 by a small circle enclosing the letter "H," will hold the last state that the system was in before leaving the composite state at an earlier time.

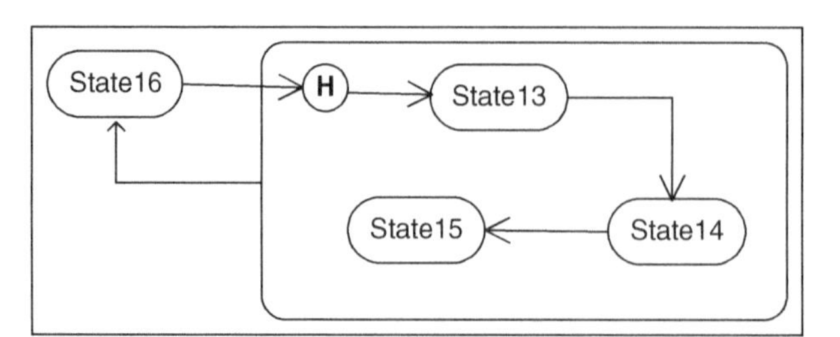

Figure B.21 Expressing a history substate in a UML state chart.

Such a state can be useful when modeling interrupt behavior or if one encounters a situation in which it is necessary to temporarily switch to another context to perform some operation prior to continuing. In either case, the present state is temporarily exited. Sometime in future, flow of control will return to that same state.

B.10.2.7 Concurrent Substates

A system may be in a composite state and also in more than one of the substates. Such is the situation in which the system may have two or more sets of substates representing parallel flows of control. When system enters a composite state with concurrent substates, it enters into initial state of both flows. Resynchronization is achieved by showing a final state for each flow as in the drawing in Figure B.22.

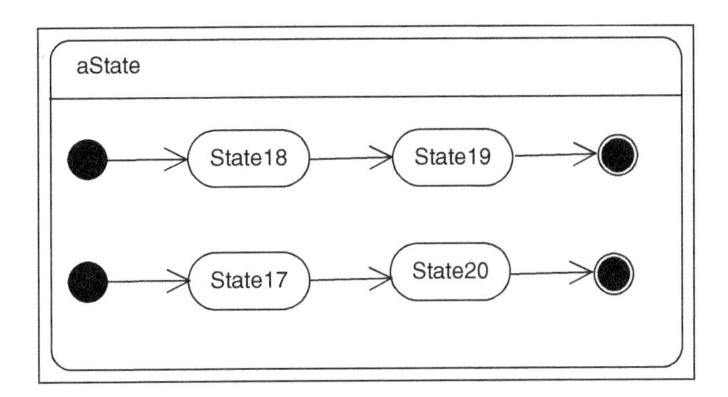

Figure B.22 Expressing concurrent substates in a UML state chart.

We have only touched on some of capabilities of the static and dynamic UML diagrams. This will be sufficient for our work. There is a vast amount of literature available for those who are interested in more detailed study.

B.10.2.8 Data Source/Sink

As the name implies, the source identifies where the data originates, for example, from an input port and a sink indicates where data goes to, for example, to an output port. The source or sink is drawn as a labeled box with an arrow to indicate direction of data flow as seen in Figure B.23. The source or sink are usually entities that are outside of the system.

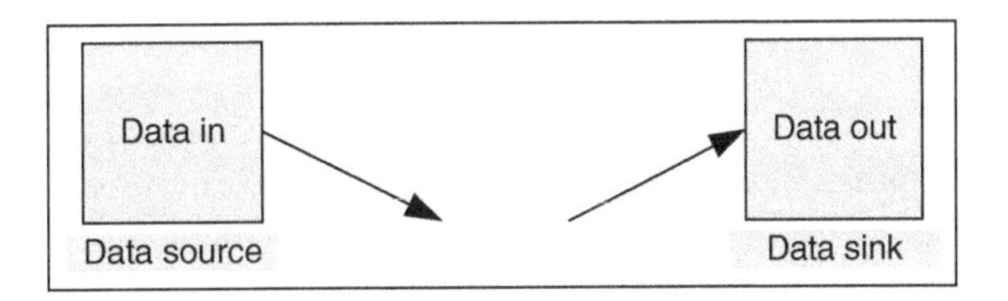

Figure B.23 Expressing a data source or sink.

B.10.2.9 Data Store

The final element is the data storage. The data store reflects the temporary storage of data or a time delayed repository of data. The data store is represented by two parallel lines or two parallel lines that are closed on the left-hand side. To our electrical engineering students, this should look just like a capacitor – and does much the same job. The graphic is accompanied by a labeled arrow to indicate the direction of the data flow as shown in Figure B.24.

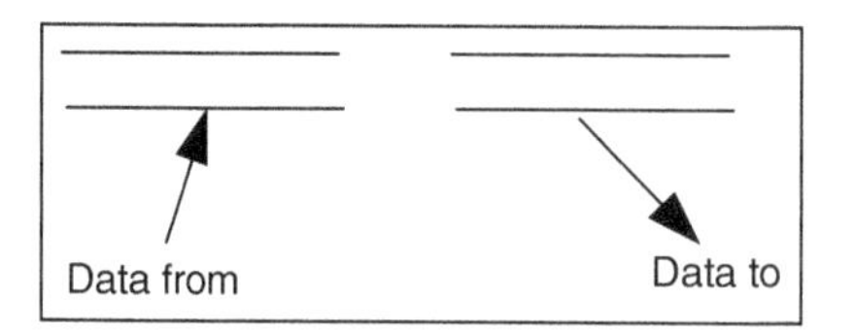

Figure B.24 Expressing a data store.

Example B.1

The diagram in Figure B.25 presents a level 0 – top level – data and control flow diagram for a system that accepts commands from a remote source; collects image data at a local site; then sends the information back to the remote site.

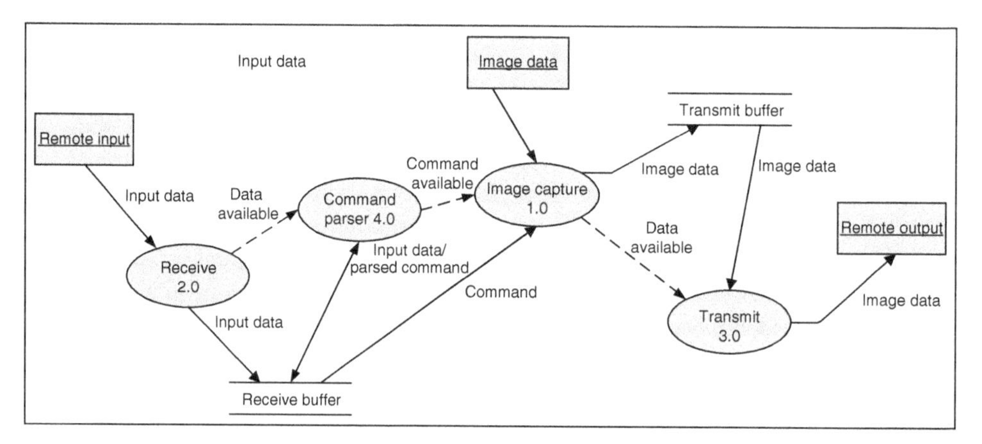

Figure B.25 Capturing the data and control flow in an imaging system.

Command data comes into the system from the remote site. This input is shown as a data source. The reception is managed by the Receive task which brings the information into the system and stores it in the Receive Buffer. Once a compete message has been accepted by the Receive task, it sends a control message to the Command Parser task which parses the data and interprets the command. When it finishes, the Command Parser writes the command back into the buffer and sends a message to the Image Capture task to execute the capture. The Image Capture task collects the data from an external source and stores it into the transmit buffer. When the capture is complete, it signals the Transmit task to send the collected data back to the remote site.

The drawing in Figure B.26 illustrates a hierarchical decomposition of a data flow diagram through three levels. At each level, greater detail is provided.

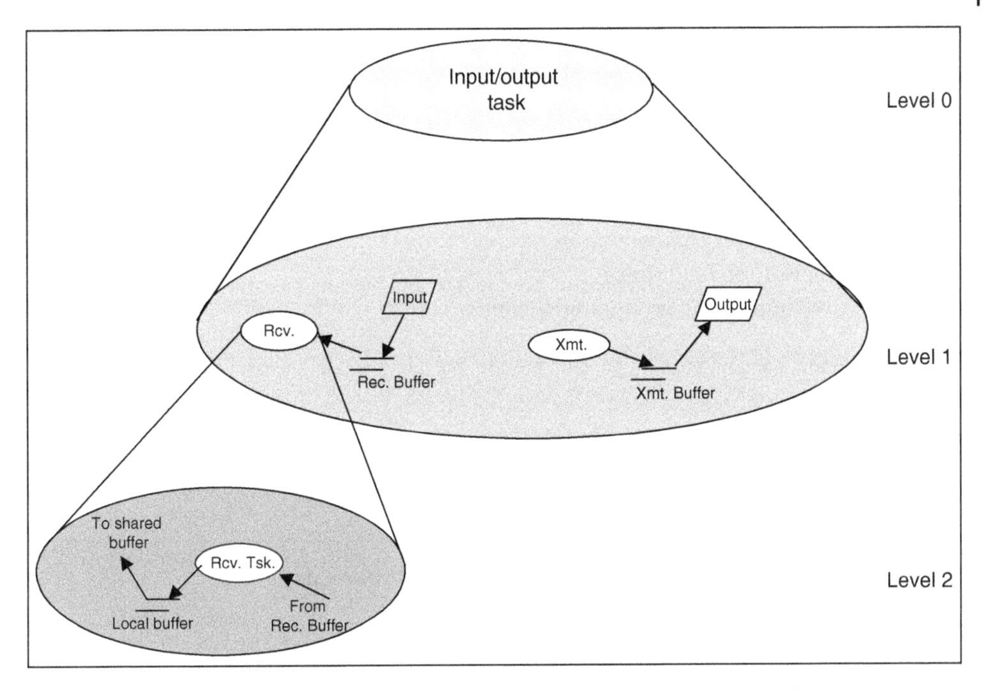

Figure B.26 A hierarchical data and control flow diagram in an input/output task.

B.11 Preparing for Test

In Appendix A, we learned to identify System Design Requirements and to create a System Design Specification. In Appendix B, we learned about UML, the universal modeling language, as a tool for aiding the design process. In Chapter 10, we learned about implementation, execution concerns, and testing of the Perceptron. In several chapters we touched on FMECA, *Failure Modes Effects*, and *Criticality Analysis* for a system. At this stage of the design and development process, we need to test our design to ensure that it meets the specified Design Requirements and that during operation that the system routinely performs in a safe and reliable manner. Volumes have been written addressing design test and the test process. The text *Embedded Systems a Contemporary Design Tool* devotes several chapters to the topic. In the next few sections, we will present some of the major highlights.

B.11.1 Thinking Test

Today, in the test system and the test process, we must focus increased attention on the following concerns:

- Transmission lines and transmission line effects
- I/O capacitance
- Conduction path inductance or impedance

- Impedance mismatches:
  Silicon to lead frame to pin to trace
  Trace to pin to lead frame to silicon
  Inner layer variations
- Signal routing:
  Trace length, impedance, signal propagation velocity
- Signal termination schemes and values
- Receiver setup and hold times
- Power distribution and grounding schemes

Other aspects of the design environment that we need to pay close attention to include:

- Signal rise and fall times
- Signal-to-noise ratio
- Signal reflection
- Race conditions and hazards
- Conduction path roughness
- Clock to Q times – Q is a storage device's signal(s) output port(s)
- Setup and hold times
- Structural Faults: stuck at, open circuit, bridging
- Propagation velocity/delay of related signals down the trace(s) from source to a common/related destination
- Printed circuit board and transmission material and structure
- Signal skew
- Path discontinuities

B.11.2 Examining the Environment

With the many excellent hardware and software tools available today, it is beyond the current scope to go into each in detail. As a start, simulation at different levels of granularity can be quite effective. With a coarse grained view, tools and techniques supported by Verilog or VHDL can be quite effective for looking at path and device delays. There are also many good, low cost or open source simulators available. For a fine grained view, tools such as PSPICE and S-Parameters can be very useful for signal and path modeling and analysis. A well-designed behavioral model of a design can be an effective aide to identify how precision waveforms scatter from ends of interconnects.

B.11.2.1 Test Equipment

There are a number of laboratory tools that can be very useful for examining the physical designs. The Time Domain Reflectometer (TDR) measures reflections resulting from signals traveling through a transmission environment. The vector network analyzer (VNA) measures network parameters of networks and is commonly used to measure s-parameters, the frequency domain scatter parameters of two port networks. Then there is the old standby; the oscilloscope remains a potent weapon in the battle to ensure and preserve signal integrity. Always have the probes on ×10 rather than ×1.

B.11.2.2 The Eye Diagram

A digital signal travels a difficult path from transmitter to receiver. As noted, a signal's integrity can be affected by external sources such as PCB traces, connectors, or cables and internal sources such as crosstalk from adjacent IC pins or PCB traces.

The simple eye diagram is a common indicator of signal quality in high-speed digital transmission. It derives its name from the diagram's shape which resembles an eye shaped pattern. It can give a quick view into a system's performance and issues leading to errors.

B.11.2.3 Generating the Eye Diagram

The eye diagram is generated by overlaying sweeps of different segments of a long data stream. The specific triggering edge not important and the displayed pulse transitions may go either way. Positive- and negative-going pulses are superimposed on each other to yield an open eye shaped pattern.

A properly constructed diagram should contain every possible bit sequence: patterns of alternating 0s and 1s, isolated 1s after long run of 0s and vice versa, or other patterns that might indicate a design weaknesses. The objective of generating the eye diagram is to gain insight into nature of signaling imperfections that can lead to errors in interpreting data bits at the receiver.

In working with the oscilloscope and probe to generate the diagram, it is very important to determine where to place probe to generate useful eye diagram and to aide in identifying source of problems. We must recognize that probe placement and grounding can affect signals being measured and can yield differing (potentially incorrect) displays based upon placement.

In the ideal signaling world, the eye diagram looks like rectangular box. In the real-world, signaling is imperfect, transmissions and communication are imperfect, and transitions do not always align perfectly. The shape of diagram depends upon the triggering: clock based, divided clock, or pattern trigger and the signal timing and amplitude which can cause the eye opening to decrease.

We illustrate the construction of an eye diagram by starting with a 3-bit signaling sequence: the transmission and then the capture of eight 3-bit patterns. The eight component patterns and the composite that make up the diagram are shown in Figure B.27.

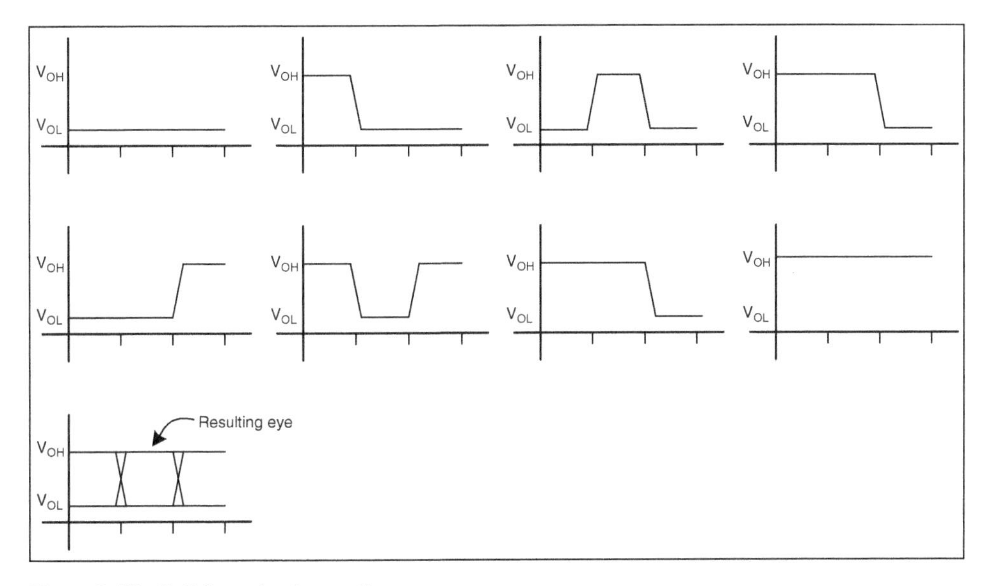

Figure B.27 Building a basic eye diagram.

B.11.2.4 Interpreting the Eye Diagram

A completed eye diagram contains large amount of parametric information. How can we interpret that information and what does it not show?

Looking first at what it will not show. The eye diagram will not reflect logic errors, protocol problems, or the wrong bit(s) sent. It will show if a logic 1 is sufficiently distorted to be interpreted as logic 0.

Let's examine the completed diagram in Figure B.25 to identify the basic metrics identified earlier. We look at three important characterizing aspects of the aggregated traces: the rise and fall times, jitter at the middle of signal crossing point, and the overshoot and undershoot.

The diagram in Figure B.28 provides an instant graphic view of data that the engineer can use to check and assess the signal integrity of the design and to identify problems early in design cycle. When used in conjunction with bit error rate and other measures, it can help to predict performance and to identify sources of problems.

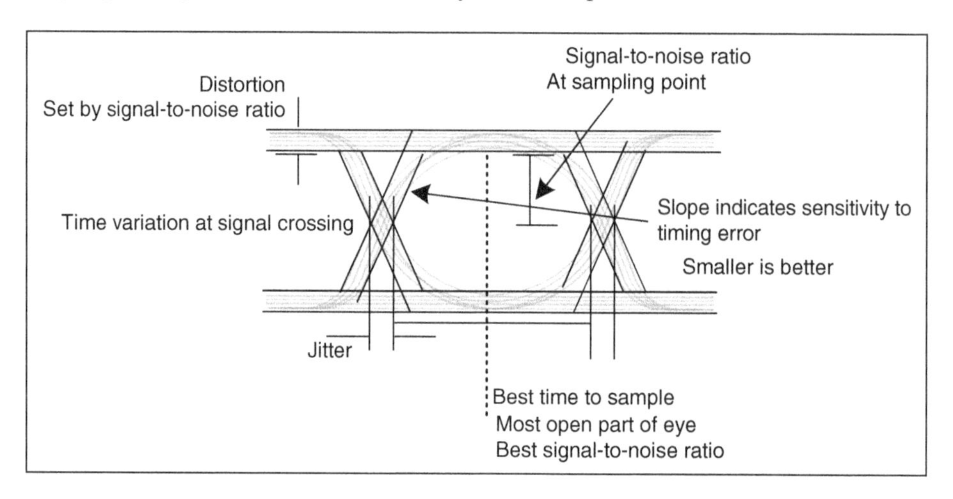

Figure B.28 A basic eye diagram.

Examining our diagram, we can:

- See the best place to sample signal.
- View the signal to noise ratio at sampling point.
- Track the amount of jitter and distortion.
- See the time variation at signal crossing.
- Assess the amount of jitter.

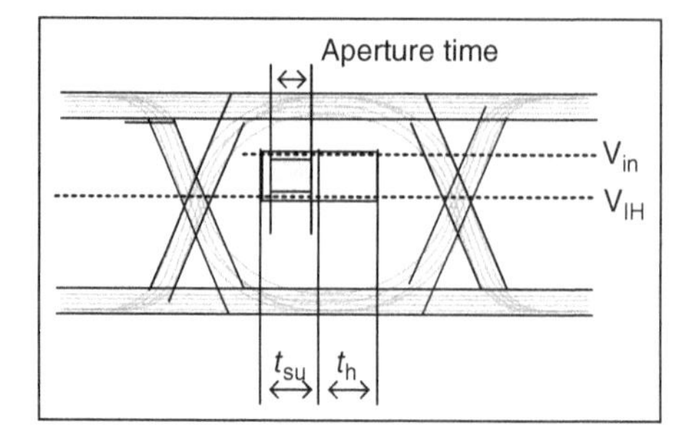

Figure B.29 Signal aperture time.

The next eye diagram in Figure B.29 illustrates the aperture time for logic 1 input.

In the diagram, the two voltages V_{IH} and V_{in} specify the minimum and typical high-level input signal levels. The times t_{SU} and t_{HOLD} specify the setup and hold times for an input signal. The light filled box width sets setup and hold times and its height sets valid input voltage levels. The gray filled box must be inside and smaller than the light filled box.

B.11.3 Back of the Envelope Examination

B.11.3.1 A First Step Check List

Let's put together a simple high-level check list as a guide for dealing with signal integrity issues. Examining the signals involved in design provides a good starting point for analysis.

- Identify and classify signals:
  Source synchronous
  Common clock
  Controls
  Clocks
  Other special category
- Determine best estimates of system variable values:
  Mean values
  Estimated maximum variations
  Example variables:
    I/O capacitance
    Trace length, prop velocity, impedance
    Inner layer impedance variations
    Buffer drive capability and edge change rates
    Termination impedances
    Receiver setup and hold times
    Interconnect skew specs
- Ensure security and safety for all critical data and systems

B.11.4 Routing and Topology

Signal routing, the associated net topology, and path material can have significant impact on system signal integrity. A key qualifier of such topology is symmetry. We must ensure that the net topology appears symmetric from driver's perspective. The length and loading should be identical for each leg in net and the impedance discontinuities at junctions minimized.

B.12 Summary

- Unified Modeling Language overview and diagrams.
- The major static modeling diagrams in UML and the utility of each.
- UML class diagrams and use cases
- The need for dynamic modeling
- Thinking system test

Bibliography

von Altrock, C. (1995). *Fuzzy Logic and NeroFuzzy Applications Explained.* Prentice Hall.

Anderson, K. (1992). *Control Systems Sample Life in the Fuzzy Lane*, 78–82. Personal Engineering & Instrumentation News.

Bezdek, J.C., Keller, J., Krisnapuram, R., and Pal, N.R. (eds.) (1992). *Fuzzy Models for Pattern Recognition – Methods that Search for Structures in Data.* IEEE Press.

Brubaker, D.I. (1992). Fuzzy-logic basics: intuitive rules replace complex math. *EDN* 18: 111–116.

Carbonell, J.G. (1984). *Paradigms for Machine Learning*", Tutorial Presented at AAAI-84. Austin, TX.

Cios, K.J., Shin, I., and Goodenday, L.S. (1991). *Using Fuzzy Sets to Diagnose Coronary Artery Stenosis*, 57–63. IEEE Computer.

Cox, E. (1992). *Fuzzy Fundamentals*, 58–61. IEEE Spectrum.

Cox, E. (1993). *Adaptive Fuzzy Systems*, 27–31. IEEE Spectrum.

Cox, E.D. (1994). *The Fuzzy Systems Handbook.* AP Professional.

Cox, E.D. (1995). *Fuzzy Logic for Business and Industry.* Rockland, MA: Charles River Media, Inc.

Davis, R. (1979). TEIRESIAS interactive transfer of expertise: acquisition of new inference rules. *Artificial Intelligence* 12 (2): 121–157.

Doyle, J. (1979). A truth maintenance system. *Artificial Intelligence* 12 (2): 231–273.

Duda, R.O. (1978). Development of the prospector consultation system for mineral exploration. *SRI Projects 5821 and 6415 final report AI Center Computer Science and Technology Division*, SRI International, Menlo Park, CA, October 1978.

Feigenbaum, E.A. and Buchanan, B.G. (1978). Meta-dendral. *Artificial Intelligence* 11: 5–24.

Carbonell, J.G. (1982). Experimental learning in analogical problem solving. In: *Proceedings of the AAAI-82*, 168–171. Pittsburgh: Pennsylvania.

Feigenbaum, E.A., Buchanan, B.G., and Sutherland, C. (1969). *Heuristic-DENDRAL*", Machine Intelligence, 4. New York, NY: American Elsevier.

Haykin, S. (1994). *Neural Networks a Comprehensive Foundation.* Macmillan Publishing Company.

Hebb, D. (1949). *The Organization of Behavior.* New York: Wiley.

Humphrys, M. *Single-layer Neural Networks (Perceptrons).* School of Computing, Dublin City University.

Introduction to Fuzzy Logic, First Edition. James K. Peckol.

Kandel, A. and Lee, S.C. *Fuzzy Switching and Automata Theory and Applications*. Crane, Russak & Company, Inc.

Kaufmann, A. and Zadah, L.A. (1975). *Introduction to the Theory of Fuzzy Subsets, Vol. I*. New York: Academic Press.

Klir, G.J. and Folger, T.A. (1988). *Fuzzy Sets, Uncertainty, and Information*. Prentice-Hall, Inc.

Klir, G.J. and Yuan, B. (1995). *Fuzzy Sets and Fuzzy Logic: Theory and Applications*. Prentice Hall, Inc.

Klir, G.J., St. Clair, U.H., and Yuan, B. (1997). *Fuzzy Set Theory: Foundations and Applications*. Prentice Hall, Inc.

Korner, S. (1967). *The Laws of Thought*. Encyclopedia of Philosophy.

Kosko, B. (1992). *Neural Networks and Fuzzy Systems*. Prentice-Hall, Inc.

Kosko, B. (1993). *Fuzzy Thinking, the New Science of Fuzzy Logic*. Copyright.

Lee, C.C. (1990). *Fuzzy Logic in Control Systems: Fuzzy Logic Controller – Part I*, 404–418. IEEE Transactions on Systems, Man, and Cybernetics.

Lee, C.C. (1990). *Fuzzy Logic in Control Systems: Fuzzy Logic Controller – Part II*, 419–435. IEEE Transactions on Systems, Man, and Cybernetics.

Leung, K.S. and Lam, W. (1988). *Fuzzy Concepts in Expert Systems*, 43–56. IEEE Computer.

McCarthy, J. (1980). Circumscription-A form of non-monotonic reasoning. *Artificial Intelligence* 13: 27–39.

McCulloch, W. and Pitts, W. (1943). *A logical calculus of the ideas immanent in nervous activity. Bulletin of Mathematical Biophysics* 52: 99–115.

McDermott, D. and Doyle, J. (1980). Non-monotonic logic I. *Artificial Intelligence* 13: 41–72.

McNeill, D. and Freiberger, P. (1993). *Fuzzy Logic*. Simon & Schuster.

Minsky, M.L. and Papert, S.A. (1969). *Perceptrons*. Cambridge, MA: MIT Press.

Mostow, D.J. (1978). *FOO (First Operational Operationalizer)*", Program accepted advice about the game of hearts, vol. 3. Handbook of Artificial Intelligence.

Norman, D. (1988). The Design of Everyday Things. *Doubleday*.

von Oech, R. (1983). *A Whack on the Side of the Head How to Unlock Your Mind for Innovation*. Warner Books.

Rao, V.B. and Rao, H. (1993). *C++, Neural Networks, and Fuzzy Logic*. MIS Press.

Reiter, R. (1980). A logic for default reasoning. *Artificial Intelligence* 13: 81–132.

Rosenblatt, F. (1957). *The Perceptron—A Perceiving and Recognizing Automaton". Report 85-460-1*. Cornell Aeronautical Laboratory.

Rosenblatt, F. (1958). The perceptron: a probabilistic model for information storage and organization in the brain. *Cornell Aeronautical Laboratory, Psychological Review* 65 (6): 386–408. https://doi.org/10.1037/h0042519.

Rosenblatt, F. (1962). *Principles of Neurodynamics*. Washington, DC: Spartan Books.

Schank, R.C. and Colby, K.M. (1973). *Computer Models of Thought and Language*. W. H. Freeman.

Shortliffe, E.H. and Buchanan, B.G. (eds.) (1984). *Rule-Based Expert Systems: The MYCIN Experiments of the Stanford Heuristic Programming Project*. Reading, MA: Addison-Wesley.

Terano, T., Asai, K., and Sugeno, M. (1989). *Applied Fuzzy Systems*. AP Professional.

Terano, T., Asai, K., and Sugeno, M. (1992). *Fuzzy Systems Theory and Its Applications*. Academic Press, Inc.

Waterman, Donald, "*Learn to Play Draw Poker*". Bulletin of Mathematical Biology Vol. 52, No. 1/2. pp. 99–115. 1990.
William of Ockham (or Occam) (1280–1348). *Occam's Razor*. Encyclopædia Britannica.
Winston, P.H. (1970). *Learning Structural Descriptions from Examples*. dspace.mit.edu.
Zadeh, L.A. (1965). *Fuzzy Sets*, 338–353. Berkeley, CA: University of California.
Zadeh, L.A. (1968). Probability measure of fuzzy events. *Journal of Mathematical Analysis and Applications* 10: 421–427.
Zadeh, L.A. (1973). *Outline of a New Approach to the Analysis of Complex Systems and Decision Processes*, 28–44. IEEE Transactions on Systems, Man, and Cybernetics.
Zadeh, L.A. (1979). A theory of approximate reasoning. In: *Machine Intelligence*, vol. 9 (eds. J.E. Hayes, D. Michie and L.I. Kulich), 149–194. New York, Wiley.
Zadeh, L.A. (1983). *Commonsense Knowledge Representation Based Upon Fuzzy Logic*, 61–65. IEEE Computer.
Zadeh, L.A. (1988). *Fuzzy Logic*, 83–93. IEEE Computer.
Zadeh, L.A. and Janusz, K. (eds.) (1992). *Fuzzy Logic*. For the Management of Uncertainty, John Wiley, Inc.
Zadeh, L.A., Fu, K.-S., Tanaka, K., and Shimura, M. (1975). *Fuzzy Sets and Their Applications to Cognitive and Decision Processes*. Academic Press, Inc.

Further Reading

HOW CAN I LEARN MORE?

There are numerous articles and books available on the subjects covered in the text. Good sources for non-mathematical treatments are these magazines and books:

BYTE
CONTROL
EDN
Electronic Design
Computer Design
Scientific American
"Fuzzy Thinking: The New Science of Fuzzy Logic" by Bart Kosko published by Warner in 1993.
"Fuzzy Logic" by Daniel McNeill and Paul Freiberger published by Simon and Schuster in 1992.

You can find a more formal coverage in journals and texts such as:

International Journal of Approximate Reasoning
Journal of Fuzzy Sets and Systems
IEEE Transactions on Fuzzy Systems
IEEE Spectrum
IEEE Edge
Communications of the ACM

Numerous books. . .Google search fuzzy texts

RECENT DEVELOPMENTS IN FUZZY LOGIC AND FUZZY SETS
Dedicated to Lotfi A. Zadeh
Shahnaz N. Shahbazova, Michio Sugeno, Janusz Kacprzyk

GUIDE TO DEEP LEARNING BASICS
Logical, Historical and Philosophical Perspectives
Sandro Skansi

DEEP NEURO-FUZZY SYSTEMS WITH PYTHON
With Case Studies and Applications from the Industry
Himanshu Singh, Yunis Ahmad Lone

DESIGN OF TWO OR THREE INPUT SINGLE OUTPUT FUZZY LOGIC CONTROL SYSTEM
Dr. Arshia Azam, Dr. Dr. Mohammad Haseeb Khan

E-LEARNER'S ASSESSMENTS WITH FUZZY LOGIC
WEB APPLICATION DESIGN AND DEVELOPMENT
Mousumi Mitra, Atanu Das

FUZZY SETS AND FUZZY LOGIC WITH APPLICATIONS
Imprecision, Uncertainty and Vagueness
M.K. Hasan

HANDBOOK OF MACHINE LEARNING
Tshilidzi Marwala

FUZZY LOGIC
Think Like a Veterinarian
J. Aaron Gruben DVM

APPLYING FUZZY LOGIC FOR THE DIGITAL ECONOMY AND SOCIETY
Andreas Meier, Edy Portmann, Luis Terán

A FIRST COURSE IN FUZZY LOGIC
Hung T. Nguyen, Carol L. Walker, Elbert A. Walker

FUZZY LOGIC THEORY AND APPLICATIONS
Part I and Part II
Lofti A. Zadeh, Rafik A. Aliev

ADVANCED FUZZY LOGIC APPROACHES IN ENGINEERING SCIENCE
Mangey Ram

FUZZY LOGIC CONTROLLER FOR REAL TIME NETWORKED CONTROL SYSTEM
B. Sharmila, K. Srinivasan

Popular Languages and Tools
- Python
- C++
- C
- Java

Matlab
AI

Index

Introduction to Fuzzy Logic, First Edition. James K. Peckol.

c